Marine fish & invertebrates

of Northern Europe

Marine fish & invertebrates
of Northern Europe
First english edition published 2004

A CIP catalogue record for this book is available from the British Library

Web page: http://www.marinbi.com

AquaPress
25 Farriers Way
Temple Farm Industrial Estate
Southend-on-Sea
Essex, SS2 5RY

First published in Norway by
KOM, PB 865, 6501 Kristiansund N, Norway
Norwegian ISBN: 82-92496-084

Graphic design: Erling Svensen
Printed by T.I.Trykk

ISBN: 0-9544060-2-8

The printing of this book is supported by:

Marine fish & invertebrates

of Northern Europe

Author
Frank Emil Moen

Photographer
Erling Svensen

Translators
Dr. Sabine Cochrane and Prof. Fredrik Pleijel

KOM 2004

Contents

Foreword

Despite the fact that I have been privileged to venture into many of the world's seas and oceans, my favorite dive sites are still the cold waters of Northern Europe. Perhaps it is because that as a terrestrial primate and a botanist I enjoy life amongst the kelp forests or maybe it is that the marine habitats of these northern climes are not quite as susceptible to the catastrophic damage now so obvious in our coral seas. A little more than 50 years ago Cousteau and Gagnan made it possible for us mere mortals to move freely in three dimensions as we began to probe the secrets of Earth's own inner space. Not long after that Hans and Lotte Hass were warning the World of the problems of over fishing, nutrient enrichment and pollution of the sea around us. This very useful book will not only speed our understanding and hence enjoyment of all we see when down beside or below the sea, but will allow us to interface with our ancestors over the last 600 million years of creative evolution. Our genesis, via the not so simple sponges through the sea squirts to the fishes and beyond are all there waiting for you to discover.

Recent research indicates that 600 million years ago the Earth was a giant snowball and all the seas were cold. The Life Giving Sea came to the rescue producing great blooms of microscopic organisms whose protective shells sequestered both calcium and carbon into long-term store. The chalk so formed acted as a safety valve smoothing the wilder fluctuations of the global greenhouse. Once this was in place evolution speeded up filling the seas with an ever-greater diversity of larger plants and animals.

Based on sound science and wreathed with real life pictures this superb book not only allows you to name their names but to understand the roles each creature plays in the balance of marine life. As we damage that balance at our own peril, the more people who see and understand the potentials and the problems the easier it will be to persuade governments to take the correct action.

I dream of the time when most of the fish and shellfish we eat will be harvested from closed circuit farms that use no chemicals nor pollute the sea in any way. That the farm stock will be fed, not on sand eels and capelin stolen from the marine food chains by industrial scale fishing but on invertebrates fed on waste material from the land. Then and only then will our fish stocks recover making sustainable fishing by local fishers a reality. Then and only then will our children's children be able to see the animals described in this gem of a book that lets you into the secrets of the wealth and the health of these cold waters.

Prof David Bellamy
OBE., BSc., PhD., Hon: FLS., DSc., D.Univ.,
C.Biol. President of The Conservation Foundation,
Coral Cay Conservation, The Marine Conservation Society of Australia
and The Galapagos Conservation Trust

Bedburn 2004

Corals - Anthozoa Page 94

Soft corals - Alcyonacida **Page 95**

Horny corals - Gorgonacida **Page 98**

Sea pens - Pennatulacida **Page 105**

Tube anemones - Ceriantharia **Page 115**

Stone corals - Scleractinida **Page 117**

Sea anemones - Actiniarida **Page 130**

Comb jellies - Ctenophora Page 145

Sipuncula Page 195

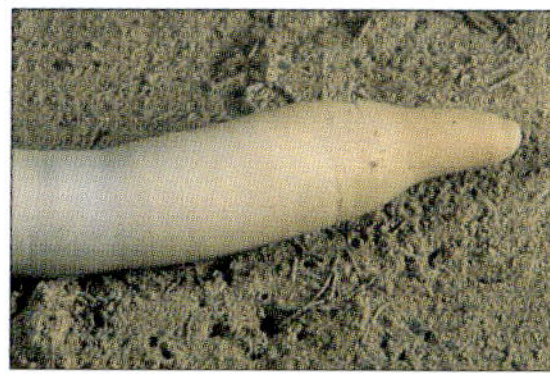

Arthropoda Page 196

Systematics

This book presents a broad selection of marine shallow-water invertebrates and fish from the coasts of northern Europe. The animals are introduced in traditional systematic order, commencing with 'simpler' animals, and continuing with the more 'complex' ones. For the non-specialist reader we here introduce some basic concepts in systematic biology.

The salmon, **Salmo salar**, *is considered a fresh-water organism, despite the fact that it spends most of its life in the sea. The fish represent the most developed group of animals discussed in this book.*

Species concepts

Specimens exhibiting internal and external similarities, and which can produce fertile offspring, are referred to as the same species. This probably represents the most commonly employed species concept. In practice, however, species identifications largely rely only on **morphological** similarity. At the introduction of a new species, the author selects a "**type specimen**" which is specified in the publication. This specimen serves as the name-bearer for the species; any specimens that are judged conspecific will carry the same name. The type specimen is usually described in detail, and is subsequently deposited in a museum. Following the so-called morphological species concept, all specimens that exhibit morphological resemblance to the type are referred to this species, where degrees of sufficient

Sponges are thought to be the oldest, or most 'primitive' of the multicellular animals

*The term **fauna** refers to animal life. In addition to the groups presented in this book, the fauna of northern Europe also includes marine mammals, such as seals and whales, as well as a variety of birds.*

resemblance generally are determined by the group specialists. Obviously, there are weaknesses with this approach, and ideally the allocations should be based on genetic analyses as well, rather than morphological similarity only.

Species usually evolve slowly through time, and species identifications are complicated by various intermediate cases. Thus, a dog and a wolf can have fertile offspring, but they are not referred to the same species, and a donkey and a horse can have offspring–the mule–although this offspring is sterile. Furthermore, there are several dog races which–due to human selection–have diverged so much from their ancestral form that they cannot reproduce with other dog races. A St. Bernhard and a Chihuahua are referred to the same species, but they cannot actually produce common offspring.

Populations and speciation

A population is "***a group of specimens living in a restricted area***". More accurately, we can state that "***populations consist of conspecific specimens connected by gene flow***". There may be significant differences among populations within the same species. A river may have a well-defined population of Atlantic salmon, where the adult specimens will tend to return to the river where they hatched, and this in spite of the fact that they spend a major part of their life span in the sea. A local salmon population will thereby adapt to a specific river, where the specimens with optimal local adaptations are more likely to reproduce successfully and thereby transmit their genes to coming generations. A population which becomes completely isolated can, over a time-span ranging from a few hundred to many thousands of years, develop important differences and evolve into a new species. A more complete understanding of these processes, however, requires technical knowledge in **genetics**.

Did you know that...

The Swedish biologist **Carl von Linné** (1707-1778) was the founder of the classification system in current use. He introduced this system in his "***Systema naturae***" of which the first edition appeared in 1735, more than one hundred years before the publication of Charles Darwin's "On the origin of species".

Classification and taxonomy is not a new science. The first systematic accounts of different species were carried out by the Greek philosopher **Aristotle** (384-322 years BC). Nevertheless, new discoveries have not yet ended. Many new species are currently being described, they are given scientific names, and they are systematised according to their relationships. The methods for analysing relationships are today in rapid development. Whereas relationships earlier were largely determined by morphological and embryological studies, new technics now allow also for biochemical analyses of DNA and proteins to assess evolutionary relationships. The actual identification of relationships is a complex issue, and systematic revisions are continously carried out as new knowledge is gained.

Classification

Living organisms are arranged (**classified**) in different groups (**taxa**), and all these groups are related to each other. Those that are closely related have a more recent common ancestor, whereas those that are more distant have a common ancestor further back in time. Recent studies indicate that Life on Earth evolved not less than 3.9 billion years ago, about 700 million years after the birth of the planet. During the first three billion years it was inhabited mainly by simple organisms such as bacteria (**prokaryotes**), and it is 'only' during the last billion years that more complex organisms have been present. We estimate that there are 15 - 30 million living species today. However, of these only a little more than one million are described at present. The study and the systematisation of groups and their relationships is a science called **taxonomy**, and the scientists working with it are referred to as **taxonomists**.

This attractive little chiton, **Callochitin septemvalvis**, *belongs to the Mollusca.*

Binomial nomenclature

The living organisms are classified and named according to their relationships. Each species is given a name consisting of two words, a naming system which is referred to as the 'binomial nomenclature'. For "***Homo sapiens***", the scientific name of man, "***Homo***" is the genus name, "***Homo sapiens***" is the species name, and "***sapiens***" represents the species epithet. Scientific names are always given in Latin. The advantage is that scientists, no matter their origins, will refer to a certain species by the same name. This separates scientific names from common names, of which there are many synonyms, even within the same language.

Male cuckoo wrasse, **Labrus bimaculatus** *(p. 538).*

Also the scientific nomenclature has its drawbacks, including changes of taxon names. In many cases we detect that species, which previously were grouped together, actually are not closely related. As a result of this the species is referred to another genus, and the generic part of the name changes, sometimes also the species part. Likewise, we may find out that specimens referred to different species actually belong together-a classical example is provided by the **cuckoo wrasse** (*Labrus bimaculatus*) which has sexual dimorphism. Today we know that they are males and females of the same species, but previously they were referred to as two different species. In such cases one of the names becomes a synonym of the other and will not be used. A more frequent situation, though, is that we discover that a single species actually consists of several, different species. When recognised, these are removed and given new names. All these changes are commonplace, and lead to situations where the same names in the literature refer to different entities. For this reason, the reader may encounter names in this book that disagree with some previous (or later) literature. We do not here provide lists of synonymies, but usually note whenever recent changes have taken place.

Apart from the species and genera in the

Did you know...

The small, beautiful blue-striped patelliform snail, which today is known as ***Ansates pellucida*** (p. 291), also is referred to as *Helcion pellucidum* and *Patina pellucida* in the current literature.

Female cuckoo wrasse, **Labrus bimaculatus,** *clearly different from the male (above).*

Did you know...

Latin species and genus names should be written in *italics* in regular text, *but in* regular font *when the remaining text is in italics.*

classification system, there are a number of other, higher, categories, which refer to more inclusive groupings. These are listed in the table below. Note that it is not necessary to employ all levels of the classifications of a given species.

The scientific name for the edible crab is **Cancer pagurus** *(p. 267)*

		EDIBLE CRAB	
ENGLISH	**LATIN**	**ENGLISH**	**LATIN**
Kingdom	**Regnum**	Animal kingdom	Animalia
Phylum	**Phylum**		Arthropoda
Sub-phylum	Subphylum		Crustacea
Supra-class	Superclassis		Malacostraca
Class	**Classis**		Eumalacostraca
Sub-class	Subclassis		
Infra-class	Infraclassis		
Supra-order	Superordo		Eucarida
Order	**Ordo**		Decapoda
Sub-order	Subordo		Pleocyematina
Infra-order	Infraordo	Crabs	Brachyura
Supra-family	Superfamilia		
Family	**Familia**		Cancridae
Sub-family	Subfamilia		
Tribe	Tribus		
Genus & species	**Genus & species**	Edible crab	*Cancer pagurus*

The shore crab **Carcinus maenus** *(p. 269) is a member of the same sub-order, Brachyura, as the edible crab, but is separated at the family level. It belongs to the family Portunidae, together with the swimming crabs.*

Today there is not one single accepted classification, which is uniformly accepted by all zoologists. Systematics is an evolving science, and as there are many thousands of active taxonomists, revisions and changes are common. We do not provide complete classifications in our book, since we saw little gain in representing the full hierarchies. Instead we provide the most commonly used names of higher taxa. Accordingly, for some groups we may give names of infra-order or superfamily, when these are generally applied, while for others we may give only the name of the phylum. On p. 592 we provide a more detailed systematic account of the included species.

Sabella sp. and *Sabella* spp.

When sp. (abbreviation of species) is given after a genus name, such as in *Sabella* sp., it denotes that the specimen(s) belongs to the genus *Sabella*, but that the actual species remains unidentified. Sometimes also spp. is used, as in *Sabella* spp., and means that several, unidentified species are involved (see p. 188).

Photographs and identifications

Many marine zoologists claim that identifications from photographs amount to no more than qualified guesses. This is due to the situation that many specialists actually pay little attention to macroscopic features. As a matter of fact, the belief that most marine biologists are keen divers with good knowledge about the specimen's ***in situ*** appearance, is erroneous, and much of the literature on marine invertebrates illustrates this lack in our knowledge. Illustrations are often copied from earlier studies, or are based on dull, formalin-fixed specimens. Collecting is usually achieved by dredging, trawling or by taking grab-samples. On the contrary, the combination of *in situ* photographs, and collection for further and more detailed study, allows us to obtain a fuller knowledge of the biology of the species, as well as adding new features for identification without the use of microscopes. The extensive photographic material in this book illustrates the amount of detail and information that can be provided in this way. However, there obviously are a large number of groups that cannot be identified from ***in situ*** photographs only. Nonetheless, it is our hope that this book can convince you that the "**habitus**" of many animals is highly characteristic when observed in their natural environment.

HABITUS

"Habitus" refers to the external appearances of an animal or plant, such as its shape and colour. The term is often employed when referring to closely related taxa, differing in features which otherwise are difficult to communicate in words.

Marine biotopes

On land we can easily observe differences between different habitats, including ***pine*** *or* ***foliiferous forests, swamps, lakes, rivers, mountains, plains, beaches,*** *and* ***islands****, including areas subject to human activity. These different biotopes harbour different fauna and flora.*
Along our coasts we can also identify a series of different biotopes, all inhabited by different animals and plants. We here present some basic information on four marine biotopes. The ***intertidal zone*** *is the area between the lowest and highest tides; the* ***kelp forest*** *appears on exposed rocky shores from 0 to approx. 20 metres depth, and is dominated by the kelp* **Laminaria hyperborea***;* ***soft bottoms*** *consist of mud and sand. Finally,* ***fjords*** *constitute the whole system within a fjord, independent of depth and bottom characteristics.*

Did you know...

The term habitat refers to the place where a species occurs. For example, the kelp *Laminaria hyperborea* is the habitat of the snail *Lacuna vincta*, whereas the open sea is the habitat of the mackerel. The term biotope refers to a special kind of area with a particular flora and fauna, such as a lake, a swamp, the intertidal zone, or a mud bottom.

Underwater life is influenced by bottom conditions, degree of exposure, water currents, salinity, depth and latitude. This picture was taken on a winter's day at Egersund in southwestern Norway.

The intertidal zone

The knobbed wrack, **Ascophyllum nodosum**, *is one of our most common algae in the tidal zone.*

Regarding physical characteristics, there are few biotopes that provide harsher conditions for the organisms than the intertidal zone, and those living there have to suffer extreme changes in environmental factors. Temperatures over the day may reach up to 20°C, and the salinity can span from 0 ‰ to over 30 ‰. Perhaps most critical, the organisms must be able to survive the drought at low tide which occurs two times a day.

Different abilities to withstand drought is one of the major causes for the characteristic intertidal zonation patterns. This is perhaps most obvious if we examine the distribution of brown algae. Tufts of **channelled wrack** (*Pelvetia canaliculata*) appear highest up on the shore, and are followed by the **flat wrack** (*Fucus spiralis*). Below these there is a wider belt of **bladder wrack** (*Fucus vesiculosus*), either mixed with or followed by the **knobbed wrack** (*Ascophyllum nodosum*). Furthest down on the shore we encounter the **serrated wrack** (*Fucus serratus*). Variations in tidal amplitude, salinity, bottom characteristics, and degrees of wave exposure, all influence the species composition and distribution in the intertidal zone.

In the inner parts of the coast, in fjords and protected inlets, the upper water layer often is low salinity brackish water. This affects the composition of algae.

TIDES

Tidal variation is due to gravitational impact by the moon and the sun on the water masses. The impact of the sun is 46% of that of the moon. The highest tidal amplitudes are reached at new and full moon, when the attraction by the sun and by the moon are reinforcing each other. In the first case the moon will be situated between the earth and the sun, and in the second case it will be furthest away from the sun. Both these situations give rise to spring tides. **Extreme high** water conditions can appear if the new or full moon is accompanied by a low-pressure situation. There are, however, some complicating factors. The moon's orbit around the earth is elliptical, and the earth's orbit around the sun also is elliptical and has both monthly and yearly cycles. The elliptical orbits mean that the distances to the moon and to the sun vary, and thereby also the gravitational impact of these. Furthermore, there are monthly and yearly variations in both the sun and the moon declination (angle relative to the equator). The most extreme astronomical (=independent of low or high pressure) tidal amplitudes are reached when the distances between earth, moon and sun are minimal, when the sun and moon declination is zero, and at new and full moon. In an area with mean amplitude of 3.5 metres, and mean spring tide amplitude of 5 metres, such situations will yield an increase to 6 metres amplitude.

Rocky shores are often dominated by the barnacle **Semibalanus balanoides** *and the limpet* **Patella vulgata.**

However, do not panic if you have a property close to the sea–the next 'high astronomical spring tide' will appear in the year 6580!

AMPHIDROMIC POINT

In an area near Egersund in southwest Norway there is no tide–a fact that is related to the presence of an amphidromic point in the area. Amphidromic points are areas where tidal waves are extinguished, owing to topographical conditions combined with the Coriolis effect (deviation of water masses due to the rotation of the earth). This can be illustrated by comparing with water movements in a bathtub: if we create a wave that follows the border of the tub, a central point without wave movement will accur. The North Sea, however, is not a perfect bath tub, and therefore three amphidromic points arise: one in the north of the English Channel, one west of Denmark, and one in the southwest of Norway.

This underwater cliff extends from the surface straight down to 50 m depth. It is fully colonised by the sea anemone **Urticina eques** *and the dead man's finger coral* **Alcyonium digitatum**. *25 m depth.*

When is the best time to study life at the sea shore?

Obviously, an excursion to the seashore is best planned at low tide during days when the amplitude is maximal, which is at new or full moon. Nonetheless, it is recommended to check local tide charts for best timing.

In shallow water, hard bottom substrates often are covered by various calcareous red algae.

Colourful bottom of a narrow, current-rich sound, dominated by filtering organisms. 10 m depth.

What kinds of organisms live on rocky shores?

A range of animals are associated with rocky shores. Below we provide a brief list of the most commonly encountered organisms on a rocky shore excursion. Perhaps most obvious are the **acorn barnacles** (*Semibalanus balanoides*) and the **periwinkles** (*Littorina* spp.). Under stones, crawling in the small pools, we will find **amphipods** of the genus *Gammarus*, often together with **nereid polychaetes**. Under larger boulders, and among brown algae, we may find an eel-like fish, the **butterfish** *Pholis gunnellus*. **Chitons** (Polyplacophora) often graze on the stones, commonly in company with sessile **jingle shells** (Anomiidae) and the polychaetes *Spirorbis* and *Pomatoceros triqueter*. More active polychaetes are the **scale worms** (super–family Aphroditoidea), which crawl under the surface of the stones.

Among the brown algae we find **isopods** of the genus *Idotea*, and growing on its surface there are different species of **bryozoans**, together with various **cnidarians**, such as **hydroids** and small **sea anemones**. Between stones, near the low water mark, live **Dahlia anemones** (*Urticina felina*) and **beadlet anemones** (*Actinia equina*). Other encountered animals near the low water mark are **nudibranchs**, the **sponge** *Halichondria*, and **patelliform snails** of the genus *Tectura*. Further common **molluscs** include **blue mussels** (*Mytilus edulis*), **dog whelks** (*Nucella lapillus*), and the large **common limpet** (*Patella vulgata*). Among echinoderms the **common starfish** (*Asterias rubens*) is frequent, and the **green sea urchin** (*Strongylocentrotus droebachiensis*) and the **small urchin** *Psammechinus miliaris* appear furthest down on the shore. Many smaller **fish** will follow the incoming tide in search of food, such as the **two-spotted goby** (*Gobiusculus flavescens*), the **goldsinny wrasse** (*Ctenolabrus rupestris*), the **stickleback** (*Gasterosteus aculeatus*), the **fifteen-spined stickleback** (*Spinachia spinachia*), and young **pollacks** (*Pollachius pollachius*). **Common prawns** (*Palaemon* spp.) live in this biotope, but will follow the water when the tide goes out. A large number of other species also occur, but this introductory list of common taxa above may be helpful for the first examinations of rocky shores.

Balanus
(p. 205)

Gammarus
(p. 216)

Gunnel
(p. 545)

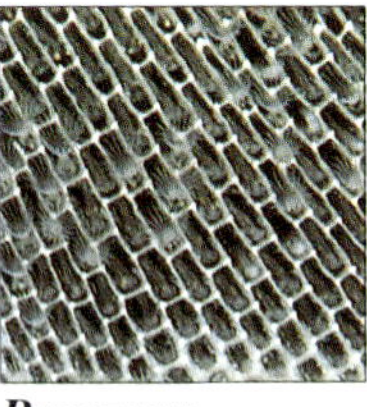

Bryozoan
(p. 398)

Dahlia anemone
(p. 131)

Nudibranch
(p. 322)

Common mussel
(p. 355)

Sea urchin
(p. 437)

Goldsinny wrasse
(p. 540)

Above: *Kelp usually does not grow on vertical substrates. Here a diver and dead man's finger coral.*
Left: *Kelp forests can extend along large areas of coastline.*

Kelp forests

Forests in the sea? Yes, this really is an apt description of the metre-high kelps (*Laminaria hyperborea*), overgrown with epiphytes on the stalks (stipes) and with waving crowns (laminae). Divers gliding over, under or between these 'tree-tops' experience a feeling of freedom comparable with the floating sensation of hang-gliding. Diving in exposed hard-bottom areas with underwater cliffs, crevasses and overhangs is a stunning experience. The kelp is less dense here, letting the light penetrate, such that the rich colours of the fauna and flora make a huge impact. The photographs speak for themselves.

The kelp forests along the coasts are important for a number of reasons. It is a biotope that serves as habitat for a large number of different groups, all of which live in complex connections to each other. Many of the species are small and inconspicuous, and occur among the kelp holdfasts. Others use the stipes and laminae as living places, and some graze directly on the kelp. The forests also play an important role as nursery areas for a number of fish species and for crabs and lobsters, all of which can use the rich associated fauna as a food resource.

The depth distribution of the kelp is constrained by the amount of accessible light. As a rule of thumb, no less than 1% of the surface

THE ENEMY OF THE KELP FOREST

The sea urchin, ***Strongylocentrotus droebachiensis*** (p. 437) probably is the worst enemy of the kelp forest. In northern Norway, it is estimated that the kelp forest is reduced to half its previous size due to grazing activities of this sea urchin.

light must be available for photosynthesis, and how deep the light penetrates depends on the amount of particles in the water. In heavily polluted areas, and in areas with high input of particulate matter from rivers, this limit may be reached already at a depth of 1 metre. In clear coastal waters, on the other hand, kelp can exceptionally be found down to 36 metres depth. Under more normal coastal conditions, though, they will disappear at around 20 metres (kelp forests are defined as areas where the area of the lamina equals or exceeds 1 m^2 per 1 m^2 bottom area).

Laminaria hyperborea may reach an age of 20 years, although length increase stops at about 10 - 11 years. The species diversity in kelp forests is very high, and even a list of the most common ones is too extensive to be given here. Throughout this book the reader will encounter pictures and text about species that belong to this biotope. Wide-angle pictures give some idea of the diversity, but note that most of the inhabitants are small and require close examination. A large number of them live among the holdfasts, including the edible crab (***Cancer pagurus***), which uses the kelp forest as a nursery ground. The highest diversity among holdfast dwellers is encountered among plants older than six years.

A diver peeps among the algae in search of a photo opportunity.

Shady areas are often exploited by sponges or, as here, the hydroid **Tubularia larynx** *(p. 71).*

KELP TRAWLING:

The kelp forests are commercially exploited for production of alginates, and are trawled in several areas in western Norway. This represents an important income in a number of coastal communities. The harvests in later years have reached 165,000 tonnes, amounting to a value of more than 200.000 Euro. However, intense kelp trawling in the same areas should be avoided–the diversity of the forests increases with age of the kelp, and a number of the associated species require plants older than 6-7 years.

Below the kelp forest, the faunal diversity decreases. In areas deeper than 20 m, with modest water currents, the bottom often can appear poor in life. Green light penetrates deepest in coastal waters but in clear oceanic water, blue light penetrates farthest. The colour of the water varies according to the amount and type of suspended particles and plankton.

Soft bottoms

Most divers regard soft bottoms as a bad choice for diving, and many would only scan larger sandy areas in search of flatfish. Such bottoms may look poor, although this is only partly correct. The number of species is fewer than on rocky bottoms, but, on the other hand, the number of specimens per species can by far exceed other biotopes.

Large stones are often found scattered on shallow sandy bottoms. These almost always are covered in algae.

A large number of species are characteristic for soft bottoms, and here we will only mention a few of the more conspicuous ones. Of fish, the most common are flatfish, such as **dab, plaice, turbot** and **flounder**, but we may also encounter **haddock, cod, angler, dragonet, grey gurnard, greater weaver** and **hooknose**.

Flatfish
(p. 577)

Dragonet
(p. 568)

Hooknose
(p. 528)

There are few cnidarians on soft bottoms; exceptions include the **tube anemone** (*Cerianthus lloydii*), and a few octocorals such as the **sea pen** (*Virgularia mirabilis*) and the beautiful **phosphorescent sea pen** (*Pennatula phosphorea*). Of polychaetes there is a large number of species, including many which live buried down in the sediment and therefore also are less accessible for *in situ* observations. The **sand worm** (*Arenicola marina*) produces characteristic casts on

Tube anemone
(p. 115)

Phosphorescent sea pen *(p.105)*

Sea cucumber *(p. 449).*

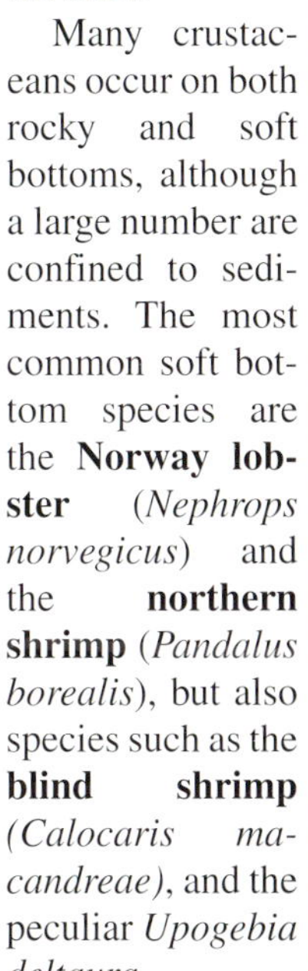

the surface in intertidal and shallow water. Whereas many have seen these traces, fewer have seen the actual animals which live far down in the sand.

Crustacean within the genus **Upogebia** *(p. 244)*

Many crustaceans occur on both rocky and soft bottoms, although a large number are confined to sediments. The most common soft bottom species are the **Norway lobster** (*Nephrops norvegicus*) and the **northern shrimp** (*Pandalus borealis*), but also species such as the **blind shrimp** *(Calocaris macandreae)*, and the peculiar *Upogebia deltaura*.

Sand gaper *(p. 379)*

Pelican's foot *(p. 304)*

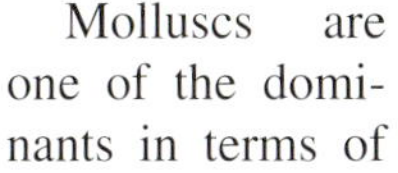

Razor shell *(p. 378)*

Molluscs are one of the dominants in terms of diversity and abundance. Many bivalves live as infauna, buried in the sediment with only the siphons communicating with the surface. The **sand gaper** (*Mya arenaria*) and the **blunt gaper** (*Mya truncata*), several **razor shells** (Solenidae), and **cockles** (Cardiidae), may serve as examples of typical members. Of gastropods we have the **pelican's foot** (*Aporrhais pespelecani*), the **auger shell** (*Turritella communis*) and the **moon snail** (*Polinices pulchella*), which are typical soft bottom inhabitants. The **common bobtail squid** (*Sepietta oweniana*) lives on sandy bottoms, where we also may observe the **octopus** *Eledone cirrhosa*.

Squid *(p. 383)*

Sand star *(p. 410)*

There are a series of echinoderms living on or in soft bottoms. The **sand star** (*Astropecten irregularis*) occurs in more sandy sediment. A series of **brittle stars** (Ophiuroidea) are associated with soft bottoms, and **irregular sea urchins** (Spatangoidea) live in the sediment, whereas the **holothurian** *Stichopus tremulus* feeds on deposited dead organic matter on the sediment surface. It is therefore evident that a diverse group of organisms are associated with soft bottoms.

Sea potato *(p. 439)*

Brittle star *(p. 428)*

Fjords

Fjords and sheltered inlets lack the kelp forests found along more exposed coasts. However, the sea anemone **Urticina eques** *(p. 132) generally grows to a larger size in sheltered areas than individuals further out on the coast.*

A fjord is characterised by the presence of a shallow sill, usually situated at the fjord opening, and any water exchange will have to pass across this sill. A shallow sill can lead to limited water exchange, and result in lack of oxygen. The composition of fauna in a fjord is characteristic and differs from that in other coastal waters. The surface water often has a reduced salinity, and this effectively excludes a number of species. Compared to coastal waters, the deeper parts of a fjord tend to be more stable, both concerning temperature

ANOXIC WATER - IS THAT POLLUTED?

We associate anoxic water with pollution (influenced by human activities). This is usually the case in fresh water, where nutrients from households and farms causes increased algal growth. Dead algae accumulate on the bottom and are decomposed by bacteria and fungi during the consumption of oxygen. In the sea it is less common that human activities lead to anoxic conditions. However, it is not uncommon that such conditions develop from natural causes, and this relates to the poor exchange of water in fjords with shallow sills. The surface water is brackish, and when the sill is shallower than the brackish water, it will isolate the deeper fjord water from the outpage and prevent exchange of full salinity water. Decomposition of organic matter consumes the available oxygen, resulting in anoxic conditions. Both bacterial activity and inorganic processes will continue, although the necessary oxygen will instead be obtained by reduction of sulfate (SO_4).This process leads to the production of hydrogen sulfide (H_2S)–the same gas which emerges from a rotten egg.

The upper water layer often is highly influenced by fresh water, which affects the algal composition.

Northern shrimp
(p. 229)

Coral reef
(p. 117)

and salinity; coastal waters are more influenced by currents, as well as by autumn and winter gales. The stable conditions in fjords can also serve as an explanation for the phenomenon that deep-sea taxa may appear in more shallow waters–indeed an added value in fjord diving! At nighttime one may encounter deep-sea fish at depths of 20-30 metres, or even shallower. The photographs of **rat-fish, velvet-belly, black-mouthed dog-fish, hake** and **argentine**–all typical deep-water species–were taken during night-dives in fjords, as were those of the **Northern shrimp** (*Pandalus borealis*). Other species living in more shallow waters in fjords than in other places are the **Norway lobster** (*Nephrops norvegicus*), several species of the **squat lobsters** (genus *Munida*), the **sea anemone** *Bolocera tuedia*, and not least colonies of the reef-building **cold-water coral** *Lophelia pertusa*, and the large **octocorals** *Paramuricea placomus*, *Paragorgia hyperborea* and *Primnoa resedaeformis*. All of these occur also outside the fjords, but then usually at depths greater than 150 metres. A number of **sponges** (Porifera), and the **basket star** (*Gorgonocephalus caputmedusae*) also appear

Rat-fish
(p. 482)

Hake
(p. 495)

Black-mouthed dog-fish
(p. 478)

Velvet-belly
(p. 475)

Argentine
(p. 493)

Norway lobster
(p. 237)

in more shallow waters in fjords than elsewhere. A symbiotic relationship, which usually only occurs in fjords, is the **hermit crab** (*Pagurus prideaux*), carrying the **sea anemone** *Adamsia palliata*.

Basket star
(p. 429)

Turnip sponge
(p. 50)

Hermit crab *(s. 262) with the shell covered by a* ***cloak anemone*** *(p. 141)*

NIGHT DIVING

A night dive in a fjord during winter (when water visibility tends to be very good) is an exciting experience. It is not unusual to encounter deep-water fishes that come up into shallow water during the night. A number of organisms also hide during the light hours, but will appear after dusk. The limited light source from the torch also increases your focus on anything that might move. Some may think that diving in the dark is unpleasant, not being able to see what is going on behind your back. On the contrary, however, this is a highly relaxing and enjoyable experience. The stillness is remarkable, and the only sound is the air hissing through the regulator valve as you breathe. Breathing frequency actually decreases during night dives, often reducing the air consumption by about one third.

PORIFERA

Phylum Porifera - sponges

Sponges (Lat. porus = pore + Lat. fero = bearing) occur in a wide variety of shapes and sizes. The smallest (Cliona) *has a height of 2 mm, whereas others have a height/ diameter of over 2 metres. The sponge body is supported by a skeleton comprising spicules of* ***lime carbonate*** *or* ***silicate*** *and/or a fibrous part made of the proteinaceous material spongin. The form of the skeleton is very characteristic and is used in species identification. Sponges are sessile organisms, and can be either solitary or colonial. Sponges have no organs as such and lack a mouth opening and nervous system. However, they do possess complex and distinct types of cells, which are grouped to perform different kinds of functions. Although exterior appearances may vary a great deal, the inner body plan is relatively simple. Sponges can be thought of as a '****beaker****' containing many pores, through which water flows in and then is distributed through a water canal system. From there, water flows into a larger cavity, where it is further transported out through one or several large exhalent openings (****osculum/oscula****). Water is moved through the canal system by means of so-called collar cells (****choanocytes****), which have whip-like structures (****flagellae****).*

Identification of sponges usually requires microscopy.

Desmonspongia haliclona. *Photographed at 20 m depth at Egersund, south-west Norway.*

Three structural sponge types can be distinguished by the positioning of the collar cells and the form of the water canal system. The **ascon-type** has a simple 'beaker' shape and each of individual water canal (**ostium**) is formed by a cylindrical cell (**porocyte**). A single porocyte extends through the entire 'beaker' wall and the inner wall is covered by collar cells. All the water filtered by the sponge passes out through a single osculum. In the somewhat more complex **sycon-type**, the pores in the wall are formed by channels between the cells. The inner wall has numerous **radial canals** lined by flagellated collar cells. The actual inner walls in this type of sponge are covered by the same type of cells as the outer walls. Sponges of the ascon- and sycon-types are only found within the calcareous sponges (Class **Calcarea**). Sponges of the **leucon-type** have the most complex structure. Here, the collar cells are located in flagellated chambers, which are rounded cavities in the water canals with the collar cells forming the inner wall. In **leucon-type** sponges, the water usually passes through several of these chambers before being expelled through several exhalent openings (**oscula**). Leucon-type sponges are found among some of the calcareous sponges and within the glass sponges (class **Hexactinel-**

Phakellia rugosa. *Photographed in Trondheimsfjorden (Skarnsundet), Norway, around 30 m depth.*

Blue is an unusual colour in nature. The blue encrusting sponge in the picture probably belongs to the genus **Hymedesmia**. *The yellow sponge on the right (with 'veins') belongs to the genus* **Clathria** *(=* Microciona *sp.).*

lida) and the class **Demospongiae** (which includes bath sponges). Bath sponges from the Mediterranean lack calcareous or siliceous skeletons; their's are entirely composed of spongin fibres.

Food particles are captured by the collar cells, by means of their flagellae. Food is ingested by phagocytosis; individual cells 'eat' the food particles, which are then digested **intracellularly** (within the cell). These cells then transfer much of the food content to another type of cell, the **amoebocytes**. These can move in an amoeboid manner and supply the remaining types of cell with nutrition. Food uptake, cellular organisation, gas exchange, reproduction and response to external stimuli are generally similar to single-celled organisms. Sponges feed by filtering out microscopic organisms such as unicellular algae and bacteria as well as dead organic material (**detritus**) and even large free organic molecules can be ingested. Their filtering capacity is impressive; studies have shown that sponges can remove around 90 % of suspended bacteria.

Most sponges are hermaphrodites, producing both male and female gametes. Egg cells are made from transformed amoebocytes, whereas spermatozoa are derived from wandering collar cells or amoebocytes. The eggs are retained within the sponge and fertilised internally. Sperm are released into the water through the exhalent openings, and fertilise the eggs of the same or another individual. However, the eggs and sperm usually mature at different times in the same individual, ensuring cross-fertilisation. Fertilised eggs develop and hatch into larvae, which are dispersed by the water masses before settling at a suitable loca-

A RARE CARNIVORE

A group of sponges have been found to have a characteristic trait that separates them from all other sponges, in terms of feeding habits. These sponges are carnivores, in the sense that they devour small crustaceans that become attached to the sponge spicules.

Above: *Water streams out through the large openings (oscula) after food is filtered from it.*

Right: **Haliclona urceolus**, *photographed at 40 m.*

Hymedesmia paupertas *usually is found deeper than 40 m. Here from Jørpeland in southwestern Norway.*

EFFECTIVE PUMPS

Sponges have a remarkable ability to pump water. Some can pump 10 000 to 20 000 times their own volume per day.

tion. Sponges also can reproduce asexually by budding and regeneration–small, torn-off pieces often grow back into a new sponge.

Up till now, more than 9000 sponge species are known, most of which are marine and all live attached to the bottom as adults. They are found at all depths and in almost all colours, but in the largest abundances and variations in shallow waters. Most occur on hard substrata, forming crusts or sometimes erect or branching structures, but a few are found on soft sediments. Some sponges have symbiotic relationships with bacteria or unicellular algae, which help them to obtain food.

Typical bottom at 65 m in the south-west of Norway. The large sponge to the right of the picture is **Geodia barretti**, *and the flat, disc-like sponges are* **Phakellia ventilabrum**. *The sponge on the left is* **Mycale lingua**.

Above: *This sponge,* **Guancho** ***sp.,*** *has a very characteristic appearance.*

Left: *This* **Mycale lingua** *was found at 45 m depth.*

Below: Halichondra panicea *occasionally grows around kelp stipes.*

Sponges have relatively few enemies, but some sea slugs (order **Nudibranchia**), chitons (class **Polyplacophora**) and starfish (class **Asteroidea**) prey on certain sponge species. In tropical waters, certain fish and sea turtles feed on sponges. On coral reefs, around half the sponges present are poisonous. In some species, the calcareous or siliceous spicules that form the skeleton stick out from the body surface as protection. Some sponges also are a hazard to humans. The Caribbean fire sponge, ***Neofibularia nolitangere*** causes an intense burning pain that lasts several hours after contact.

Sponges provide habitat for a whole range of different marine species. The tube sponge, ***Shenciospongia vesparia***, which can be up to two metres high, can contain more than 16 000 different species!

Sponge systematics are not fully understood. The order is usually divided into three classes; calcareous sponges–**Calcarea**; glass sponges–**Hexactinellida**, and the **Demospongiae**, which include bath sponges. In earlier works, a fourth class, the coral sponges Sclerospongiae, is mentioned. Species within this class are now transferred to the Demospongiae. Recent studies suggest that the glass sponges that usually live in deep waters are so distinct from the other two groups that they

The Trondheimsfjord is host to a rich array of sponges. Here are two different species in front of **Paramuricea placomus**. *From 40 m depth.*

should be separated, into either their own sub-order or order. In Scandinavia, there are only a few known deep-living glass sponges.

Species determination of sponges is difficult because the external construction and colour seldom give enough information. For positive identification, it is necessary to examine the internal construction of the sponge, especially the skeletal structure. In recent times, sponge classification is based on larval development (**embryology**), cellular organisation (**histology**) and construction (**cytology**), as well as biochemical make-up. Because of the difficulties in identification, the distribution of many species is unknown. Along the Norwegian coast, a total of 263 species currently are recorded, many of which are found in deep waters.

PHAGOCYTOSIS AND INTRACELLULAR DIGESTION

Single-celled, or unicellular organisms, and simple multicellular organisms such as sponges carry out what is known as intracellular digestion. This means that a food item, for example a bacteria, is taken up into a cell by phagocytosis; a process where a small part of the cell membrane encircles the food item, becomes detached and so enters the body of the cell. A food vacuole is formed, within which enzymes dissolve the food material. The released nutritional matter is further metabolised, releasing energy, which can be used to form new cell structures or for growth. Some of our own white blood cells are so-called phagocytes (eating-cells) that do exactly the same–they 'eat' foreign particles or bacteria that enter the bloodstream, and break them down by intracellular digestion.

Class Calcarea - calcareous sponges

The calcareous sponges (Lat. calcareous = of chalk, calcified) possess an inner skeleton of calcareous spicules formed of ***calcium carbonate****. These spicules will therefore dissolve in a weakly acidic solution, a property useful for classification. The spicules are either simple, or equipped with 3 - 4 rays, but are almost never joined to each other. Calcareous sponges are all marine and contain members of the ascon-, sycon- and leucon-types. The group is divided into two sub-classes. Most members are relatively small (5 - 120 mm), and found in shallow water down to approx. 40 metres. They are usually found in low-light conditions; under stones, cliffs, in cracks or between algae.*

Clathrina sp.

Clathrina *sp. forms characteristic white lattice-like structures in shallow water.*

Distribution: Entire Norwegian coast and large parts of the Atlantic and Pacific. There are 8 - 10 species in Norway. All coasts of the British Isles.
Description: Belonging to the ascon-type, with a simple body plan and up to 30 mm thick. Small individuals of around 1 mm high have the classic 'beaker'-shape. One variant forms a white network of tubes (chalk-lattice), but other colour variations also exist (grey, light pink or yellow). Despite the calcareous spicules, the sponge is soft. Oscula (exhalent openings) are visible as elevations, whereas the ostia (inhalent canals) are invisible to the naked eye. Note that some 'moss animals' (order **Bryozoa**) have a similar form, but these are harder.
Habitat: On hard substrata from the shore-line, usually not deeper than approx. 10 metres, but some specimens have been found as deep as 650 metres. Usually attached to the underside of stones, or kelp holdfasts. Can also attach to mollusc shells.
Biology: Asexual reproduction during summer. Fragmented parts either grow into new sponges, or may even form larvae asexually, which swim in the water before settling. These sponges die when the water temperature exceeds 19°C, and seldom live longer than one year.

A gammarid amphipod sitting on **Clathrina** *sp.*

Sycon quadrangulatum

Note the ring of spicules surrounding the single exhalent opening.

Distribution: Entire northeast Atlantic, Mediterranean. Very common.

Description: Characteristically vase-shaped. Only one large osculum (exhalent opening), often surrounded by a crown of spicules. Height up to 90 mm. White, silver-grey, brown-grey or brown. Solitary or in clusters.

Habitat: On the shore and in shallow water on hard substrates or brown algae.

Biology: Generally annual, with larvae released in springtime, after which the adults die.

Grantia compressa

This sponge can form dense aggregations in shady spots in shallowwater.

Distribution: Atlantic and Mediterranean, but most abundant from the English Channel, along the Norwegian coast north to the Arctic. A common species on all coasts of the British Isles.
Description: The form of the 'beaker' is characteristic, being more or less compressed, with constrictions at either end. Usual proportions are 10 - 20 mm high, 5 - 10 mm wide and 2.5 mm thick. Up to 100 mm high. Colour usually yellowish-white to pale brown-grey.
Habitat: The species is common in shallow water, often in clusters attached to kelp. Also found on rock surfaces and under overhangs in low light conditions.
Biology: Annual cycle, with overwintering individuals. Larvae are released in early spring, after which the adults die. The larvae settle on the bottom after a short planktonic stage. The earliest released larvae can reach sexual maturity during the summer, in which case they can reproduce before winter. However, most will first reproduce during the following spring.

SCIENTIFIC NAMES AND FAMOUS PEOPLE

The genus *Grantia* was named after the Scottish biologist Robert Edmond Grant (1793 - 1874), who amongst others worked with marine invertebrates. In 1825 - 27 he published several articles on sponges, including the family Grantiidae. He was one of Charles Darwin's teachers, and gave Darwin an excellent foundation in marine biology before he left on his world-famous five-year long excursion on the sailing vessel "Beagle". Grant (who was a keen proponent of Lamarckism) was most likely the first biologist to express the heretical belief that all life has a common origin.

Class Demospongiae

More than 90% of all the sponges currently known belong to the Demospongiae and all are of the leucon-type. Representatives are found at all depths and some inhabit fresh water. The skeleton consists of either siliceous spicules only, siliceous spicules and spongin or only spongin. Some species lack a skeleton altogether. Identification of species within this class is complicated and external characteristics are not sufficient.

Sponges often grow on top of other organisms. Here covering the shell of a brachiopod.

Geodia barretti

Distribution: Northern Atlantic, north coast of Svalbard, coast of Greenland and south to the coast of northern Spain.

Description: Large, solid and relatively inflexible. Juveniles more or less rounded, but becoming more irregular in shape as they grow. Colour light brown or grey. Height/diameter more than 30 cm, up to 50 cm. Weight up to 24 kg! Surface even and smooth, with the exception of a few marked depressions (oscula). The small pores (ostia) form a mesh-like structure. The upward-facing part of the sponge is generally darker in colour, because sinking particles tend to become embedded in its surface. A similar sponge is ***Geodia macandrewii***, which can be at least equally large, but which usually occurs in deeper water (> 150 m). Its shape is more symmetrical than ***G. bar-***

Geodia barretti *is often attached to vertical rock walls.*

The surface of **Geodia barretti** *is often dark in colour due to sinking particulate material.*

At a certain site in Lysefjorden in Norway, we found some 10-20 hairy individuals, which otherwise look like **Geodia barretti**. *However, we are unsure of this identification.*

retti, and oscula are usually aggregated on a flattened peak.

Habitat: Along the Norwegian coast, seldom found shallower than 50 m, but can in certain fjords be found as shallow as 20 m. Occurs down to approx. 1000 m. Attaches to rock, often cliff projections or mounds where water exchange is greatest.

Tethya norvegica (= *T. aurantium*) - Orange ball sponge

The orange ball sponge has a characteristic shape, allowing it to be identified without a microscope. The colour varies, but it is often orange.

Distribution: Northeastern Atlantic and Mediterranean. Entire Norwegian coast. Common on western and southern coasts of the British Isles.

Description: This sponge resembles an orange in shape, but is more yellow in colour. Diameter up to 70 mm. Uneven surface, often covered by a thin layer of sediment and with a small root-like attachment to the substratum. Can be confused with other members of the genus ***Suberites***, but these have an even, smooth surface. Another species, ***Tetilla cranium***, has a similar shape and surface, but is usually white.

Habitat: Shallow waters in low light conditions, such as under outcrops and large rocks and in cracks. Some finds have been made at 930 m. Individuals usually cluster together to form small colonies. Small individuals often are very abundant on kelp stipes.

Polymastia mammillaris

Individuals show large variations in shape, but the colour is usually in tones of yellow and green.

Distribution: Entire northern Atlantic, Arctic, Mediterranean, northern Pacific. Common all along the Norwegian coast.
Description: Forms a relatively thin, often somewhat irregular pillow/disc with a number of erect papillae, of which most have small pores (ostia). A few of the papillae are larger and have a terminal (positioned at the end) osculum. Shape, length and number of papillae greatly varies with age and condition/environment. As a result, two individuals are seldom alike. Diameter up to 130 mm. Excluding papillae 10 - 30 mm thick. Papillae often approx. 155 mm high and 1 - 4 mm thick. The surface of the base is rough, with smooth papillae. Colour usually pale grey, less commonly yellow or pink. The inside of the sponge is orange, as seen if the basal part is cut. Papillae characteristically paler (often transparent) than the base of the sponge. Can be confused with ***P. robusta*** (see below), but distinguished, amongst others, by the shape and size of the papillae. *P. robusta* is characteristically more pillow-shaped. The taxonomic status is currently uncertain, and different variants may belong to different species.
Habitat: Around the British Isles, the optimum depth interval appears to be 5 - 15 m. Found as deep as 2300 m. Usually located in the transition between soft sediments and bedrock, with the disc partially covered in sediment. This contrasts with *P. robusta*, which usually attaches to open rock surfaces.

DID YOU KNOW...

The scientific name ***Polymastia mammilaris*** is inspired by the latin and greek names for breast (Lat: Polys = many + mastos = breast; Gr. mamma, dimin. mammalia = breast).

This sponge is distinguished from its close relative P. robusta *by having slightly transparent papillae.*

Polymastia robusta (= P. boletiformis)

Distribution: Large parts of the northeastern Atlantic, east coast of Canada. In Europe, south to Portugal and north to Greenland and the Barents Sea. Recorded as far north as Troms, northern Norway. A common species on western and southern coasts of the British Isles.

Description: Colour ranging from orange, ochre, green to dark grey. Pillow-shaped, with a number of close-sitting papillae with ostia and an osculum. Diameter up to 120 mm, with the papillae reaching up to 50 mm in length and 10 mm thick at their bases. Both the bases and the papillae have smooth surfaces. The osculum is small, and is located at the end of some of the papillae. At the tips of the remaining papillae are very small pores. The papillae are not transparent, and have the same colour as the rest of the sponge. Can be confused with the ***P. mammilaris***, but is distinguishable by the form, length and colour of the papillae. The papillae in the latter species are partly transparent. Two other species within the genus *Polymastia* is found along the Norwegian coast. In southern Norway, these are found deeper than 60 m (***P. bursa***) and 120 m (***P. uberrima***), respectively. The taxonomic status is currently uncertain.

Habitat: Usually from 20 m or deeper, but can be found up to surface waters. Records exist down to approx. 2300 m. In southwest Norway, the optimal depth interval is between 20 - 40 m. Usually found on partially sediment-covered, relatively flat rock surfaces, but can also attach to shelled organisms.

Characteristic pillow-like shape with 'warts'. Relatively common from 20 m and below.

Suberites ficus - Hermit crab sponge

This sponge usually encloses the snail shell house of hermit crabs. It typically has a single, or only few relatively large oscula.

The colour often is yellow or orange.

Distribution: Widely distributed in the north-Atlantic. Recorded from the Oslofjord to Troms in northern Norway. A common species on western and southern coasts of the British Isles.

Description: Often covers the snail shells of hermit crabs, or the surface of certain scallops (***Chlamys* sp.**). Usually greyish in colour, but white, blue and some green-speckled variants exist, the latter in shallow water as a result of symbiosis (co-operative living) with green algae. The surface is smooth but the tongue sticks to it if you try to lick it. One or several large oscula (exhalent opening) usually located in small mounds. Individual sponges can be up to 200 mm in diameter. The species can be confused with ***Pseudosuberites carnosus***, whose surface feels more like velvet when licked. The latter contracts dramatically to about a quarter of its original size when brought to the surface, whereas *S. ficus* only slightly contracts, to about three quarters of its original size. Recent British studies of the genetic make-up of different specimens of *S. ficus* suggest that it comprises several species.

Habitat: Attached to rock, shipwrecks or partially or completely surrounding hermit crab shells. As both the crab and the sponge grow, the sponge gradually breaks down the shell such that in the end the hermit crab is covered only by the sponge. Also covers the auger shell (***Turritella communis***). Observed from 0 - 1300 m.

Cliona sp. - Boring sponge

Cliona *sp. 'drills' irregular channels in calcareous material, here in the bivalve* Arctica islandica.

Distribution: Atlantic and Mediterranean. Common on most coasts of the British Isles but apparently absent from the North Sea.
Description: Green-yellow when alive but turns almost black when brought on land. If one is lucky enough to see it in its natural element, it is a beautiful sight with papillae stretching out of the drilled holes in the shell. The papillae retract into the holes upon touching the surface of the shell. These boring sponges cannot be identified to species based on external characteristics alone, because there are several very similar species in the same genus (***Cliona celata, C. lobata, C. vastifica***). Under favourable conditions, *C. celata* can grow out of the shell it bores into, forming large characteristic 'crusts' up to 1 x 0.5 m, whereas the other two species are only found boring into calcareous shells.
Habitat: Boring sponges drill irregular holes into snail or bivalve shells, calcareous red algae or limestone. A common beach find is old perforated shells that have been attacked by this sponge. Such shells have several holes beside each other, with a diameter of 1 - 3 mm. However, it is uncommon to find the actual sponge. We have often come across boring sponges in the icelandic syprine (***Arctica islandica***), but it is also common on oysters (***Ostrea edulis***) and the shell of the common whelk (***Buccinum undatum***). *C. celata* appears to prefer thicker shells than the other two species. Depth range approx. 3 - 600 m.
Biology: The yellow-green colour of shallow-living individuals is due to symbiotic green algae that photosynthesise in the sponge's outer tissues, providing a food supplement.

Halichondria panicea - Breadcrumb sponge

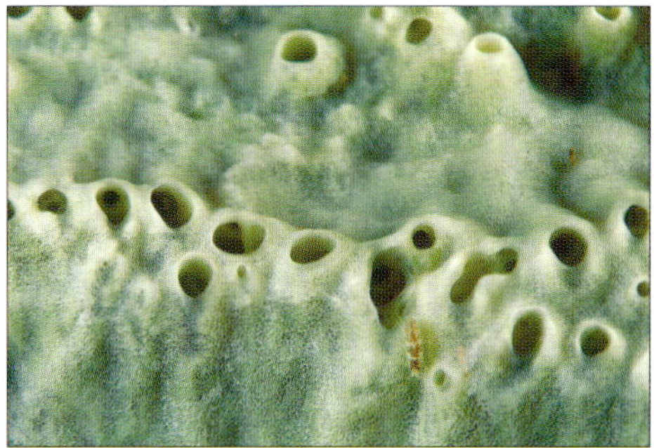

Halicondria panicea *sometimes spreads out over large areas, up to 1 m².*

Distribution: Found in most seas. A common species on all coasts of the British Isles and Norway.

Description: In common with many others, sponges within this genus form a carpet-like layer on hard substrata. Very variable in colour and shape, but usually yellow, green or orange. In strong light conditions, it is usually green due to symbiotic algal growth, whereas individuals in low light areas are usually creamy-yellow. Oscula often forming prominent craters. This sponge fragments easily and gives off a strong smell when taken to the surface. This distinguishes it from ***H. bowerbanki*** (named after the British J. S. Bowerbank who, in the mid-1800s described not less than 40 of the near 230 sponges found in our waters). Usually pale yellow - yellow-white, and often branches bush-like from the substratum. Species determination from external characteristics is difficult. Seven species within this genus are recorded along the Norwegian coast, of which most live in relatively deep waters.

Habitat: Found from the intertidal and upper sub-littoral zone. Very common in exposed areas where large waves give good water exchange. In such conditions, this sponge can form carpets on rocky walls up to several metres long. Often attached to kelp stipes, especially tangle or cuvie (***Laminaria hyperborea***).

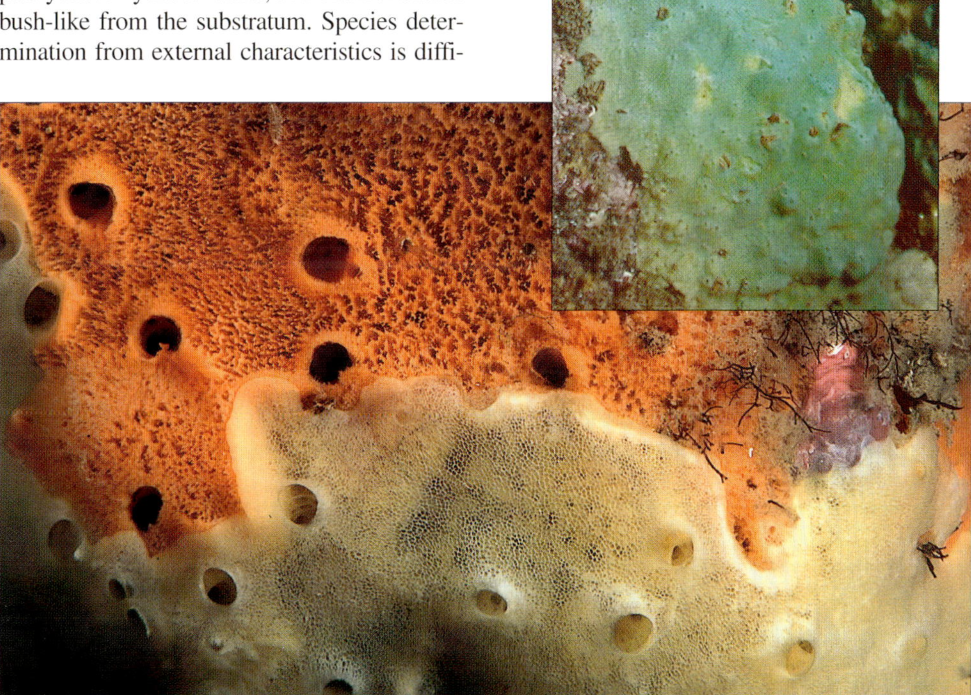

Here a yellow Halichondra *meets a myxillid sponge (orange).*

Phakellia ventilabrum - Chalice sponge

The chalice sponge usually forms thin fan-like structures with a characteristic pattern of 'veins'. On this sponge is the sea lily **Antedon petasus** *and two sabellid fanworms at the base of the sponge.*

Distribution: Arctic, northeastern Atlantic coast and northern Pacific. Recorded all along the Norwegian coast except east of the North Cape. Some localities on the western and northern coasts of Ireland and also the west coast of Scotland.

Description: Very characteristic exterior structure making it easy to identify. Up to 500 mm high, 5 mm thick. Colour pale yellow. Usually forms large thin fan-like structures, with elevated nerve-like lines radiating out from the short stem. The fan may also be more or less curved to form a funnel. These can be distinguished from the funnel sponge, ***Axinella infundibuliformis***, because the chalice sponge has much thinner edges (see below). Another species, ***Phakellia robusta***, also forms a fan, but is more robust and usually occurs deeper than 80 m. However, we have observed ***P. robusta*** at 30 m depth in the Trondheimsfjord.

Habitat: Usually positioned with the fan at 90 degrees to the main current direction, to optimise filtering. Usually found deeper than 20 m, in current-rich areas.

Axinella infundibuliformis - Funnel sponge

Distribution: West and southern coasts of the British Isles, Atlantic coast of France, entire Norwegian coast.

Description: Relatively easy to identify from external features. Usually forms a funnel-like construction up to 250 mm diameter with a short stem. May also resemble a fan. Distinguished from chalice sponges (***Phakellia*** sp.), which also can be funnel-shaped, by the thick rounded edges of the former. Colour off-white, to yellow-brownish or buff.

Its very characteristic shape makes the funnel sponge one of the few sponges that can be correctly identified from a photograph.

Habitat: Often attached to rock, usually deeper than 20 m. Recorded down to 630 m.

Here are two sponges, which confusingly resemble the previously described species **Phakellia ventilabrum** *(largest) and* **Axinella infundibuliformis** *(smallest). These were photographed at the entrance to Egersund at around 50 m depth. On the bottom is a member of the starfish genus* **Henricia** *and on the nearest sponge a common starfish* **Asterias rubens.**

Mycale lingua

Distribution: Widely distributed from the French Atlantic coast and along the entire Norwegian coast. West coast of Scotland.

Description: Characteristic surface broken up into deep cracks. Exhalent openings large. Much variation in shape, but mostly oval without branches. Most of the individuals we have observed are relatively large, approx. 30 - 40 cm length/height. Colour usually white-yellow to grey, occasionally orange. Skeleton of both spicules and spongin fibres. Very elastic.

A horizontal individual from 40 m depth.

This sponge often occurs together with the stone coral **Lophelia pertusa.** *From 55 m in Trondheimsfjorden, Norway.*

Habitat: Common on ***Lophelia*** reefs, in fjords and near the coast, less common further out on the continental shelf and slope. Specimens on reefs are usually 'upright' such that they are considerably higher than wide. However, individuals have also been observed 'lying down'. From approx. 45 m - 2500 m depth.

Biology: Individuals of this species often are inhabited by a range of small crustaceans, especially amphipods that live in the sponge's pore system.

Myxilla fimbriata/incrustans

Distribution: From the Arctic south to the British Isles. Recorded sporadically along the Norwegian coast from Rogaland north to Troms. The two species are often found together and are difficult to separate.
Description: Bright orange to brown-red surface, yellowish inside. Forms a well-defined carpet usually 6 - 7 mm, but up to 30 mm thick, with a diameter of approx. 20 -30 mm. The oscula (exhalent openings) are irregularly distributed and flush with the sponge surface.
Habitat: On hard substrates, shallow water up to the tidal zone, but usually deeper than 35 m.

The sponge in this picture has the brittle star **Ophiotrix fragilis** *nestling on it. Brittle stars often exploit the extra water currents created by sponges. Uppermost is the sea squirt* **Ascidia mentula.**

Esperiopsis fucorum (= *Amphilectus fucorum*)

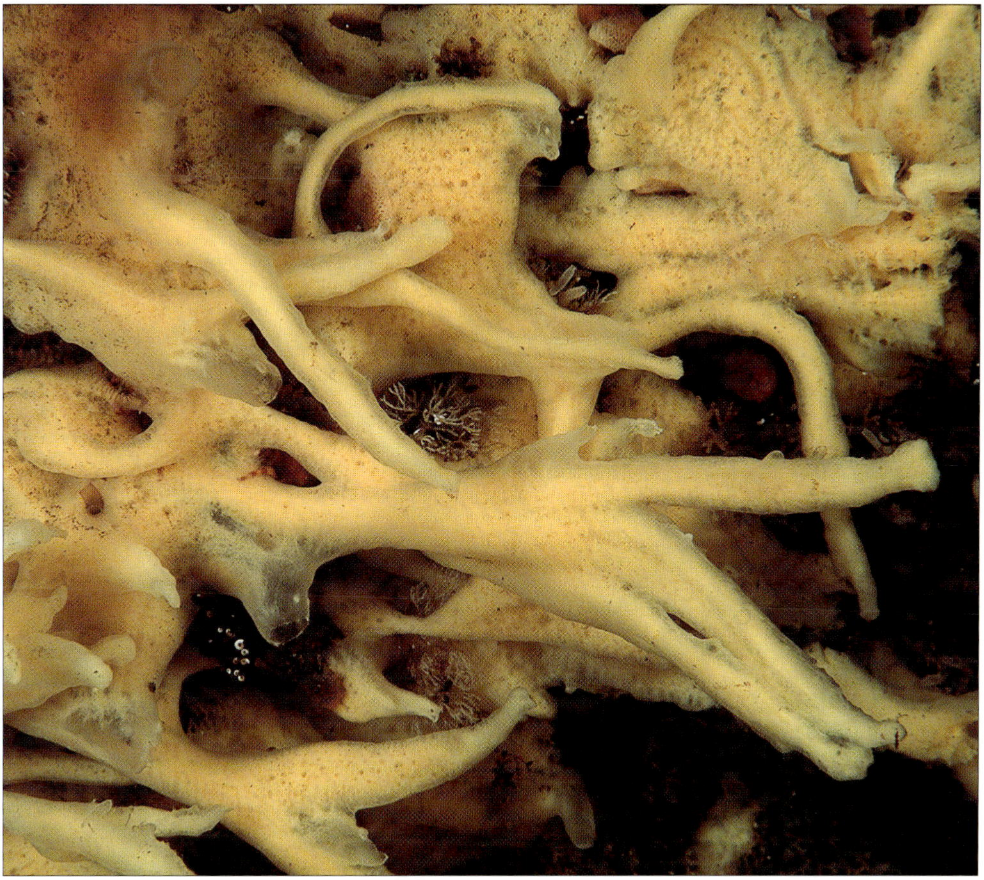

This close-up shows hair-like growths characteristic for the Esperiopsis fucorum.

Distribution: Northeast Atlantic from France to Norway. Recorded north to Troms, northern Norway. Common and widespread around the British Isles.

Description: Extremely variable in shape. Forms crust-like layers, 'pillows' and tubes. With or without prominent 'hair' around the exhalent openings or oscula. Colour usually orange, yellow or brown-grey. Individuals living in deep waters sometimes occur without pigment, Soft and very porous. Orange individuals can lose colour after mechanical pressure. If taken to the surface, they give off an intense pungent smell.

Habitat: Shallow water and down to approx. 200 metres, preferentially in areas of good water exchange. Also grows on kelp stalks (stipes), often without the otherwise prominent 'hair', or on stalks of the hydroid ***Tubularia*** sp.

Haliclona urceolus

Distribution: A common species on all western coasts of the British Isles and most of the Norwegian coast. Some records from the North Sea coasts.

Description: White-grey to grey-yellow. Tubular, either as a single tube or joined with others, each with a terminal osculum. The sponge is soft and flexible and the individual tubes are attached to the substratum by a hard but flexible stalk. The actual tubes make up around half the height of the sponge. Up to 150 mm long. Extremely slimy if brought ashore.

Habitat: Grows on rock, shipwrecks, and fragments of shells or other hard substrates. From approx. 3 - 1000 m depth.

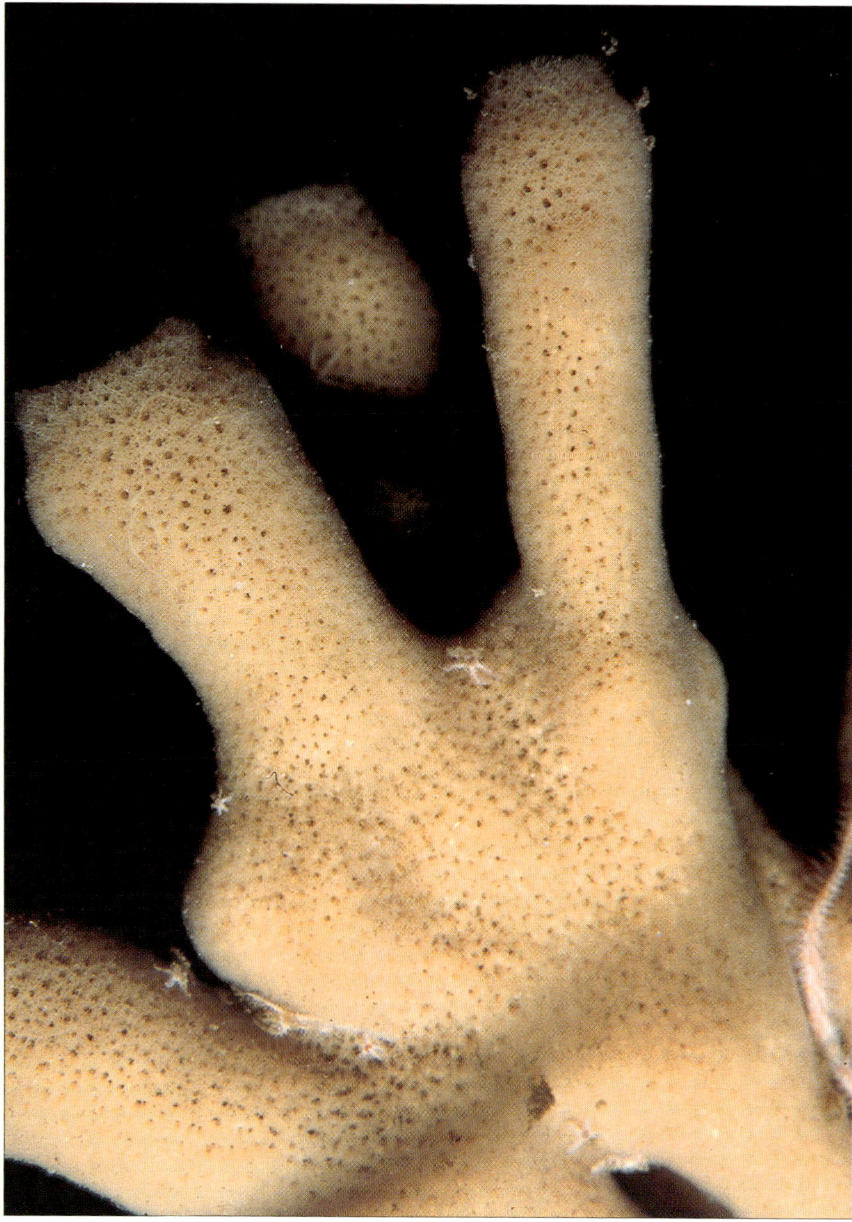

This sponge forms tube-like extensions, each of which bears a terminal osculum, or exhalent opening, as well as numerous ostia, or water canals.

Antho dichotoma

Distribution: North Atlantic from sub-Arctic areas south to Ireland. Entire Norwegian coast north to Lofoten, but also recorded from Kola.

Description: This species belongs to a large genus, i.e. containing many species. Identification generally depends on microscopic examination of the spicules. *Antho dichotoma* forms characteristic tree-like sponges up to 35 cm high, with branches around 3 - 5 mm in diameter. The grey-yellow branches have a hairy appearance due to the protruding spicules.

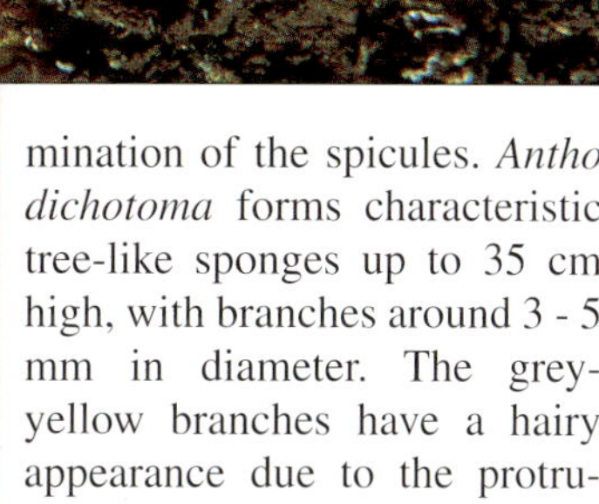

Habitat: Usually at depths greater than 50 metres, on hard substrates, usually in areas with some current. The literature cites the depth distribution as between 70 - 800 metres, but it also has been observed at 25 metres.

We have observed **Antho dichoma** *at 35 m depth, but the really large colonies occur from 50 m and below.*

The sponge **Ciocalypta penicillus**, *which has a southerly distribution in Norway. It is relatively common around south-west Norway and occurs at scattered localities on the western coasts of the British Isles.*

This spectacular large unidentified sponge was photographed at 52 metres depth in Trondheimsfjorden. It grows out from coral reefs and the deep-water coral **Lophelia pertusa.** *Also visible on the photograph is the remains of a hermit crab that has left both its former skeleton and shell.*

CNIDARIA

Phylum Cnidaria

Cnidaria (Gr. knide = nettle + Lat. -aria = similar) is a large and diverse group, including jellyfish, sea anemones and corals, but also lesser-known taxa, such as hydroids, sea pens, siphonophores, and zoanthids. The animals have a simple body construction, consisting of three layers; an inner layer lining the gut, an outer layer of epidermis, and between these the mesoglea. Cnidarians exhibit radial symmetry, such that many similar sections are symmetrically arranged around the central axis through the gut. There are about 9000 living species within the group.

These beautiful cnidarians are hydroids within the genus **Corymorpha**. *Note the sea slug sitting on the hydroids (see also enlargement).*

Cnidarians are more complex than sponges. They have different and specialised cells that form tissues, many can swim, they react to outside stimuli, and they can capture and swallow prey. They also have a specialised cell type, **cnidae**, which include the poisonous **nematocysts**. However, they lack respiratory organs, excretory organs, and anus. There is but one single opening that is used both for eating and defecating. There is no brain or well developed nervous system, and the nerve cells are dispersed as an irregular net in the animals. One feature of special interest in medusae and jellyfishes is the presence of statocysts, or balance organs. These consist of a small fluid-filled chamber filled with sensory cilia and some small calcareous grains, and this organ allows the animals to

MANY MILES LONG

Cnidarians vary in size from microscopic, as is the case with some polyps, to several metres in diameter and with tentacles up to 50 metres, as in some jellyfish. Corals can produce colonies that are several miles long.

Some cnidarians live their whole life as plankton. Here the colonial **Apolemia uvaria**.

orientate themselves in the water column. Colonial cnidarians often have different polyps that are specialised for various purposes, such that some are involved in reproduction, some eat, and some protect the colony. The reproductive polyps bud off small, transparent medusae, which usually are planktonic and contain the sexual products (gametes). The fertilised eggs develop into free-swimming larvae, usually in form of planula-larvae, which are characterised by being covered by cilia. Following a short planktonic phase, the larvae settle and produce a polyp, which then multiplies asexually and founds a new colony. Some hydrozoans lack polypoid stages, and the planula instead develops directly into a medusa. In others there is no medusa stage, and the gametes are formed directly from the polyps. Two features have clearly been important in the evolution of the group. One is the ability of the polyps to reproduce asexually, and the other is the alteration between a benthic polypoid stage and a pelagic medusa. Most cnidarians are marine, although some taxa also occur in fresh water. The vast majority are predators, although a number of corals live in symbiosis with photosynthesising unicellular algae, zooxanthellae.

Few organisms eat cnidarians. However, various sea-slugs have overcome the defence mechanism of cnidaria. Here, the sea slug **Tritonia hombergi** *is feeding on the dead man's finger coral* **Alcyonium digitatum.**

Cnidaria is currently subdivided into Medusozoa, which includes hydrozoans (Hydrozoa) and jellyfish (Scyphozoa), and Anthozoa, which include octocorals (Octocorallia) and zoantharians (Hexacorallia).

Superclass Hydrozoa - Hydroids

*Hydrozoans (Gr. hydro = water / hydra = myth. many-headed water serpent + Gr. zoon = animal) include hydroids (**Hydroida**), which form feather- or bush-like colonies, and the colonial medusae (**Siphonophora**). The colonial medusae, or siphonophorans, are pelagic, and some float on the surface of the sea. The colonies actually consist of a combination of polypoid and medusoid stages. They are provided with a high diversity of specialised polyp cells, which are connected to a gas-filled medusa. They tend to be more common in warmer waters.*

Above: *Unidentified hydromedusan.*

Left: *This hydromedusan is called* **Tima bairdii.** *It has very long tentacles, and a robust, funnel-shaped mouth opening. Annual.*

Below: **Aequorea vitrina**

Class Hydroida - hydroids

Most hydroids have colonial, benthic polyps, alternating with free-swimming ***medusae*** *(see previous page). The medusa stage is the sexual generation, and produces gametes which unite and develop into the planktonic, ciliated* ***planula larvae****. These larvae will settle on the bottom and found new colonies, or just grow up into a single adult specimen in the case of non-colonial polyps. In some species, however, a free-swimming medusa is lacking. In these, the polyps instead have so-called* ***gonophores****, which produce the gametes. Sea spiders (****Pycnogonida****) and nudibranchs (****Nudibranchia****) are the most common predators on hydroids. Identification of hydroids from photographs can be difficult; many are small and require microscopical studies, and this may also be the case for similar-looking larger species, which can be separated only by study of finer details. Most hydroids along our coasts belong either to* ***Thecata****, with naked polyps, or to* ***Athecata****, where the polyps are protected by a hard extension from the exoskeleton of the stalk, called the theca. Other hydroid groups can also be found in northern European waters, including the* ***Limnomedusae*** *(p 82).*

LOOKING LIKE PLANTS

In spite of this, all hydroids are predators. They use a toxin to immobilise their prey, and transport it to the mouth of the feeding polyps.

This characteristic hydroid is attached to an algal frond. Unidentified.

Subclass Athecata

Members of Athecata (Gr. a- = not = Gr. thece = case/covering) are identified by their naked, unprotected polyps. This is due to the situation that the ***perisarc****, which provides external support and protection for the colony, ends at the base of the polyps, and does not cover the distal-most part with the tentacles (the hydranth).*

Corymorpha nutans

Hydroids within this group are reminiscent of sea anemones. They have a very thin, almost transparent stalk.

Distribution: Widely distributed in north Atlantic coastal waters; in Europe it occurs from the Mediterranean to northern Norway. Common throughout Britain and Ireland.

Description: This large and solitary hydroid, which from its size can be mistaken for a sea anemone, attains a height of 100 mm. The stalk is cylindrical and stouter than in most hydroids. The polyp is white to pink. Around the mouth a ring of short tentacles, and below the mouth there is a second ring with much longer and thicker tentacles. Between these rings are the so-called gonophores, which produce the medusae; they are usually pink in colour. Three other and less common species of *Corymorpha* occur in the region.

Habitat: Unusual for hydroids, which mostly occur on hard substrata, this species lives on bottoms with sand and gravel. Down to approx. 100 metres.

Tubularia larynx

Distribution: Present along most northeast Atlantic coasts; in Europe from the Mediterranean to Iceland. Common all around the British Isles.

Description: It is not surprising that these organisms originally were classified as plants by early taxonomists. The colony is bush-like, with comparatively large, naked polyps, reaching a height of about 45 mm. The stalk is branching, the polyps are pink and have two rings of tentacles. The outer ring carries approx. 20 tentacles, whereas the inner ring, which surrounds the mouth opening, carries 14 - 20 shorter tentacles. It is similar to ***Tubularia indivisa***, but can be separated by the tentacle number, the shape of the colony, and the colour of the stalk. In ***T. larynx*** the stalk is pale yellow, whereas in ***T. indivisa*** it is orange-yellow.

Habitat: Usually attached to stones and rock, but can also occur on other hydroids and on algae. From the intertidal down to 100 metres.

This hydroid is known as "shore flower" in some languages, because of the way it forms colonies in shallow water that look like bunches of flowers. It has relatively large, naked polyps with numerous outer, and short inner tentacles.

Tubularia indivisa
- Oaten pipes hydroid

Long thin outer tentacles and an inner ring of short mouth tentacles.

Distribution: Present along most north-east Atlantic coasts. Not in the Mediterranean. Common on all British Isles.
Description: Similar to ***T. larynx***, but with straighter and less branching stalks, which connect basally on the colony. Fewer polyps than the previous species. Furthermore, the branches are stiff, and usually orange-yellow. Polyps large, light pink, with 20-30 tentacles in the outer ring, and approx. 40 in the inner. Stalks usually 30-150 mm, but can attain a length of 400 mm.
Habitat: Attached to hard substrata. Often on shady places, growing upside down.

Biology: Lacks free-swimming medusae, but has sessile medusae on the polyps that produce gametes.

Colonies have orange-coloured, upright stems, distinguishing them from the previous species.

Clava multicornis - Club-headed hydroid

This hydroid is very common on brown algae in shallow water along many European coastlines.

Distribution: Widely distributed along European coasts. Found around the north, west and southern coasts of the British Isles.
Description: Creeping colonies, with erect, orange polyps. Each polyp with approx. 20 slender, white tentacles, dispersed over the outer part. Collections of grape-like gonophores situated below the tentacles. The outer part of the polyp, the hydranth, can be extended up to 10 - 15 mm, or withdrawn into a small lump. Stalks and polyps are light pink, whereas the mature gonophores are blue.
Habitat: Intertidal down to a few metres depth. Usually attached to knobbed wrack (*Ascophyllum nodosum*), serrated wrack (*Fucus serratus)*, bladder wrack (*Fucus vesiculosus*), or mussel shells.
Biology: Planula larvae are released from the gonophores in late spring and summer. Once settled, these will grow up into primary polyps, which, by means of budding, will form a new colony.

Hydractinia echinata

The hermit crab **Pagurus bernhardus**, *protected by its snail shell house and the hydroid* **Hydractinia echinata.**

CRAWLING LARVAE

An unusual feature of this species is its lack of planktonic larvae. Larvae are released from specialised polyps (gonozoids) in the colony, and can distinguish shells occupied by hermit crabs which moves fast from those of living snails. The larvae select the shells occupied by hermit crabs.

Distribution: From Morocco to the Arctic in the northeast Atlantic. Not in the Mediterranean. Found all around the British Isles.
Description: Occurring as approx. 3 mm thick, often light pink crusts on shells occupied by hermit crabs. Polyps emerging from this crust attain a length of 13 mm. There are several kinds of polyps. The longest, which are the feeding ones, have two rings with eight longer and eight shorter tentacles, with the former situated outside the latter. It can be confused with another species, ***Podocoryne carnea***, which, among other places, also occurs on shells containing hermit crabs. It can be separated by the single tentacular ring, which has tentacles that alternate in length.
Habitat: Characteristically occurring with hermit crabs, although not always. It occurs from the surface down to approx. 30 metres depth.
Biology: Usually lives on shells containing hermit crabs, and represents a classical textbook example of symbiotic relationships. The hydroid benefits from access to a substratum, and from the food spill of the hermit crab. The benefits for the crab are less obvious, but it may be protected by the defence-specialised polyps (dactylozoids), which can be activated by means of chemical stimuli from the crab.

Sarsia tubulosa

Distribution: In European waters from the west coast of France in the south to the Barents Sea in the north. Circumpolar. Around the British Isles.

Description: Colonies usually creeping, richly branched, with individual polyps arising approx. 10 mm from the base. Some, less branching, colonies can occasionally reach a height of 30 mm. The stalks and the polyps are usually orange to light brownish, and the club-shaped polyps carry up to 20 (usually 12-16) dispersed, rather long tentacles that terminate in a knob. Two-three gonophores with short stalks are situated below the tentacle region in mature specimens.

Habitat: From the intertidal down to a few metres depth. Medusae are common in spring, down to approx. 60 metres depth.

Biology: This colonial hydroid produces only few gonophores. These release medusae during the spring, characterised by being bell-shaped and provided with four tentacles.

This hydroid usually occurs in exposed parts of the coast.

The colony usually spreads out over a relatively large area.

Eudendrium rameum

Distribution: Cosmopolitan, from tropical to Arctic waters. Common around the British Isles.

Description: Large tree-shaped colonies, up to 250 mm high. Usually strongly and irregularly branching, with a thick stem that decreases evenly in diameter towards the distal parts. Stalk basally usually brown, whereas distal branches are lighter in colour. The polyps are pink, and provided with 20 - 40 tentacles in a ring. They are short and have a spherical mouth region when closed, but trumpet-shaped when open–a unique feature for this genus. There are a number of species within the genus, but only this one has tree-shaped colonies.

We have observed the commensal amphipod Stenopleustes latipes *several times on this hydroid.*

Habitat: The colonies are found on various hard substrata, from about 5 - 100 metres depth. We have often found it on ***Lophelia pertusa***. All observed colonies harboured the amphipod ***Stenopleustes latipes***.

Such hydroid colonies also occur on coral reefs of **Lophelia pertusa**. *This photograph is taken in Trondheimsfjord, Norway.*

Leuckartiara octona

Distribution: Common in European waters, especially in the northern parts up to Iceland. Occurs also in the Mediterranean. Common on British and Irish coasts.

Description: The medusa has a classic bell-shape, approx. 20 mm high and less in width. The aboral, upper part is transparent and slightly pointed, and can be difficult to observe from photographs. When extended, the tentacles are rather long, whereas they are spiralling in their contracted state. They are usually 16 in number, although large specimens may have up to 28. Newly hatched medusae have only a few tentacles, and add them in a pair-wise fashion. A characteristic feature is the appearance of small, new tentacles between the fully-grown ones, which separates it from the less common *L. nobilis and L. breviconis.*

Habitat: Occurs both in surface and deeper waters, usually during the autumn.

Biology: A special feature of this, and of a number of other medusae, is the association with a number of small crustaceans. Hyperiid amphipods are common, sitting on the outer, aboral side. This species belongs to a group of hydromedusae with a short, benthic polypoid stage with asexual reproduction.

Two individuals on a collision course.

Subclass Thecata

Polyps are partly or fully covered by the theca, formed from the perisarc

Halecium halecinum

Distribution: Widely distributed in European waters, rare in the Arctic. Very common around the British Isles.
Description: Stout basal stem and main branches. Basal part often composed of a number of joined branches. The branches are straight and grow in a single plane (the British 18th century zoologist Ellis called this hydroid the "herring-bone coral"). The branches are alternating along the stem, separated by equal distances. The polyps cannot be entirely withdrawn within the thecae. The colonies grow up to a height of 250 mm. The genus includes a number of species, and identification usually requires microscopy.
Habitat: Usually at 20 - 100 metres depth, but may occur from a few down to 1300 metres depth. Hard substrata. Very common.

This hydroid forms characteristic tree-shaped colonies. There are polyps on both sides of the branches.

Colonies also can have branching of the main stem.

Obelia geniculata

Distribution. Recorded from virtually all oceans.
Description: The species forms characteristic zigzag-shaped colonies on brown algae, up to approx. 50 mm high. The polyps alternate on the stem, and each is covered by a large theca. May be confused with ***Laomedea flexuosa***, but has a less pronounced zigzag-shape of the stalks, and lacks extensions between the polyps on one side of the stalk.
Habitat: Very common on cuvie (*Laminaria hyperborea*), oar weed (*Laminaria digitata*) and serrated wrack (*Fucus serratus*). Intertidal down to a few metres depth.
Biology: Releases medusae during summer.

The stem of this hydroid is characteristically zigzag shaped.

This hydroid usually grows on seaweed or kelp laminae.

Kirchenpaueria pinnata

This hydroid forms erect, thin colonies with the individual polyps on the upper side of the side branches.

Distribution: Widely distributed in European waters, from the Mediterranean to Iceland and northern Norway. Common around the British Isles.

Description: Fine, graceful colony, usually about 20 - 30 mm high, but may reach 100 mm. Colony simple feather-shaped, with a non-branching and rather thin main stem. Each articulation of the stem carries one or more branches, with the polyps and their bowl-shaped thecae located on the upper side. The thecae are as wide as high. The branches are alternating along the main stem, which is straight or weakly zigzag-shaped.

Habitat: Growing on stones or algae in shallow water; rarely deeper than 100 metres.

Nemertesia ramosa

Distribution: Northern Europe to Iceland, Barents Sea and west Greenland; north-west Africa. Common around the British Isles.
Description: Similar to ***N. antennina*** (not illustrated), from which it differs by having a branching stem. The yellow to orange stem consists of a segmented tube, and is irregularly branching. Segmentation less distinct in basal parts of the colony. Up to 250 mm high.
Habitat: Attached to stones or shells, or anchored with a tassel-shaped stolon in sand and gravel. Usually at depths exceeding 20 metres.

The robust stems are characteristic for hydroid colonies in this group.

The colonies are relatively sturdy and heavily branched, usually yellow-orange in colour.

Subclass Limnomedusae

As implied by the scientific name (Gr. limne = lake), most species within this group occur in fresh water. There are a few exceptions, including the hydromedusa described below.

Gonionemus murbachii

A beautiful hydromedusan. The species is recognised by the characteristic bend near the tip of each tentacle.

Distribution: Both sides of the North Atlantic and the Mediterranean.
Description: Medusa with shape like a half sphere, up to 20 mm in diameter. Tentacles 60-80, thin and relatively long. The outermost parts of the tentacles are characteristically bent and slightly thickened and pigmented. These distal parts serve in attaching to algae and other objects. The nematocysts are visible as distinct bands all along the tentacles. Gonads are present as orange to brown, strongly folded bands, extending from the central part of the medusa to the edge. Based on these unique features, it cannot be confused with other species.
Habitat: In shallow water, either pelagic or attached with the tentacles to algae and other objects.
Biology: Under calm conditions it will swim with rhythmic contractions towards the surface. The contractions stop upon contact with the surface, the animal turns upside down and sinks passively towards the bottom while spreading the tentacles. Any animal of suitable size that comes in its way will be caught and eaten. By means of small suckers on the

tentacles, it will attach to any algae that it may encounter, and, still upside down, remain there some time waiting for prey.

The four orange parts visible in this hydroid are the gonads.

Class Siphonophora

(Gr. siphon = tube = Gr. phoreus = carrier) These colonial medusae are all planktonic. The colonies are complex and consist of highly polymorphic 'individuals', all with different specialisations. There actually is a mixture of medusoid and polypoid individuals, which all develop by budding from a larval polyp, producing a medusoid gas-filled sac (serving as a float in many species), swimming bells, feeding polyps, and sensory polyps. There are also attached reproductive medusae, and these usually are of both sexes, such that the colony as a whole is hermaphroditic. Many siphonophorans have very bright colours and striking morphologies, and are among the most beautiful sea-living creatures. Some of them, on the other hand, also have large number of highly potent nematocysts, and an encounter can be very painful for humans and occasionally even lethal.

Apolemia uvaria. *This colonial jellyfish 'invaded' large parts of the Norwegian coast in autumn 1997, causing mortality of large quantities of farmed salmon.*

Superclass Scyphozoa - Jellyfish

The benthic, sessile stage in scyphozoans (Gr. scyphos = cup + Gr. zoon = animal) is either completely absent, or present only as very small polypoids. There are approx. 200 species, all of which are marine. Although often referred to as primitive, a series of features in their biology and morphology are highly specialised. A number of species are provided both with 'eyes' and statocysts. Most are pelagic, with the notable exception of the stauromedusae. The medusae are usually large, and the mouth has four extensions called oral arms. The gelatinous mesoglea, situated between the outer ectoderm and the inner endoderm, actually makes up the most part of the wet weight of the animal. The mouth opens into the gut that forms a distinct cavity, and is where the extracellular digestion takes place. Undigested food remains are expelled through the mouth. There are four orders of jellyfish occurring in European waters. Among these the Stauromedusae are sessile. Of the order Coronatae there are eight species in the North Atlantic, but these occur mainly in deep water, and are not included here. The dominant group is Semaeostomaeida, to which order most of our well-known jellyfish belong. Of the last order–Rhizostomae–we have one southerly species in the area.

EXTRACELLULAR DIGESTION

Jellyfish have what biologists call extracellular digestion. Enzymes are produced in the gut cavity and start the digestion of the proteins. The prey, say a small fish, is transported through the mouth opening into the gut. Enzymatic gland cells, situated in the gut wall, initialise the decomposition. The nutrients are then taken up intracellularly, and subsequently distributed to other cells. Undigested food remains are expelled through the mouth.

This red stinging jellyfish **Cyanea capillata**, *together with Moon's jellyfish, is probably our most common large jellyfish. Large individuals are truly an impressive sight.*

Order Stauromedusae

Small, sessile medusae (Gr. stauros = cross + L. myth. medusa = Phorky's daughter, who charmed Poseidon with her golden hair, and thereafter gave birth to, among others, Pegasus) that develop directly from a benthic planula-larvae. A polypoid stage is completely absent. Body goblet shaped with a moveable peduncle ending in an adhesive disc. Most species have eight arms. The short, thin tentacles are numerous, and are situated in clusters at the tip of the arms. The animals are usually attached hanging upside-down, commonly on algae or sea-grass in shallow water. Most species have a northernly distribution.

DID YOU KNOW...

Even though stauromedusae are sessile, they actually have three ways of locomotion: 1) creeping across the surface by motion of the adhesive disc; 2) walking upside down by with the sticky tentacles; and 3) moving like a leech by bending the arms down to the surface, letting go of the adhesive disc, attaching the tentacles, bending the stalk and re-attaching the disc, etc.

Haliclystus salpinx

Distribution: Northern parts of the North Atlantic. Less common. Not yet recorded in British waters.

Description: Length of stalk about half the length of whole animal. Height and width up to approx. 25 mm. Colour usually reddish brown, with a tint of violet. The eight arms are evenly distributed, and each arm carries approx. 60-70 (up to -100) tentacles. In between each pair of arms there is a so-called anchor, which has a large, trumpet-shaped adhesive disc. Another congener, ***H. auricola***, is of the same size, but can be separated by the anchors, which are smaller and kidney-shaped.

Photographed near Kristiansund in southwest Norway. Note the large 'anchors' between the arms.

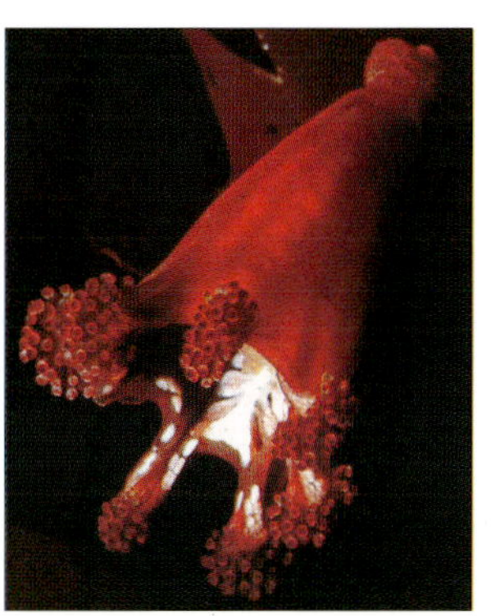

Craterolophus convolulus. *This jellyfish also has relatively short arms and stalk, but lacks the 'anchors' between the arms.*

Habitat: Similar to other stauromedusae, mainly on algae and eel-grass (*Zostera*) in shallow waters.

Lucernaria quadricornis

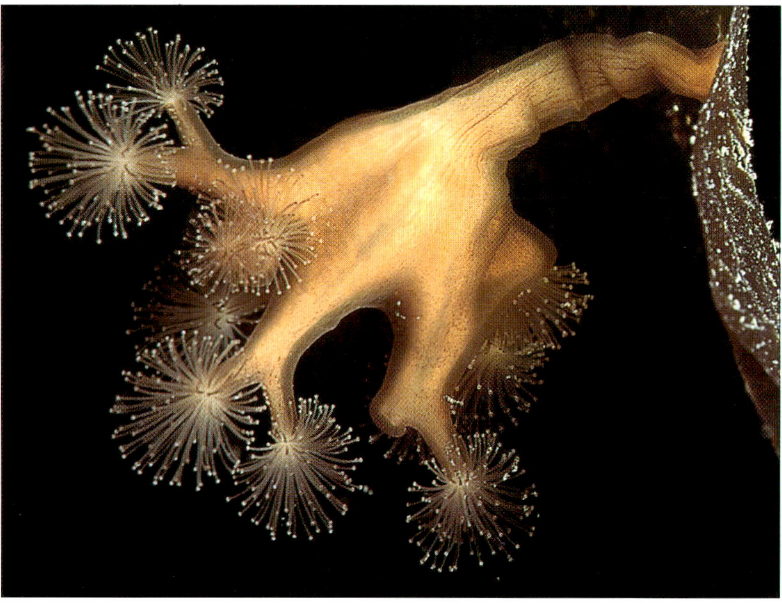

At the ends of the arms are clusters of thin tentacles, with knob-shaped tips.

Distribution: Northern parts of the British Isles north to Spitsbergen and Greenland, and the northwestern parts of the Atlantic. Common.
Description: Our only stauromedusae with pairwise arranged arms. Comparatively large, up to 70 mm high. The stalk is slightly longer than the cup and the eight arms. Each arm carries approx. 120 short and thin tentacles. Distal knob-shaped part of tentacles with large amount of nematocysts. Colour highly variable, including green, brown, brownish red, and grey. The genus lacks the anchors present in ***Haliclystus***.
Habitat: Usually occurring on various algae in shallow water. Recorded from 83 metres depth on Greenland.
Biology: Both asexual and sexual reproduction, with high capacity for regeneration. The animals appear to have a life cycle of one year. The favourite food consists of small snails (*Lacuna* and *Rissoa*) living on algae, and various amphipods.

The stalk often is longer than the body and the arms.

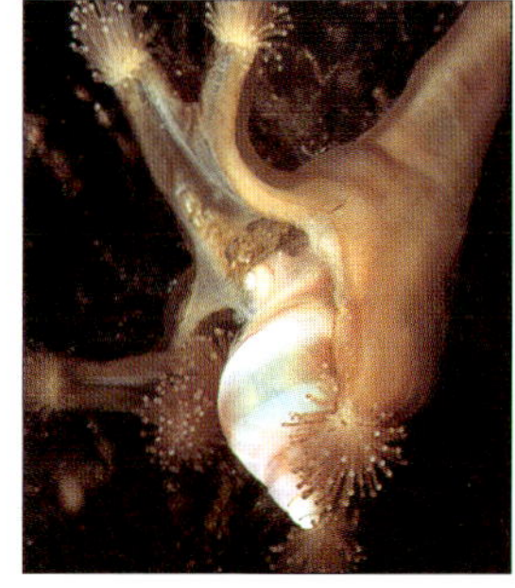

The prey organisms are huge in relation to the size of the jellyfish. Here pictured with its favourite meal, the small snail **Lacuna vincta.**

Order Semaeostomaeida

Within this group (Gr. semaia = banner + Gr. stoma = mouth) we find the largest and the most common jellyfish in the area. Most have a size of 50 - 500 mm in diameter, although some species may reach up to two metres. They have typical medusoid characteristics, and drift more or less passively as plankton in the free water, although vertical movements through the water are achieved through rhythmic contractions of the bell. Some feed on small particles and plankton, whereas others catch larger prey, such as fish. There are eight species in the north-east Atlantic, and we here include the most common five from coastal areas.

Periphylla periphylla *is a deep-water jellyfish, which in wintertime may occur near the surface. It occurs in all oceans except in the Arctic. Egg fertilisation occurs in deep waters and the development takes several months– this jellyfish is thought to reach an age of up to 30 years. It has 12 tentacles, and the bell can reach a height of 20 cm.*

Cyanea capillata - Lion's mane jellyfish

The tentacles can stretch to over 30 m in length.

Distribution: Circumpolar in the northern hemisphere, from Bay of Biscay in the south to the Arctic in the north. Very common.
Description: May reach up to 2 metres in diameter, with tentacles up to 30 metres long in Arctic waters, and is the largest existing jellyfish. Colour yellow to yellow-brown.
Habitat: Occurring in surface waters.
Biology: Life span is about one year. Small specimens appear in April-May, and the growth rate is usually high, depending on available food. The sexes are separate. Fertilisation occurs in July-August, with sperm from the male swimming through the mouth into the gastric cavity of the female. The early development of the planula-larvae takes place within the female, after which they emerge and settle on stones, algae, etc, in water, growing into small polypoids (**scyphistomae**). These multiply asexually, and each scyphistoma will produce three to five small medusae (**ephyrae**) during March and April. The ephyrae subsequently grow into new, sexually mature jellyfish. The number of medusae decreases during autumn, and they are usually absent during winter, although a second strobilation sometimes takes place in the autumn. These medusae may actually survive the winter, although many will perish in the winter storms. Scyphistoma sitting on stones or algae may reach an age of several years, and strobilate on several occasions. The jellyfish feeds primarily on plankton, although also small fish are caught.

UNPLEASANT POISON

The poison from the nematocysts may also affect humans. In less serious cases they burn and cause erythema, but can sometimes also lead to cramps, breathing and cardiac problems, agony, and influensa-like symptoms. Danmarks Fauna (1937) suggests the following treatment: "*Caustic soda or diluted salammoniac spiritus is applied as a cure and rubbed into the affected areas, and a glass of cognac will prevent any further feeling of feebleness*"-Enjoy!

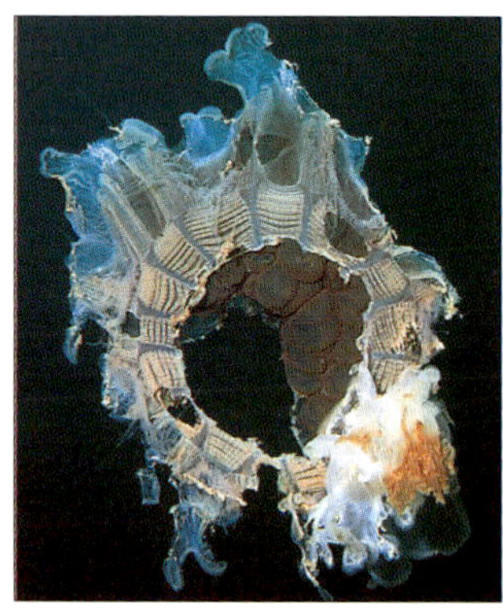

This stinging jellyfish is almost fully decomposed, but it was still alive. It most likely was exposed to a strong autumn storm. Photographed in December.

Cyanea lamarckii

This blue stinging jellyfish regularly occurs in the south-western parts on Norway.

Distribution: In European waters from the Bay of Biscay to Iceland. Common in the south-western parts of the British Isles and in the southern North Sea.
Description: Similar to the lion's mane jellyfish (***Cyanea capillata***), from which it is distinguished by colour and smaller size. Maturity may be reached already at a size of 40 mm in diameter, whereas fully-grown individuals rarely exceed 200 mm. The bell also tends to be more arched. Nevertheless, the colour is the main diagnostic difference, with the gastric cavity being sky blue. It has a smaller number of tentacles, and also fewer nematocysts, but still the same ability to inflict pain.
Habitat: In open and coastal surface waters. Appears in May-September.

Aurelia aurita - Moon's jellyfish

Distribution: Moon's jellyfish may occur in huge densities all over the world.
Description: This transparent jellyfish is familiar to most who have been in contact with the sea. Its colour and shape are highly characteristic, and cannot be misidentified. The diameter of large specimens may exceptionally reach· 400 mm, although they rarely exceed 250 mm. The flat bell has eight small incisions, where the sense organs (**rhopalia**) are situated. The aboral surface appears smooth, although it is actually covered by fine warts with nematocysts. The edge is densely provided with small tentacles, interrupted only by the sense organs. The four horseshoe shaped gonads are clearly visible

Moon's jellyfish has short tentacles with relatively small amounts of poison. The almost circular gonads are characteristic.

LICKING UP THE FOOD

The food of the moon's jellyfish consists mainly of microscopical plankton. Following the rhythmic contractions of the bell, the tentacles are bent over the edge and the aboral surface, thereby cleaning the upper side from any organisms that may have settled there. Small specimens may also capture fish, whereas the larger, mature ones generally have less nematocysts, and live mainly from microscopic organisms and suspended particles.

Close-up of the gonads.

Moon's jellyfish contains more than 98% seawater, making it a relatively unattractive prey item.

through the gelatinous mesoglea, and tend to be more closed in shape in larger specimens. The four short mouth-arms, together with the tentacles and the gonads, have a faint tint of violet, sometimes also of red and yellow. Remaining parts are colourless. The animal consists of 98% seawater.

Habitat: Very common in coastal, surface waters.

Biology: Sexes are separate, and fertilisation takes place by the emitted sperm from the male entering the mouth of the female. The larvae develop in brood-pouches on the female's mouth-arms. The fully-grown larvae subsequently leave the female and settle on the bottom, where the polypoids multiply asexually. These scyphistomas will overwinter, and bud off 12-14 small medusae (ephyrae) next spring. The scyphistomas are very abundant in shallow waters on stones, algae, mussels, etc, during the winter season. The medusae die during late autumn and winter, and specimens that are falling apart into pieces may then be encountered. The nematocysts of moon's jellyfish normally do not affect humans, although one may occasionally feel a weak burn on places with more delicate skin, such as on the lips.

Chrysaora hysoscella - Compass jellyfish

Distribution: Common in European waters in the Mediterranean and the European west-coasts up to Helgoland. It is rare around the British Isles and further north.

Description: This beautiful jellyfish has a characteristic colour pattern on the aboral side. The bell is weakly arched, up to 300 mm wide, faint yellow to white in colour, and with highly distinct brownish-red stripes, which are V-shaped and radiate from the centre. The edge is brown and consists of 32 rounded lappets. It has 24 long tentacles, arranged in eight groups with three in each. The four mouth-arms are elongated, pointed and strongly folded. The lengths of the mouth-arms are commonly twice the length of the diameter of the bell, although in undamaged specimens they can be much longer.

Habitat: Coastal, shallow waters.

Biology: This species is unusual in being **hermaphroditic**, although not simultaneous. Male gonads develop first, such that small specimens will be males. In medium-sized specimens both male and female gonads are present, whereas in the larger ones only the female gonads remain. Reproduction, however, appears to occur between different kinds of specimens, even in the medium-sized ones, due to the fact that eggs and sperm do not reach maturity at the same time. Furthermore, some specimens appear to be only males or females all through their life. Fertilisation and early development takes place within the female gonads. Similar to the moon's jellyfish and the lion's mane, the scyphistomae overwinter in shallow water.

Detail shots of this richly pigmented jellyfish.

The compass jellyfish is a rare guest in our waters. Very characteristic.

Order Rhizostomae

Species within this order (Gr. rhiza = root + Gr. stoma = mouth) have branching mouth-arms, provided with a series of folds containing "secondary mouths".

Like the compass jellyfish, this species is a rare late-summer visitor in our waters.

Rhizostoma octopus

A very attractive and characteristic jellyfish, with an eye-catching violet-blue bell edge.

Distribution: Common in European waters from the Mediterranean up to the English Channel, occasionally further north during late summer and autumn.

Description: A beautiful and highly characteristic jellyfish, up to 900 mm in diameter. Shape of bell strongly arched. The colour is milky white with transparent bluish (males) or brownish (females) gonads. Edge lappets are deep blue to violet. The outer-most parts of the sense organs are orange, and located in the incisions separating the edge lappets. Tentacles are absent. The aboral surface appears smooth, but is covered by fine warts with nematocysts. The consistency of the animals is tough, almost cartilaginous. The lengths of the mouth-arms equal the diameter of the bell.

Habitat: In coastal, shallow waters.

Biology: Separate sexes. Sessile polypoid stage unknown.

This photograph shows colonies of the soft coral dead man's finger, **Alcyonium digitatum**, *and a number of sea anemones,* **Urticina felina**. *Picture taken at Saltstraumen in northern Norway on a winter's day.*

Subphylum Anthozoa - Corals

Most people would associate the term "coral" (Gr. anthos = flower + Gr. zoon = animal) with exotic reefs in tropical seas. However, it has a much wider application, and includes both ***Octocorallia*** *and* ***Hexacorallia****.*

What, if anything, is a coral?

The term "coral" is not strictly defined, and is often applied only to a subgroup of Anthozoa. Sea anemones, for example, are closely related to the reef-building corals, but are not usually included.

Paragorgia arborea, *photographed at 50 m depth in Trondheimsfjorden. On the left is a small colony of the coral* **Paramuricea placomus**. *On the coral are basket stars,* **Gorgonocephalus caputmedusae**, *and a colony of the reef-building fanworm* **Filograna implexa**.

Class Octocorallia

The polyps of octocorals have eight arms.

*Octocorals (Gr. octo = eight + Gr. korallion = coral) are named for the number of tentacles on the polyps. The more familiar ones include the three orders gorgonians or horny corals (**Gorgonacida**), soft corals (**Alcyonacida**), and sea pens (**Pennatulacida**).*

Order Alcyonacida - Soft corals

*Most members of this group (Gr. alkyon = sea foam and Gr. alkyonion = name of a specific mushroom) occur in the tropical parts of the Pacific among reef-building corals. In European waters there are but a few members of the families **Alcyoniidae** and **Nephtheidae**.*

Alcyonium digitatum - Dead man's finger

Distribution: Very common in European waters from the Bay of Biscay in the south to Iceland and northern Norway.

Description: Characteristic, leather-like, orange colonies. The English name relates to specimens brought up to the surface–when seen *in situ* the polyps are usually extended, giving the colonies a hairy appearance. The leather-like 'skin' is usually orange, although white specimens also occur. The polyps–like other octocorals–have eight tentacles. The colonies reach 200 mm in height. Young colonies form crusts on hard surfaces.

Habitat: Suspension-feeder, and thereby depending on currents and good water exchange. Colonies appear from low water mark, often in darker places without competition from algae. May appear in high abundance in areas with strong currents; also in wave-exposed places. Occurs down to 700 metres depth, common down to 100 metres.

Biology: Most colonies have separate sexes, although some are hermaphrodites. Spawning of eggs and sperm takes place during winter, and there is a planktonic planula-larva.

Two colour variations, white and orange.

Gersemia rubiformis

Distribution: Mainly Arctic distribution, also northern parts of the Pacific and Atlantic coasts.

Description: Can be similar to dead man's finger (***Alcyonium digitatum***), but has a distinct stem with branches that can be repeatedly subdivided, whereas dead man's finger never produces bush- or treelike colonies. The polyps also are more prominent and give the colony a less 'orderly' and more hairy impression. Similarly to dead man's finger, but in contrast to the nephteid genus *Capnella*, it has fully retractile polyps. Colour is variable, including reddish, orange, pink, greyish, yellowish, and white. The colonies are usually 10 - 15 cm high, although they may reach 25 cm.

Habitat: In the northern-most parts of its distribution, the colonies occur from 6 metres depth, whereas they appear deeper in more southerly waters, all the way down to 3600 metres.

This coral can be found in shallow water at high latitudes. Here photographed at Svalbard.

Drifa glomerata

Right corner: *A sea spider 'strolling' on the coral. Due to the appearance of the branches, this coral is known as the cauliflower coral in some languages.*

Distribution: Northern parts of the Atlantic, Davies Strait, Baffin Bay, Norwegian Sea and Barents Sea.

Description: Bush- or tree-shaped colonies, relatively large, usually with distinct longitudinal furrow along stem. Densely distributed polyps on the numerous, fine branches, usually appearing in groups of three. The polyps are not fully retractable. Colour white, brownish or reddish. Another species, ***Capnella glomerata***, occurs in the area, and can be difficult to separate from *D. glomerata*.

Habitat: In areas exposed to currents, usually deeper than 40 metres. Recorded down to 1500 metres, but also up to 15 metres.

Biology: Many of the colonies occurring in northern European waters have a very slow growth-rate, and should be treated with care by divers.

The branches look like cauliflower.

Order Gorgonacida - Horny corals - Gorgonians

Gorgonian (Gr. myth. Gorgo, genit, Gorgonis = female monster) colonies have a central stem that is attached to a hard substratum. In many species the branches subdivide in one plane only, giving the colony a fan-like appearance.

Paramuricea placomus

Paramuricea placomus *with the polyps retracted around the stem. Also a colony of the reef-building fanworm,* **Filograna implexa**.

Distribution: Widely distributed in the North Atlantic.

Description: This orange-yellow octocoral may reach a height of 1 metre. The colony has a short stem, followed by numerous branches of various sizes, which all subdivide in a single plane. The densely distributed polyps are approx. 3 mm high. The colonies are a

DID YOU KNOW THAT...

Coral colonies may have very slow growth rates. *Paramuricea clavata,* which occurs in the Mediterranean, grows maximally 1 cm per year, suggesting that a one-metre high colony is 100 years old! Obviously, divers should treat such colonies with respect.

Paramuricea placomus, *here with the polyps extended. In the centre of the colony is a colony of the reef-building fanworm,* **Filograna implexa**. *Photographed at 50 m depth in the Trondheimsfjord, Norway.*

Various crustaceans often live in between the polyps. Here the isopod **Astacilla longicornis.**

beautiful underwater sight indeed, but lose their colour and turn black if conserved.

Habitat: Lives in waters with strong currents, and turns one of the flat sides against the water direction in order to maximise food capture. Colonies are often associated with reefs of *Lophelia pertusa*. Common at depths of 150-1600 metres, but is known from 18 metres in the Trondheimsfjord.

Biology: Some of the colonies have a rich epifauna, notably amphipods and isopods. One colony was shown to harbour no less than 6499 crustaceans belonging to eight species.

Paragorgia arborea

Distribution: Northerly distributed in European waters, probably not south of the Norwegian coast. Also south and north Pacific.

Description: Often called "sea-tree" by fishermen who catch it in trawls, since it reminds of a tree, or parts thereof, when landed on deck. The impact is even more stunning when seen while diving. Recent observations by means of a ROV (Remotely Operated Vehicle) show colonies up to 6 metres high, although a more common size is 1-2 metres. Branching is usually in a single plane. There are two colour morphs, white, and variants of reddish pink, of which the latter is more common. The surface of the colony is completely smooth when the polyps are retracted.

Habitat: Excepting a few shallow occurrences in Norwegian fjords, this species lives mainly at

Paragorgia arborea, *seen from the side away from the direction of current.*

Paragorgia arborea *photographed at 40 m depth in Trondheimsfjord in Norway, the worlds' shallowest known locality for this coral.*

depths of 200-1300 metres. Often associated with reefs of *Lophelia pertusa*, and other places exposed to currents. The shallowest known locality is the Trondheimsfjord at 39 metres depth.

Biology: A series of animals are associated with this octocoral. The ophiuroid basket star, ***Gorgonocephalus caputmedusae*** is common, as are various polychaetes and small crustaceans.

THE TRONDHEIMS-FJORD

These photographs of *Paragorgia arborea* and *Lophelia pertusa* are probably the only *in situ* observations by divers of these corals. For uncertain reasons, the Trondheimsfjord is unique for these shallow occurrences, but it may be related to stable conditions in temperature and salinity, combined with good water circulation. Furthermore, the corals may actually benefit from the low water visibility and light penetration in the fjord, which is due to high input of particulate matter from surrounding rivers.

Polyps out feeding! See also the picture on p. 94. Like the whole colony of **Paragorgia arborea**, *the polyps are very large.*

Primnoa resedaeformis

Primnoa resedaeformis, *photographed at 55 m depth.*

Distribution: Widely distributed in the North Atlantic and northernmost parts of the Pacific.

Description: Colonies may reach 1 metre in height. They are similar to those of ***Paramuricea placomus***, but the growth pattern is more bush-like, branching in many directions. Colour is faint to bright orange. Each of the densely distributed knobs on the colony represents a bell-shaped polyp, which under undisturbed

This coral has a somewhat untidy, strongly branched shape. Here pictured together with the author at 50 m depth in Trondheimsfjord, Norway.

conditions will extend the eight tentacles.

Habitat: The shallowest record is 32 metres in the Trondheimsfjord in Norway, whereas it is known from 100-1000 metres depth from other areas. Marine biologists collect them when dredging on or nearby ***Lophelia*** reefs.

Biology: This species does not appear to have a rich epifauna; our observations include the isopod ***Astacilla longicornis,*** and the northern stone crab ***Lithodes maja.***

The species has relatively small, bell-shaped polyps.

DID YOU KNOW...

The term **epifauna** refers to animals living on other organisms or on hard substrata. Conversely, the term **epiphytes** refers to plants living on other organisms, usually larger algae or sea-grass.

Swiftia rosea

This attractive red horny coral has a very limited distribution.

Distribution: Known from the Azores and the Canary Islands to Trondheim, mid-Norway, although older records indicate a previous distribution north to northern Norway.
Description: Colonies up to 150 - 200 mm high. Colour deep red, with white or almost uncoloured polyps. Colonies usually consist of a main stem with few branches occurring in a single plane. Basally, the branches appear pair-wise, whereas they are more irregular in the more distal parts. The polyps cannot be withdrawn. There is a white colour variety found in western Norway, at the Swedish west coast, and around the British Isles, which is known as ***Swiftia rosea pallida***. However, it probably represents a distinct species, ***Swiftia pallida*** (Madsen 1970).
Habitat: This gorgonian requires specific conditions to survive, and has a restricted distribution. It is found on rocky bot-

NECKLACE OF CORALS

Coral gems are often made from the Mediterranean red precious coral (*Corallium rubrum*), which not only has a red surface tissue, but also a red skeleton. Pieces of this skeleton are polished and sold as gems.

toms, usually at 30 - 400 metres depth. Furthermore, it requires full salinity, low temperatures and water currents. Large numbers of colonies may be found under favourable conditions. Although it is tempting for divers to collect this beautiful coral, it will soon–like most of our other gorgonians–lose its colour. The surface tissue decays rapidly and the colony turns black, which is the colour of the skeleton.

Above: *The horny coral,* **Swiftia rosea**, *closely resembles the previous species. At the bottom of the picture are the bryozoans* **Cellepora ramosa** *and* **Reteporella beaniana**.
Right: *Polyps.*

Order Pennatulacida - Sea pens

The colony within this group (L. pennatulus = wing carrier) develops from a primary polyp that produces a lower part, the ***pedunculus****, without polyps, and an upper part, the* ***rachis****, with polyps. The lower part has a terminal extension that anchors the colony in soft bottoms. The colony is held upright by the internal skeleton, which consists of the protein* ***pennatulin****. Pennatulids have separate sexes, and many of them are bioluminescent.*

Pennatula phosphorea - Phosphorescent sea pen

Detail showing the rows of polyps.

Distribution: Cosmopolitan and widely distributed in the North Atlantic. Recently located in the northern parts of the British Isles.

Description: Beautiful, usually deep red. Consists of an upright, thick and fleshy stem, which terminates in a foot extension sitting in the sediment. The branches grow pair-wise, and carry rows of polyps. The colonies reach a height of approx. 250 mm. Two other, less common, species occur in the area, ***P. aculeata*** and ***P. grandis***, of which the latter is a deep-water form and may reach a height of 1 m.

Habitat: Mud bottoms, usually below 15 metres.

Biology: A striking feature for this species is its ability to glow if touched. The light, which is produced by symbiotic bacteria, is first emitted from the touched area, and then spreads across the whole colony. Pennatulids, in contrast to other octocorals, are not fixed to a surface, and can therefore live on soft bottoms.

The colonies are very characteristic and cannot be confused with other species in our waters. Seen only rarely, because it usually occurs on muddy bottoms in relatively deep water. However, we have seen it at 15 m depth at Sognefjorden in southwestern Norway.
***Left:** Sea pen with clusters of gametes (reproductive cells) ready for spawning.*

Under unfavourable conditions they also have the ability to move, due to relatively well-developed muscular and nervous systems.

Reproduction in cnidarians

A brief account of cnidarian reproduction, with interesting observations from Sognefjord in western Norway.

by Alf Jacob Nilsen

Phylum Cnidaria is divided into the three classes Hydrozoa (hydroids), Scyphozoa (jellyfish), and Anthozoa (corals and sea anemones). The reproductive patterns in these groups are highly diverse, and some further information is therefore provided here.

Patterns of reproduction

Two basic, different kinds of reproduction can be identified; **sexual** and **asexual**. In sexual reproduction, the eggs are fertilised by sperm, leading to a combination of female and male genes. The offspring thereby becomes genetically different from the parents, and the accumulation of such differences may lead to evolutionary differentiation. The fertilised eggs usually develop into larvae with the ability to disperse, in cnidarians represented by ciliated **planula-larvae.** In asexual reproduction there is no fertilisation or combination of genes, and the offspring is genetically identical to the parent; a clone. Some organisms alternate between the two basic modes of reproduction. During a specific time-span, for example a year, a species can accordingly have two different generations, one sexual and one asexual. This leads both to the formation of a large number of genetically identical individuals that can spread over large areas, and genetic recombination following the sexual phase. These two modes are encountered within the hydroids and the scyphozoans. The Moon's jellyfish (*Aurelia aurita*) is a classical example, but there is a lot of variation present within the two groups. Corals and sea anemones exhibit both sexual and asexual reproduction, but never with two distinct generations.

This plumose anemone has produced small individuals exactly like itself, by budding off parts of the foot, or pedal disc. This process, called basal laceration, is effectively a cloning of the parent individual.

***Opposit page:** Sea pens with gametes. Picture from Sognefjord in western Norway at full moon. The gametes are released by the polyps, and our observations indicate common spawning for various corals during nights at full moon. Most corals observed on this night were ready to spawn.*

Polyps and medusae

Cnidarians have two basic forms, polyps and

medusae. Both of these have the same body structure, with an inner and an outer lining, and a tissue between these with motile, gelatinous cells (mesoglea). At the centre is the mouth surrounded by tentacles. The medusae are planktonic, whereas the polyps are sessile, and have a basal point of attachment, a stem, and a mouth surrounded by tentacles at the top.

Within the hydroids we encounter both polyps and medusae, although the polyps are the dominant stage. They often form colonies with differentiated polyps, where some are specialised for food intake, and others for reproduction. The latter bud off small medusae, which as adults will reproduce with eggs and sperm. The fertilised eggs develop into larvae, settle and found new colonies. However, there is a large amount of variability of this basic pattern.

Whereas scyphozoans also have polyps and medusae, here it is the the medusa-stage that is dominant. We are all familiar with the large jellyfish that appear during the summer. The cosmopolitan Moon's jellyfish (see page 89) constitutes a classic illustration of alteration between polyp and medusa stages. Medusa stages are completely absent within the anthozoans.

Eggs and sperm

It is usually not possible to determine the sex of a coral or a gorgonian without microscopical examination of the gonads of the animal. In many species the gonads are large and clearly visible, such as in the moon's jellyfish, but it still is not possible to state whether they are females or males before they have reached full maturity.

Cnidarian eggs tend to be large, up to a few mm in diameter, faint yellow or orange in colour, and clearly visible. The sperm cells, in contrast, are microscopical in size, and can only been seen as a milky substance when emitted. Fertilisation usually takes place in the free water, although in some cnidarians the fertilised eggs develop within the female.

Tropical corals

The vast majority of cnidarians in our waters have separate sexes. The colourful, tropical corals, however, are generally, but not always, hermaphroditic. Some have separate sexes in some localities, whereas they are hermaphroditic in other, and some are hermaphroditic but develop eggs and sperm at different times. Nevertheless, the majority are simultaneous hermaphrodites, and have eggs and sperm at the same time. Tropical cnidarians, similar to our ones, generally shed eggs and sperm, and the fertilisation takes place in the surrounding water. However, a number of exceptions exist, where the fertilised eggs stay and develop within the parent, and there are also examples where genetically identical planula-larvae develop asexually within the parent. The variation in reproduction among tropical cnidarians is overwhelming and still far from fully known; we recommend Fosså & Nilsen (1998) and Harrison & Wallace (1990) for further reading.

Synchronised mass spawning

In Japanese history it was observed that the sea turned red once a year in coral areas, and this was explained as when the Princess of the Underwater Dragon Palace had her periods. Coral reefs in Japan are less abundant today, but we know today now that the red colouration is due to synchronised mass reproduction of these reef-building corals–one of the most spectacular events nature has to offer.

Reproductive growth and maturity in tro-

Opposite page:
The tall sea pen, **Funiculina quadrangularis**, *photographed in Sognefjorden in south-western Norway the night before the full moon in August 1999. The gametes are clustered at the bases of the polyps on the main stem and have just started to migrate out into the polyps. The day after the full moon, the spawning was over.*

pical corals takes place as the temperature increases after the tropical 'winter'. In waters around the Great Barrier Reef in Australia and in the central Pacific, this culminates during three to six nights, following a full moon and when water temperatures approach their maxima. This leads to a virtual explosion of sexual activity. It is onset by the first colonies emitting so called 'egg-sperm bundles', together with a hormone (Estradiol 17b), which probably is involved in triggering the spawning of other corals. Furthermore, this reproduction is followed, not only by other corals, but also by many other invertebrates, including bivalves, holothurians and polychaetes. Large epitokes of the Palolo worm (a polychaete) swim at the surface among the coral eggs and sperm, and the sea turns into a red-coloured soup of gametes. The mass spawning is synchronised by water temperatures and the moon and tidal cycles, and is so precise that we can predict the actual night when it will start. Each year it is studied by a large number of scientists, and there still remain many unresolved mysteries associated with this tropical phenomenon.

Clearly visible gamete clusters ready for spawning.

Is there a mass synchronisation among cold-water cnidarians?

While diving the day before full moon in Sognefjord in western Norway in August 1999, the authors of this book observed large amounts of eggs in pennatulid colonies, mainly of *Pennatula phosphorea* and *Funiculina quadrangularis* (see pp. 102 and 112, respectively). Apart from the dense occurrence of pennatulids, we also observed filtering holothurians, haddock, megrim, and long rough dab. The bottom was mud, sloping from 15 down to 35 metres. When diving again the day after the full moon, all gametes were gone. This observation can indicate that spawning is synchronised to the moon, although this can only be confirmed by further observations. And are other organisms involved in the synchronisation? And does it occur on a yearly basis, or are there longer intervals between the events?

There are certainly many highly interesting topics within cnidarian research, and many questions still await their answers.

REFERENCES

FOSSÅ, S.A. & A.J. NILSEN 1998. Reproduction in the Cnidarians. In Fosså, S.A. & A. J. Nilsen, The Modern Coral reef Aquarium, volume 2. Birgit Schmettkamp Verlag, Bornheim, Germany, pp. 34-63.

HARRISON, P.L. and C.C. WALLACE. 1990. Reproduction, dispersal and Recruitment of Scleractinian Corals. In: Dubinsky, Z. (ed.) Ecosystems of the World, vol. 25 Coral Reefs. Elsevier, N.Y. pp 133-208.

Funiculina quadrangularis - Tall sea pen

Distribution: Widely distributed in the North Atlantic, also recorded from the Indian and Pacific Ocean. On the west coasts of the British Isles.

Description: This pennatulid has a unique shape and adult colonies can hardly be misidentified. It is very tall, recorded up to 1.7 metres. The basal part of the stem is rounded and without polyps, with a slightly inflated ending. The part carrying the polyps (the rachis) is 4 - 9 times longer than the basal part, and is quadrangular (*quadrangularis* L. quadrus = four and L. angulus = corner/angle), rather than rounded. In contrast to ***Virgularia mirabilis***, the stem is not branching. The polyps sit directly on the stem, and are encapsulated by a cup with eight pointed teeth. In early stages of the colonies, the polyps are inserted pair-wise, but are more irregularly positioned on larger ones. The colour of the colonies is a faint reddish yellow, sometimes brownish with violet polyps.

Habitat: Occurs on soft bottoms, as with other pennatulids, predominantly on fine mud from 20 down to 2600 metres depth, with the shallowest occurrences in fjords.

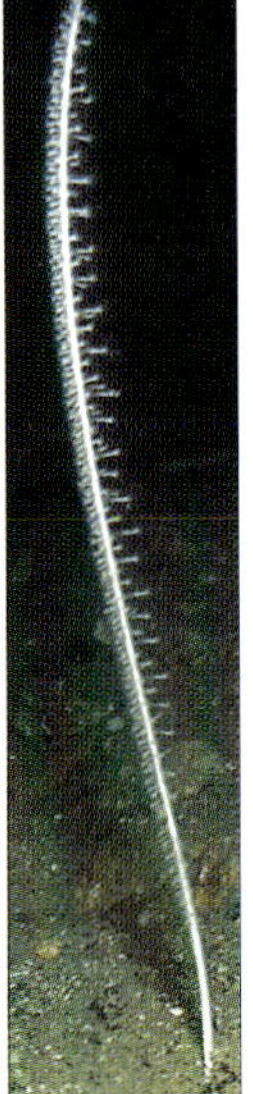

The tall sea pen can reach up to 1.7 m in height.

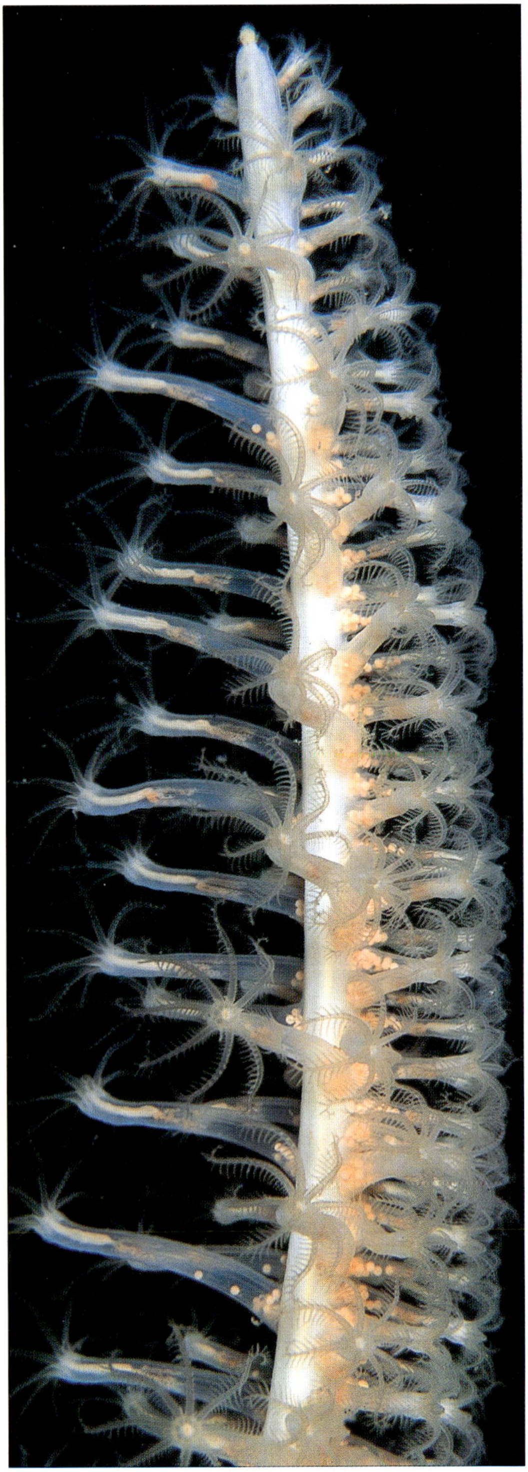

The tall sea pen with gametes ready to spawn. The individual on this picture was over 1.5 m high.

Virgularia mirabilis

Distribution: Widely distributed in the North Atlantic. In European waters from the Mediterranean to northern Norway. All around the British Isles.
Description: Long and thin colonies, with short, pair-wise arranged branches; lower parts without polyps. Polyps occurring in clusters of up to 16 together, and can be withdrawn. Up to 60 cm high. Colour yellow-orange to light pink. In contrast to ***Funiculina quadrangularis,*** the stem is rounded in cross section, it is branching, and it ends in a tip without polyps. A close relative, ***V. tuberculata***, is distinguished in having only 3 - 6 (8) polyps per branch, whereas *V. mirabilis* has up to 16.
Habitat: The colonies grow upright on soft bottoms from 10 to approx. 360 metres depth. If disturbed, they can–despite their length–withdraw into the sediment.
Biology: Like other pennatulids, the colony can move with the aid of the muscular peduncle. The species is oviparous.

Virgularia tuberculata *is distinguished from* **V. mirabilis** *by having less polyps per branch, and by the generally larger size of both the polyps and the branches.*

Right: **Virgularia mirabilis** *has a more rectangular form than the tall sea pen* **Funiculina quadrangularis** *and the smaller* **Virgularia tuberculata** *(above).*

Kophobelemnon stelliferum

Of all our octocorals, this one has the largest polyps. Photographed at Sognefjord in southwestern Norway at around 50 m depth.

Distribution: North Atlantic and parts of the Pacific.
Description: Colonies are highly characteristic. In contrast to many other pennatulids, it has polyps that are inserted directly on the stem (rachis), a feature that it shares with ***Funiculina quadrangularis***. It can be separated from the latter by the lack of encapsulation of the large polyps, which are not arranged in longitudinal rows. The presence of the large but few polyps is the reason behind the scientific species epithet *stelliferum* (Lat. stella = star). It may attain a height of 75 cm, of which a major part is buried in the sediment. The polyps usually are yellowish brown to violet, wheras the stem is faint yellow to brown.
Habitat: Occurs on soft bottoms from approx. 40 metres depth, down to over 3500 metres.

Class Hexacorallia

Polyps with six tentacles (Gr. hex = six + Gr. korallion = coral), or multiples of six (some exceptions exist). The class is subdivided into two subclasses, ***Ceriantipatharia*** *and* ***Zoantharia****. It is within Zoantharia that we find the stony corals,* ***Scleractinida****, and the sea anemones,* ***Actiniarida****. In European waters there are only a few stony corals, but a large number of sea anemones. The subclass Ceriantipatharia is represented by a few species only, belonging to the order Ceriantharia.*

A small, unidentified sea anemone

Order Ceriantharia

All species within this group are solitary. They have a long cylindrical body and live in soft bottoms. The animals construct a tube that is buried in the sediment, and only the tentacles and the mouth opening are visible. They may be confused with digging sea anemones, but these never build tubes.

Cerianthus lloydii

This individual is caught in the act of retracting its tentacles. The outer tentacles are long, whereas the inner tentacles are very short.

Distribution: From the Bay of Biscay in the south north to Svalbard and the Barents Sea. Locally abundant on all coasts of the British Isles and Ireland.
Description: Only the tentacles and the upper-most part of the tube visible when observed *in situ*. The body is up to 210 mm long and yellowish white. Up to 70 outer tentacles, up to 50 mm long; inner tentacles much shorter. The colour of the outer tentacles is usually uniform reddish brown in larger specimens, with a white ring near the basis and a brown area below that. Smaller specimens often have yellowish-white outer tentacles with brown, transverse stripes. The shorter inner tentacles are brown towards the mouth, and paler towards the outside. There is a rare but similar species, ***Pachycerianthus multiplicatus***, that is described on the next page. There are a series of digging sea anemones within **Actiniarida**, although most of these have much fewer tentacles.
Habitat: Common on soft bottoms, usually with fine sand, at depths of 2 - 100 metres.
Biology: The body wall is heavily muscular-

ised, and it can withdraw quickly down into the sediment. The tube is made of mucus secreted from the body wall, mixed with sediment. The animals are hermaphrodites, and shed the gonads into the water. The larvae are pelagic.

This individual has very long outer tentacles, a characteristic trait of tube anemones.

Pachycerianthus multiplicatus

This tube anemone is very large, with extremely long outer tentacles. The 5 - 10 reddish-brown bands on each of the tentacles are typical for adults of this species.

Distribution: Limited distribution in the North Atlantic, from Kattegatt and Skagerrak to western Norway. Recently located in western Ireland and Scotland.

Description: Considerably larger than *Cerianthus lloydii*, and a beautiful animal when seen in its habitat. The body wall is usually flesh-coloured or pink. There are up to 180 outer tentacles, which in adult specimens carry 5 - 10 brown transverse bands, otherwise light grey or pink, sometimes with a hinge of violet. The much shorter inner tentacles are uniform greyish brown, occasionally with greenish tips. The oral disc has yellowish grey and yellowish brown radiating stripes. Total length, including the buried parts, is up to 30 cm, with a column diameter of up to 5 cm. Cannot be confused with other species in the region.

Habitat: Occurs slightly deeper than *Cerianthus lloydii*, 15 - 200 metres.

Order Scleractinida - Stony corals

The stony corals (Gr. skleros = hard + Gr. aktinos = ray) include the highly familiar reef-building corals of the tropics. However, it is less well known that some reef-building corals also occur in deeper water in our region.

The stony coral **Lophelia pertusa**. *The outer edge of the colony tends to be evenly rounded. On the bottom are fragments, which can grow further and create new structures. The colonies often are occupied by large sponges, such as* **Mycale lingua**, *depicted here. The Norway redfish,* **Sebastes viviparus**, *is common around* **Lophelia** *reefs.*

Lophelia pertusa

The polyp arms are translucent, with clearly visible nematocysts (stinging cells).

Distribution: Widely distributed in the North Atlantic, including the Mediterranean, but also in the Pacific and the Indian Ocean.

Description: The colonies consist of polyps that secrete a hard skeleton of calcium carbonate. The largest reefs may attain a length of 5 km, a width of 200 metres, and a height of 35 metres. The calcareous cups, which are secreted by the polyps, are about 5 - 10 mm in diameter, and the polyps carry 60 or more transparent tentacles. The colour of the colonies is usually white, although light red forms appear in more shallow waters.

The polyps extend from a cup made of calcium carbonate.

It can be confused with the smaller ***Madrepora oculata.*** Both often occur together, but the latter has shorter cups with a diameter of 4 - 5 mm, and polyps growing in alternating rows.

Habitat: Found in places exposed to currents, and with cold and stable water conditions. Normal depth distribution is 150 -2000 metres. The shallowest known

GOOD FISHING SPOTS

It is commonly known that *Lophelia* reefs provide excellent fishing, and fishermen using lines and nets have exploited these biotopes for a long time. The complex reef structures serve as habitat for a rich fauna, and are important locally for the productivity, not least for commercially caught fishes, such as red-fish, cod, ling, blue ling, and saithe. In recent years this commercial fishing has increased, and vessels with heavier gear are actually fishing within the reefs. There have been accusations of systematic crushing of reefs in order to be able to trawl the area afterwards. Reefs, which are thousands of years old, can for short-term profits become wasted in a few days. However, from March 1999, selected areas in Norway are now protected (see further information in the note on p. 127).

The book's photographer took a photograph of the author. Diving was carried out with a triple valve and tank set, and an 80% oxygen mixture (EAN 80) was used during decompression. Be aware that diving at such depths (50 - 60 m) demands detailed preparations and a high level of experience.

Lophelia, *together with the sponge* **Mycale lingua,** *being photographed by a keen diver.*

locality is 39 metres in the Trondheimsfjord. Many new reefs have been detected following the introduction of ROV for inspection of pipes and bottom characteristics prior to underwater installations. These are situated on the continental shelves and slopes at depths of 200-400 metres.

Biology: *Lophelia* reefs constitute large and beautiful biotopes with a diverse fauna. Around 300 species have been recorded from sampling by triangular dredge around the Faeroes, and altogether over 700 species have been reported associated with *Lophelia*. The majority are suspension-feeders (sponges, bryozoans, hydroids, brachiopods, bivalves and polychaetes), but there also are a number of crustaceans (galatheoids, amphipods, shrimps) living among the branches, and the redfish also is highly typical. Investigations indicate a growth of the coral branches of about 4 - 25 mm per year, with an average of about 6 mm. Although this may not sound very impressive, it actually exceeds the growth rates of many tropical stone corals. Northern European waters, in contrast to tropical waters, are rich in nutrients, and *Lophelia* does not obtain additional nutrients from microscopical symbiotic algae, as is the case with the tropical reef-building corals. *Lophelia* reefs are found in areas with stable water temperatures of 4-8°C, and full salinity.

Deep-water coral reefs in Norway

by Alf Jacob Nilsen

The word "coral reef" certainly has exotic connotations, and tends to be associated with images of coconut palms, white beaches, and azure-blue water. And this is not without reason. Reef-building corals thrive in the tropical clear waters where water the temperature never fall below 18°C.

These corals live in symbiosis with unicellular photosynthesising algae (zooxanthellae), and an important part of the nutrititional needs of the corals are supplied by the symbionts. The northern European reef-building corals, in contrast, lack symbionts.

The deep water coral **Lophelia pertusa** *occurs in two colour variations; orange and white.*

Stone corals may either occur as solitary individuals or as colonies. The Devonshire cup-coral (*Caryophyllia smithii*, see p. 129) is an example of a solitary coral, which is common in our waters. Of reef-builders we only have two species, *Lophelia pertusa* and *Madrepora oculata*, and it is mainly *L. pertusa* that forms the larger reefs. *Madrepora oculata* is closely related to it, but more rare. Further out in Atlantic waters we also encounter *Solenosmila variabilis*, and as we proceed south the number of deep-water reef building corals increases.

History

Two 18th century Norwegian bishops showed interest in *Lophelia*. It is mentioned for the first time by Erik Pontoppidan's study from 1752, whereas his colleague Johan C. Gunnerus in 1768 published a study together with a beautiful illustration, where he mentions it as *Madrepora pertusa*.

Left: The world's shallowest known **Lophelia** *reef, at 39 m in Trondheimsfjord in Norway. On the picture are also the horny corals* **Paragorgia arborea** *and* **Paramuricea placomus**.

The coral grows densely, forming a jungle that hides many other small animals.

Gunnerus probably obtained his colonies from Rødberg in the outer parts of the Trondheimsfjord. Subsequently, several Norwegian zoologists have studied the reefs, including Michael Sars and Carl Dons, and the latter published a series of papers on them (Dons, 1932, 1933, 1934, 1935 and 1944). Until the Second World War *L. pertusa* had been recorded from a number of localities along the Norwegian coast, and Carl Dons wrote in his study *"Norske Korallrev"* [Norwegian coral reefs] that *"we now know of about 100 coral reefs in Norwegian waters, including both dead and live ones"*.

New discoveries

Enter the Norwegian *'oil-age'*, accompanied by extensive mapping of the sea floors. And it was during such mapping in the Barents Sea that Statoil in 1982 discovered a coral reef at Fugløya 70°N and 21°E at the Askeladden field. By means of sonar it was first recorded as an unknown 15-metre high structure at 280 metres depth-larger than an ordinary house (Hovland & Mortensen, 1999). Sediment samples together with ROV (Remotely Operated Vehicle) studies disclosed a coral reef.
Extensive mapping off mid-Norway at the Heidrun field was carried out in 1982, and a series of similar structures were encountered, and was an indication that coral reefs may be far more abundant on the continental shelf than previously thought. Positive identifications were carried out by means of ROVs. Within a 200 km long and 3 km wide area 57 reefs were recorded at 250-350 metres depth. The surface of these reefs varies from 1 300 to 12 000m^2, and the three largest ones are up to 140 metres long, 120 metres wide, and 25 metres high.

There also are a number of more near-shore reefs. Many of these have been long known, because fishermen get nets and lines caught in the reefs, and therefore have paid attention to their positions. One such example

Rudolf Svensen prepares for a dive on the **Lophelia** *reef in Beitstadfjorden, at the mouth of Skarnsundet in Trondheimsfjord in Norway.*

is a reef west of the island Fejde in Hordaland, western Norway. The Trondheimsfjord has a large number of reefs, some of which are known since the times of Gunnerus, and include the shallowest known one, 39 metres deep and situated between Tautra and Fosen peninsula. There are also reefs known from Agdenes at the entrance of the fjord, at Tautra, and in Beitstadfjorden in the inner parts of the Trondheimsfjord.

For those who read Norwegian the book "Norske Korallrev og prosesser i Havbunnen" by Martin Hovland and Pål B. Mortensen (1999), is a rich source of information for distribution, geological history, and biology of the coral reefs in Norway.

Structure and associated fauna of the coral reefs

The depth distribution of the *Lophelia* reefs makes them difficult to access, and in most cases one has to rely on examination with ROVs and on grab and core samples. Nevertheless with the use of echo-sounders and sonar, we are, now starting to understand how the reefs are structured.

In the upper part we find the white, live coral colonies, usually rounded structures with a diameter approaching 2 metres. Above that size, colonies will tend to break and fall down, due to their own weight. When this happens, they will end up in a zone labelled the "intermediary zone", consisting of dead reef pieces, overgrown by *Paragorgia arborea, Paramuricea placomus, Primnoa resedaeformis* (see pp. 98-103), and many sponges. In contrast to live *Lophelia*, these pieces now turn grey or brown. Below, we find the zone termed *Lophelia* gravel, which consists of sediment mixed with broken reef pieces, and a series of sponges.

These reefs, just like the tropical ones, are highly diverse. There have been 614 associated species recorded from Norwegian waters, and 860 from the north-east Atlantic (Hovland & Mortensen, 1999), and we find filter-feeders, herbivores, carnivores, and detritus feeders.

Among this associated fauna, we encounter the gorgonians *Paragorgia arborea, Para-*

muricea placomus and *Primnoa resedaeformis*, groups of the up to 15 cm long bivalve *Acesta excavata*, the galatheoids *Munidopsis serricornis* and *Munida serricornis*, the former in the coral, and the latter in the sediment among the reef gravel. Other crustaceans include a number of isopods and copepods, which also serve as prey for *Lopelia*. A large number of the associated fauna relies on plankton and the water currents, such as the sponges *Geodia baretti* and *Mycale lingua*. *Ophiuroids* of the genera *Ophiacantha* and *Ophiactis* are common, as is the bright yellow sea star *Henricia sanguinolenta* (p. 418).

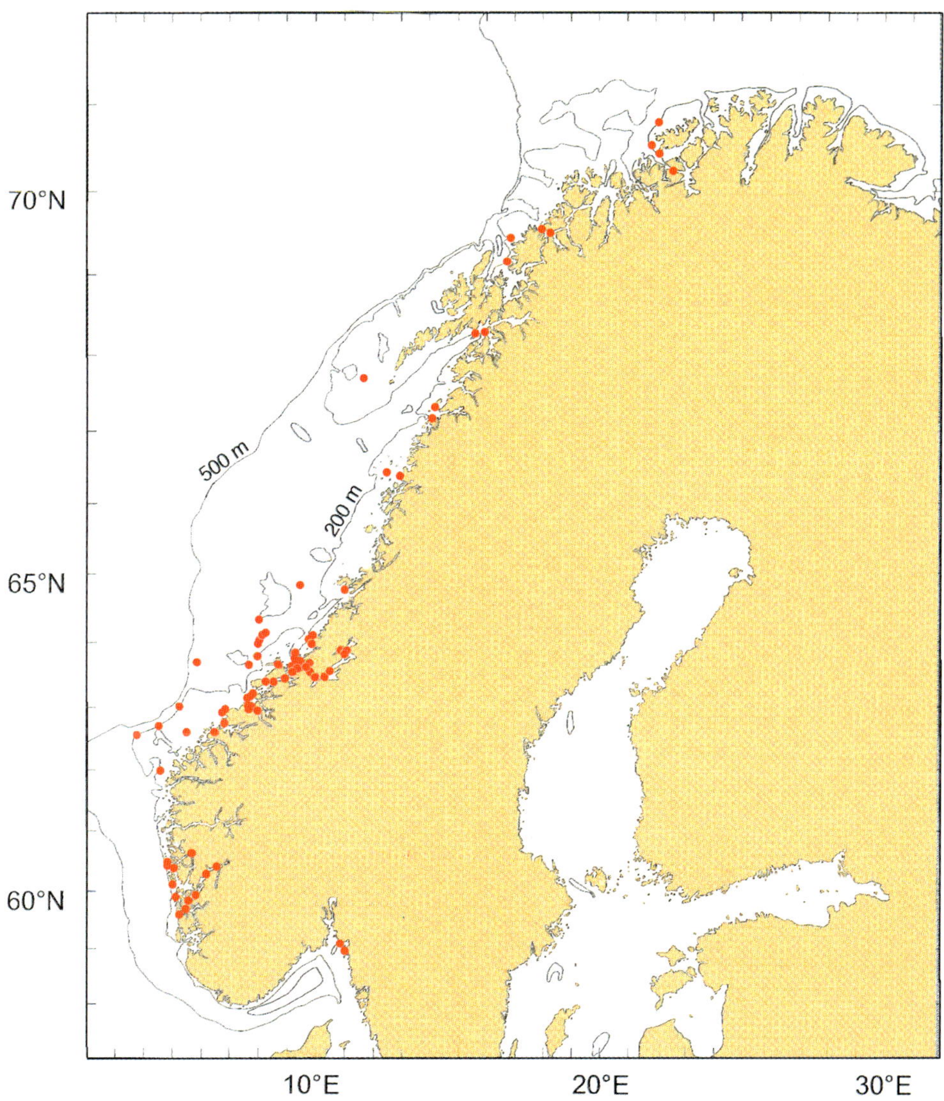

Lophelia *reefs are found along most of the Norwegian coast. They thrive inside the fjords as well as out on the continental shelf. The red dots show scientifically verified finds. The Norwegian Institute of Marine Research has in addition many records from fishermen of occurrences between 200-400 m depth at fishing banks and along the edge of the continental shelf. Map from the Norwegian Institute of Marine Research.*

Skarnsundet in Trondheimsfjord in Norway, seen from Skallen, the location of the Lophelia *reef we dived on.*

Of worms we find the long ribbon worm *Lineus longissimus* (p. 155), and the odd echiuran *Bonellia viridis* (p. 194). The polychaete *Eunice norvegica* is highly characteristic and well adapted to the reef, and is present on all reefs where *Lophelia pertusa* and *Madrepora oculata* occur. Among fish, two examples of the most common ones are Norway redfish (*Sebastes viviparus*) and tusk (*Brosme brosme*).

Chemosynthesis

Chemosynthesis is the process of synthesising organic molecules from simple, inorganic compounds by means of chemical energy. In areas rich of oil and gas, there is often a leakage of hydro carbonates, such as methane gas, from the sea floor, often referred to as seeps. Chemosynthesising organisms (bacteria and bacterial symbionts) can utilise the chemical energy of these as a resource. Deep-sea communities based on chemosynthesis were discovered for the first time off the Galapagos in 1977. This produced major headlines since it was the first finding of biological communities that were independent of photosynthesis. The basis of this food web is the chemosynthesising bacteria that oxidise hydrogen sulphides. They obtain energy from this reaction, which they use to build organic matter from carbon dioxide. Of special interest were the 'giant vent worms' (previously referred to phylum Vestimentifera, but now considered a polychaete family, Siboglinidae), which has no guts but lives in symbiosis with chemosynthesising bacteria.

However, chemosynthesising is found also along our coasts, including siboglinids and several bivalves, especially in some places in the Skagerrak. In the Mexican Gulf such seeps have provided new insights into to the processes relating to communities based on chemosynthesis (see Sassen et al., 1994). One current hypothesis under evaluation is whether *Lophelia* reefs, to a larger or smaller extent, rely on leakage of water rich in hydrocarbonates (see Hovland et al., 1987, 1998; Hovland & Thomsen, 1989, 1997). Quite apart, the reefs are likely to be important as a source of unknown chemical compounds that can be employed within medicine, including against cancer. We are only at the starting point of the exploration of the reefs, and can look forward to new and exciting findings.

Old reefs

Corals construct their calcareous skeletons through complicated chemical processes, involving uptake of calcium and the synthesis of aragonite (one crystallisation form of calcium carbonate). Whereas many of the tropical corals can have rapid growth rates, *Lophelia* only has an average growth of branch length of 6 mm per year. As a branch approaches about 5 cm in length, it splits into two or more branches, and thereby obtains a tree-like structure. Just like in branching tropical corals, it is only the distal parts of the colo-

The northern stone crab, **Lithodes maja**, *has taken refuge in the living part of the coral.*

nies that are living. In the middle of a *Lophelia* colony, we find only dead branches, lacking the characteristic white and glossy tissue that covers the live ones. Growing branches may join when they meet with other branches, and thereby strengthen the whole colony–another common feature shared with many of their tropical relatives.

It is self-evident that a 25-metre high reef, with a growth rate of 6 mm per year, must be very old. Based on dating with radioactive carbon (^{14}C), the oldest reefs are estimated at over 8500 years old, and these then actually go back to times when humans colonised Scandinavia after the last ice-age. See Hovland & Mortensen (1999) and Hovland et al. (1998) for further information.

Reef destruction

Humans are the worst enemy of the reefs! While the reefs may take thousand of years to grow, their destruction can be accomplished within a few hours. The biggest threat today is fishing. Whereas fishermen in earlier days avoided them in order not to ruin their gear, today's trawling gear is much sturdier, and a number of reefs have been destroyed in this way. And when the reefs disappear, so do the fish (for example, see Larsen 1997). In the Norwegian newspaper Stavanger Aftenblad, 21 August 1998, it was stated:

"*The reefs appear to constitute important breeding areas for fishes and other important organisms, and in spite of this they have been destroyed by fishermen using heavy gear in order to be able to trawl the areas afterwards. However, marine biologist Jan Helge Fosså at the Institute of Marine Research in Bergen does not wish to criticise the fishermen–they did not know different. And he leaves it to the appropriate authorities to decide on what restrictions should apply in these areas along the coast, especially from Møre and Romsdal and further north*".

Norway thereby became the first country to introduce laws to protect deep-sea coral reefs. On the 11 March 1999, both the Sula reef and the Haltenpipe reefs were protected, an area covering 970 km^2. **§1** in "**Regulations for coral reef protection**" states: "*The intention of this regulation is to protect the coral reefs from destruction by fishery activities...*"

Let us hope that the regulation is respected and that the mapping and investigation of these amazing underwater sculptures continues, such that also coming generations can enjoy these ancient communities.

SOURCES

DONS, C. 1932. Om Nord-Norges korallsamfunn. Det Kongelige Norske Videnskabers Selskabs Forhandlinger 5(4):13-16

DONS, C. 1933. Om Beitstadfjordens korallrev. Ibid. 5(37):143-146

DONS, C. 1934. Über die nordlichsten Korallenriffe der Welt. Ibid. 6(55):206-209

DONS, C. 1935. Die Verbreitung von Steinkorallen in West-Finnmark. Ibid. 8(18):57-60

DONS, C. 1944. Norges Korallrev. Ibid 16: 37-82

GUNNERUS, J. C. 1768. Om nogle Norske Coraller. Kgl. Norske Vidensk. Selsk. Skr. 4: 38-73.

HEINRICH, R., A. FREIWALD and SHGIPPBOARD PARTY. 1997. The *Lophelia* reef on Sula ridge, mid-Norwegian shelf. Cruise report No. 228/97, Bremerhaven, 12 pp.

HOVLAND, M. and E. THOMSEN. 1989. Hydrocarbon-based communities in the North Sea? Sarsia 74:29-42

HOVLAND, M. and E. THOMSEN. 1997. Cold-water corals - are they hydrocarbon seep related? Marine Geology 137:159-164.

HOVLAND, M., M.R. TALBOT, H. QVALE, S. OLAUSSEN and L. AASBERG. 1987. Methane-related carbonate cements in pockmarks of the North Sea floor. Journ. of petrology. 57:881-892.

HOVLAND, M., P. B. MORTENSEN, E. THOMSEN and T. BRATTEGARD. 1997. Substratum-related Ahermatypic Coral Banks on the Norwegian Continental Shelf. Proc. 8th International Coral reef Symposium, 2:1203-1206.

HOVLAND, M., P.B. MORTENSEN, T. BRATTEGARD, P. STASS and K. ROKOENGEN.
1998. Ahermatypic coral banks off mid.-Norway: evidence for a link with seepage of light hydrocarbons. Palaios 13:189-200.

HOVLAND, M. and P. B. MORTENSEN. 1999. Norske Korallrev og Prosesser i Havbunnen. John Grieg Forlag. 155 pp. ISBN 82-533-0288-6

LARSEN, P.M. 1997. Korallrevene tråles bort. Fiskets Gang 11/12: 5-8.

MORTENSEN, P.B., M. HOVLAND, T. BRATTEGARD and R: FARSTVEIT. 1995. Deep water bioherms of the scleractinian coral *Lophelia pertusa* (L.) at 64°N on the Norwegian shelf: Structure and associated megafauna. Sarsia 80:145-158.

MORTENSEN, P. B. Og M. HOVLAND. 1996. Korallrev i Mørket. Naturen 5:247-254

SASSEN, R., I.R. MacDONALD, A.G. REQUEJO, N.L. GUINASSO Jr., M.C.
KENNICUTT II, S.T. SWEET and J.M. BROOKS. 1994. Organic geochemistry of sediments from chemosynthetic communities, Gulf of Mexico slope. Geo-Marine Letters 14:110-119.

Caryophyllia smithii - Devonshire cup-coral

The polyps have a conspicuous knob-like structure.

Members of the species are usually found deeper than 30 m.

Distribution: In Europe from the Mediterranean to mid-Norway, possibly to Lofoten. Common on west coasts of the British Isles, on the east coast of Scotland and in the North Sea .

Description: Solitary coral that does not form colonies. The calcareous cup has a diameter of 15-20 mm; the height is usually slightly less. There are up to about 80 more or less transparent tentacles, which end in a characteristic knob. They are arranged in three rings. The knobs are usually paler than other parts of the tentacles. Colour is variable, and the mouth area can be red, violet, brown, yellow or orange, whereas the tentacles tend to be weak brownish or orange. The soft parts can be withdrawn within the cup, and the calcified parts are white with 12-20 radiating and sharp ridges.

The Devonshire cup-coral is a solitary coral, but it occasionally forms two-headed, even three-headed individuals. This happens when two larvae settle on an already-settled individual. The specimens in the picture were first observed 27 years ago, and they are still alive. This picture taken in December 2002.

Habitat: Attached to rock or stones. It rarely occurs in dense assemblages, although large numbers of individuals may occasionally occur. Around the British Isles it appears from the intertidal, whereas in Norwegian waters it usually does not occur above 20 metres. Found down to approx. 130 metres.

Biology: Hermaphrodite, reproducing during spring; no asexual reproduction. Although it is a solitary coral, two or even three individuals can sometimes be observed attached to each other, which is the result of larvae settling on other specimens. Repeated diving at the same spot for a number of years indicate that these corals grow old–the 'three-headed' coral on the picture was observed for the first time well over 20 years ago, and all three specimens are still thriving.

Order Actiniarida - Sea anemones

Sea anemones (Gr. aktinos = ray) are common in the intertidal, and therefore tend to be familiar to most. There is a large number of species, also in deeper water, and many are highly colourful and beautiful. They lack a calcareous skeleton, and are therefore soft. Most are attached to hard surfaces with a pedal disc, and have a limited motility.

Urticina eques.

Urticina eques *near Østerøya, south-east Norway.*

Sea anemones can be difficult to identify. This is probably Sagartia troglodytes *(p.136).*

Urticina felina - Dahlia anemone

Distribution: In Europe from the Bay of Biscay to the Arctic. Very common. All around the British Isles.

Description: This beautiful anemone occurs in several morphs. It has up to 160 fleshy tentacles, often with violet spots. The diameter is usually larger than the height, and the animal is attached by the pedal disc that serves a powerful sucker. The mouth opening is red or reddish brown. Characteristic papillae or warts cover the outside, and it often has mussel shells, algae, etc. attached. ***Urticina eques*** (see below) was earlier regarded as a variety, called ***U. felina lofotensis***, but is now treated as a distinct species. It is larger, lacks the warts, and usually does not live in shallow water.

The dahlia anemone often has shell particles attached to the papillose, or warty, body wall.

Habitat: Common in the intertidal, attached to stones, on rocks and in crevices. Down to 200 metres depth. Several morphs exist, occurring on different depths, and the specimens that are largest in size appear in deeper water.

Biology: Dahlia anemones have high regeneration capacities. If a piece of the pedal disc is removed, it will produce a new specimen. Most varieties shed the gametes without brooding (oviparity). They are predators which feed on plankton, but also catch various crustaceans, sea spiders, and even fish.

It is possible that this individual may represent a different species, distinguished by the patches on the body wall.

Urticina eques

It is possible that this individual may represent a different species, distinguished by the patches on the body wall.

Be aware that the colour of the tentacles can vary - not all have violet patches.

Distribution: Common all along the north-east Atlantic coasts.
Description: Similar to the **dahlia anemone**, but attains a much larger size, lacks well developed warts, and never have particles attached to the body wall. It also has longer tentacles, spanning up to 300 mm. It tends to be paler in colour, white to light orange with red or orange spots. It was earlier referred to as ***U. felina lofotensis.***
Habitat: Occurs up to the low water line, but usually deeper; recorded down to 400 metres.
Biology: In contrast to the dahlia anemone, the eggs and the larvae develop within the parent (viviparity).

Actinia equina - Beadlet anemone

OLD AND TOUGH

Beadlet anemones are tough animals, and can tolerate large temperature ranges. Some have been kept alive in aquaria for over 66 years.

The beadlet anemone is usually found at shallow locations, and often under stones on the lower part of the shore.

Distribution: Widely distributed in the north-east Atlantic, from West Africa to the Arctic. Very common around the British Isles.

Description: Body short, smooth, with wide pedal disc, height, when extended, up to 60 mm. Diameter of oral disc approx. 50 mm, pedal disc up to 70 mm. Adult specimens with 92 tentacles arranged in six rings; each tentacle approx. 2 cm long and relatively stout, fully retractile. Colouration usually deep red, but green, violet or brown forms occur. Tentacles and body usually without bands or spots, although the mouth is usually surrounded by a blue ring.

Habitat: A shallow-water species, usually intertidal but occurs down to approx. 10 metres depth. Together with the Dahlia anemone, the most common intertidal species.

Biology: Similar to many other sea anemones, the beadlet anemone is viviparous, and there are findings of parents having juveniles with up to 80 tentacles. The reproduction is otherwise poorly known. Both males and females produce offspring by parthenogenesis; sexual reproduction is unknown. Being a tough animal, it has been extensively studied in aquaria, and has been shown to be highly aggressive towards other sea anemones. Like most of these, it has internal, filamentous organs termed acontia, which contain large amount of nematocysts. The acontia are always blue in the beadlet anemone. On contact with other anemones, it will bend towards these and extend the acontia, a process that takes about 5 - 10 minutes. The nematocysts are discharged when the acontium comes into contact with the other animal, usually resulting in a retreat of the latter.

Bolocera tuediae

The northern stone crab, **Lithodes maja***, is often seen carrying this sea anemone, which gives it protection.*

Distribution: In the major parts of the North Atlantic. All around the British Isles, but rare in the south.
Description: A large sea anemone, which can reach a diameter of 300 mm when extended. The body shape is cylindrical, longer than wide, with a smooth body wall without colour patterns. There are up to 200 tentacles, distributed in five-six rings, in numbers that divide by six. They are long and sturdy and provided with very large nematocysts, and cannot be fully withdrawn. The body is usually flesh-coloured or pink, rarely darker red; the tentacles have similar colouration but often slightly darker. Can be confused with large specimens of ***Urticina eques***, although this has small warts on the body wall, the tentacle numbers divide by ten, it rarely has uniform colour, and it can withdraw the tentacles completely.
Habitat: Tends to be attached to solid objects on soft bottoms. Usually 40 - 2000 metres depth, but may appear already from 10 metres. Can also live on vertical cliffs.
Biology: If irritated it will discharge tentacles, which subsequently are regenerated. Shrimps of the genus ***Spirontocaris*** (p. 235) are often associated with this anemone.

BURNING

Bolocera tuediae has potent nematocysts, comparable to those of the lion's mane jellyfish.

Metridium senile - Plumose anemone

Close-up of the mouth, or oral disc.

Asexual reproduction leads to large aggregations of the same species.

Distribution: From the Bay of Biscay to the Kola peninsula. All around the British Isles. Very common.

Description: Highly characteristic: the large number of short, thin tentacles gives a fluffy appearance. Body wall smooth, pedal disc often widened and irregular. The colour may be orange, yellowish orange, or white. There are no pigmentation patterns. Our biggest sea anemone, up to 50 cm high.

Habitat: Occurs just below low water mark, attached to stones, cliffs, bridge poles, etc, also in brackish waters. Down to 100 metres.

Biology: It differs from many of the other sea anemones described here in feeding on plankton. Good ability to reproduce asexually.

The colour varies between orange, white and yellowish-brown. The plumose anemone has very numerous thin tentacles.

Sagartia troglodytes

Sagartia troglodytes. *The genera* **Sagartia** *and* **Sagartiogeton** *are difficult to distinguish from each other.*

Distribution: Widely distributed in the north-east Atlantic, including the Mediterranean. All coasts of the British Isles, but rare in the English Channel.
Description: The pedal disc is wider than the column. Height up to 120 mm, with up to 200 tentacles. Body wall usually pale; oral disc and tentacles highly variable in colour. A common feature for most specimens is a B-shaped mark near the basis of each tentacle, and some have longitudinally striped tentacles. Upper part of body wall often with attached foreign particles. Difficult to separate from ***Sagartia ornata*** (p. 140), which has slightly longer tentacles, and is more transparent and greenish in appearance.
Habitat: Often partly buried in sand where they are attached to shells or stones, sometimes also to the holdfasts of the cuvie (*Laminaria hyperborea*).
Biology: The species reproduces sexually, and identical specimens are therefore uncommon.

This is probably **Sagartia troglodytes**, *but it could also be* **S. ornata** *(see picture on p. 140).*

Sagartia elegans

Distribution: In the North Atlantic from the Mediterranean and the Bay of Biscay to western Norway and Iceland. All around the British Isles.

Description: Tall, thinnest medially, pedal disc up to 40 mm wide, upper part up to 45 mm wide. Body brick red or yellowish brown with colourless or white warts; tentacles about 200. Warts small, mainly on upper part of column. Usually without attached fore-ign particles. Occurs in five different colour morphs, *nivea* with white oral disc and tentacles, *venusta* with orange oral disc and white tentacles, *rosea* with pink or magenta red tentacles, *miniata* with spotted oral disc and tenta-cles, and *aurantiacea* with orange tentacles. The dominant form in Scandinavia is *mine-ata*.

Habitat: Often on cliffs, sitting in crevi-ces.

Biology: Reproduces asexually by deta-ching parts of the pedal disc (basal lace-ration), or sexually by shedding eggs and sperm. As a result of the asexual reproduc-

The diameter of the mouth, or oral disc is larger than the foot, or pedal disc. Clearly visible are pale tubercles on the generally orange body wall. In other aspects, the colour is very variable. Variant **miniata.**

This aggregation of asexually formed individuals are growing on the pier pilings at the biological field station at Trondheim. Variant **mineata**.

tion, large aggregations encounters groups of genetically identical specimens (clones).

Sagartia elegans *var.* **venusta** *occurs relatively rarely in our waters.*

REPRODUCTION

Wheras sexual reproduction is usually achieved by shedding eggs and sperm into the water, many sea anemones can also reproduce asexually by detaching parts of the pedal disc, a process called basal laceration. This kind of reproduction is the reason why we often find many virtually identical individuals together. Such groups, which all emanate from one single ancestor and have identical genomes, are called **clones**.

Sagartiogeton viduatus

Distribution: Limited distribution from Skagerrak to the Trondheimsfjord, mid-Norway.
Description: Provided with 190 long tentacles that exceed the body length when extended. Pedal disc, like in its congeners, wider than oral disc. Body wall grey, greyish green, sometimes bluish green, with distinct moss green longitudinal stripes. Tentacles greyish transparent with longitudinal green or brown stripes (lateral stripes), together with a weak median one. Lateral stripes wider near basis. Size of fully extended animal up to approx. 50 mm long and 15 mm wide. Can be confused with ***Sagartiogeton undatus***, but is almost always attached to eelgrass (***Zostera marina***), whereas the latter lives partly buried in sediment.
Habitat: Associated with plants in shallow water, less than 6 metres depth, and usually sits on eelgrass,

The habitat of this sea anemone is almost always associated with the eel-grass **Zostera marina.**

sometimes also on sea lettuce (***Ulva lactuca***) and stones.

Biology: The species sheds eggs and sperm into the water during summer. When observed in aquaria without aeration, they have been observed to let go of the substrate, float up to the surface, turn upside down, and 'hang' with the pedal disc in the surface. They may even 'crawl' in this way.

The foot, or pedal disc is compatible with the diameter of eel-grass.

Sagartiogeton undatus

In contrast to **Sagartiogeton laceratus***, this species lives on soft bottoms, with the body partially buried in sand or gravel.*

Distribution: From the Mediterranean in the south to the Trondheimsfjord, mid-Norway in the north. All around the British Isles.

Description: Up to 120 mm high and 20 mm wide. Oral disc slightly wider than in the previous species. Up to 200 tentacles that may attain a length of 30 mm. Body wall smooth. Pedal disc usually flesh-coloured, body flesh coloured to pink with darker longitudinal stripes. Tentacles light grey or uncoloured with characteristic, green-brown bands. It is similar to ***S. laceratus***, but this species lacks the stripes on the tentacles. Asexual reproduction is unknown, and it lacks the lumps on the pedal disc of *S. laceratus*, which are involved in basal laceration.

Habitat: From the intertidal down to 100 metres depth. Often partly buried in sand or gravel where it attaches to a stone or a shell.

Sagartiogeton laceratus

Distribution: Bay of Biscay to Norway and Iceland.

Description: A small, slender sea anemone with up to 150 tentacles. Height up to 30 mm, width to 22 mm. Pedal disc wide, usually with characteristic uneven, undulating shape due to asexual reproduction by fragmentation. Body wall tends to be smooth. Colour variable, body wall white or salmon-coloured, with brownish red or, more commonly, brick red, longitudinal rows of irregular spots. Upper part of body wall and oral disc often with weak elements of orange. The tentacles are greyish with grey, white or light yellow bands, sometimes with a brown longitudinal stripe on the oral side. Bases of tentacles with typical brown, more or less B-shaped, mark. This species, when exhibiting the undulating shape of the pedal disc, cannot be confused with others.

Probably **Sagartia ornata.** *Species within* **Sagartia** *and* **Sagartiogeton** *are very similar and difficult to identify from photographs alone.* **Sagartia ornata** *has tentacles of a similar length to* **Sagartiogeton**, *but lacks the characteristic stripes.*

Habitat: Often sitting on auger shells (*Turritella communis*), but also on other objects, usually partly covered by loose particles. From 10-100 metres depth.

Biology: Asexual reproduction by laceration, and many identical specimens often occur together.

This sea anemone is usually found attached to whelk snail shells. Most often seen in more northern waters.

Hormathia digitata

Distribution: Low Arctic species, from Labrador to Barents Sea, down to Kattegatt and Skagerrak. Previously frequent on the North Sea coast of Britain, present status in Britain unknown.

Description: About 100 fully retractile tentacles of varying reddish colouration. Oral disc flesh-coloured, light red or orange. Mouth opening orange or pale yellow-red. Pedal disc wide and flattened, up to approx. 80 mm in diameter. Body thick and wrinkled.

Habitat: Often attached to live or dead mollusc shells, such as ***Neptunea despecta*** and the common whelk (***Buccinum undatum***), and in the case of dead shells these are usually inhabited by the hermit crab ***Pagurus bernhardus***. There are also records from spider crabs (***Hyas***). Occurs from 25-660 metres depth.

This sea anemone is usually found attached to whelk snail shells. Most often seen in more northern waters.

Adamsia palliata (= A. carciniopodos)

The violet-coloured stinging cells are activated if the crab/anemone couple are disturbed.

Distribution: Widely distributed in the North Atlantic, from the Mediterranean to the Norwegian coast north to Lofoten. Occurs all around the British Isles.

Description: This sea anemone is well known for its symbiotic relationship with the hermit crab ***Pagurus prideaux***. The body wall is very thin, very light brown or orange, almost white, with violet to pink spots. Up to 500 thin, short tentacles. Oral disc and tentacles positioned under the hermit crab. Body wall with row of violet acontia, usually retracted, with

For unknown reasons, this crab is leaving its anemone.

Left: *This individual is attached to the rock. We have seen this on some localities in the fjords.*

large number of nematocysts.

Habitat: Present on soft bottoms with sand and mud, from a few metres depth down to approx. 200 metres; in deeper water during summer. Very common in fjords.

Biology: The pedal disc of this anemone is attached to the shell of the hermit crab, and the position under the hermit crab provides efficient uptake of food spill from the host. The crab, in turn, is protected by the nematocysts of the anemone, and the pedal disc of the anemone will also provide an enlargement of the shell opening, such that the growing pagurid will not have to shift shell. If disturbed, the anemone will discharge the long acontia, thereby protecting both itself and its host.

ARTILLERY

The hermit crab, if threatened, communicates with the sea anemone. If a diver or a fish comes too close, it will shake its abdomen, and this serves as a signal for the anemone to discharge the acontia.

Protanthea simplex

Photographed at 50 m depth in Høgsfjorden in southwest Norway.

Distribution: Boreal distribution. Recently located on the west coast of Scotland.

Description: Small, beautiful sea anemone, up to 20 mm high, tentacles excluded. Body salmon-coloured, tentacles slightly lighter. Tentacles up to 200, cannot be fully retracted. Body wall with 24 distinct longitudinal furrows. Pedal disc up to about 10 mm. Cannot be confused with other species in the area.

Habitat: Found as shallow as 15 metres in the Gullmarsfjord in Norway, attached to sea squirts or polychaete tubes. Otherwise common on *Lophelia* reefs, from 50 - 500 metres depth.

This young specimen was photographed on the dead part of a **Lophelia** *coral reef, where it occurs commonly.*

Halcampoides abyssorum

Distribution: The distribution is poorly known, mostly based on old collections and dating back as far as 1877, when several specimens were collected in the Norwegian Sea and the Barents Sea at depths of 640-1134 metres. There are records from divers from the northern parts of the British Isles, and we have seen it in Egersund in south-western Norway.

Description: Very thin and contractile sea anemone. There are twelve pointed tentacles, and the body wall is transparent with longitudinal furrows. Width up to 10 mm, total length up to at least 10 cm.

Habitat: Normally in deep water, although we have observed it during night dives on mixed sand and gravel at 20 metres depth. Very sensitive to light, and withdraws into the sediment within a few seconds when exposed to torch-light.

We observed this anemone over a few weeks, but it disappeared after a storm. Photograph taken during a night dive at Dyrnes near Egersund, south-western Norway.

COMB JELLIES

Phylum Ctenophora - Comb jellies

Here **Beroe cucumis** *has swallowed a sea gooseberry, visible through the body.*

All comb jellies (Gr. ktenos = comb + Gr. phoreus = carry) in northern European waters are planktonic, gelatinous animals that swim by means of ***cilia****. There are a few, more southerly species, which are benthic. They exhibit bilateral symmetry, and have a mouth opening situated in one end and a sense organ in the other.* ***Eight ciliated rows*** *situated on so-called comb plates run between these two 'poles'. The alimentary system consists of the mouth, leading via a small chamber into a system of canals.*

In contrast to cnidarians, comb jellies swim with their mouth first. Many have a pair of **tentacles** for catching prey. These tentacles, however, do not carry nematocysts, but have two kinds of cells, **sensory cells** and **colloblasts**. The former are used to localise the prey, and the latter to catch it. The prey is carried to the mouth by means of the tentacles that are retractile, and then transported into the gut system for **extracellular digestion.** Comb jellies may sometimes even catch prey that is larger than themselves, and it is actually possible to observe their gut content, the body being fully transparent. They tend to live near the surface, but move into deeper water under rough or very bright conditions. Comb jellies do not have different generations, and the larva largely resembles the adult. The gonads are positioned in longitudinal canals; the animals are **hermaphrodites** and reproduce all year around. The regeneration capacity is well-developed.

Ctenophorans do not include many species, and are divided into two classes. **Tentaculata** are provided with tentacles, whereas **Atentaculata** lack them. This latter group includes one single family, **Beroidae**, with ***Beroe cucumis*** being common in our waters.

Class Tentaculata

Tentacle-carrying comb jellies.

Pleurobrachia pileus - Sea gooseberry

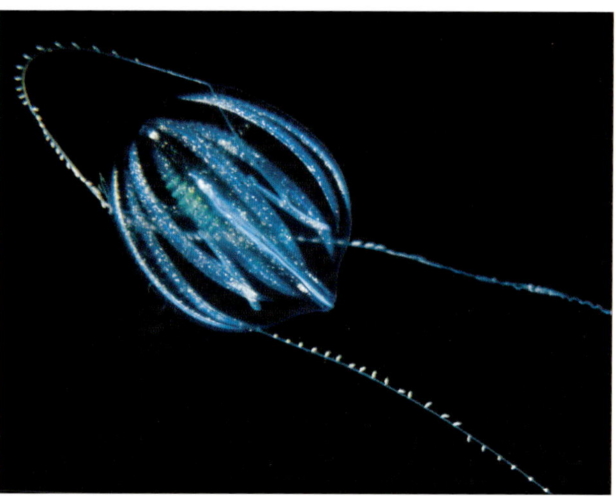

Comb jellies typically have very long tentacles.

Distribution: Widely distributed in the Atlantic, from the Mediterranean to Svalbard.

Description: Highly characteristic. Body near spherical, slightly longer than wide, length approx. 17-25 mm. The eight rows of comb plates are equidistant. Extended tentacles about 15-20 times longer than body, inserted in lateral pockets into which they can withdraw completely. Body fully transparent, anterior gut region and tentacles white, pale yellow or orange.

Habitat: May periodically be very abundant in shallow water along the coasts. Occurs all year round. Tends to be most common in summer, but may also appear in large numbers during some winters.

Biology: Hermaphrodite, with regeneration capacities, as in other comb jellies. Swimming is achieved mainly by coordinated movements of the ciliated plates, starting with the posterior ones, and continued by the more anterior ones, yielding a beautiful moving show of colour effects. Additionally, movement also is aided by musculature in the body wall. All ctenophores are carnivores, although sea gooseberries catch their prey in an unusual fashion. The two tentacles have rows of small side-branches and sweep through the water. They are provided with **colloblast cells**, which attach to the encountered prey–often small crustaceans or fish larvae–with an adhesive mucoid substance, where after the tentacles bring the prey to the mouth opening. The animals have glandular organs in the internal canals that emit a greenish light.

BALANCE ORGANS

The swimming speed and direction is controlled by a sense organ–the statocyst–situated at the aboral end. It is richly innervated and highly sensitive for movement. It consists of a ciliated invagination with a small calcareous particle, the statolith. Tilting of the animal causes the statolith to move, and this is registered by the sensory cilia. They transmit a nerve signal to the ciliary plates, which regulate their activity accordingly. Basically, this organ functions like the balance system situated in our internal ear.

Bolinopsis infundibulum

Note the cilia located along the comb rows.

Distribution: Widely distributed in the North Atlantic, including the Mediterranean and the Baltic.
Description: Relatively large, up to 150 mm long. Near transparent. Body shape almost ovoid, anteriorly with two large mouth lobes. There are two tentacles, and four shorter and four longer comb rows. Extremely delicate, and nearly impossible to catch intact using traditional methods. Larger specimens cannot be confused with other species. ***Beroe cucumis*** (below) lacks the mouth lobes.
Habitat: Planktonic, near the surface.
Biology: Feeds exclusively on other zooplankton.

Class Atentaculta

Beroe cucumis

Distribution: Widely distributed in European waters.
Description: Differs from ***Bolinopsis infundibulum*** in lacking mouth lobes. Characteristic body shape, and specimens longer than 3 cm cannot be misidentified. Large, up to 160 mm long. Like in other ctenophorans the body is virtually transparent. The eight comb rows are of equal length, and produce light interference effects.
Habitat: Similarly to other ctenophorans, it is planktonic and occurs in shallow waters. Can be found in very high abundances.
Biology: Feeds exclusively on other comb jellies, mainly on ***Bolinopsis infundibulum***, but also on ***Pleurobrachia pileus***.

F L A T W O R M S

Turbellarians, like leeches and tapeworms, belong to the flatworms.

Phylum Plathelminthes - flatworms

Plathelminthes-flatworms (Gr. platy, plato = flat, wide + helminth = worm) include about 20 000 free-living and parasitic species. Most belong to groups that are mainly parasitic, such as tapeworms and flukes. It is within the flatworms that we first encounter animals that are bisymmetrical and with distinct anterior and posterior ends; most are also distinctly flattened. They have a head-region with sense organs, and a central nervous system. They are not segmented. Most are hermaphrodites, with male and female gonads present in the same animal. The regeneration capacity is well developed, and a turbellarian, which is cut in two, will develop new posterior and anterior ends of the two pieces.

This attractive flatworm belongs to the family ***Pseudocerotidae****.*

Free-living representatives in the marine environments belong to **Turbellaria**, a group, however, which today is considered as artificial. It includes approx. 4 500 species. Within **Polycladida** we find the most commonly encountered ones, characterised by a thin, foliose body. Many are brightly coloured, some are large, and the highest diversity is found in tropical waters.

Prostheceraeus vittatus

Distribution: Northeast Atlantic and the Mediterranean.
Description: Very thin, up to 50 mm long and 25 mm wide. Fragments easily if touched. Highly characteristic pigmentation; ivory-white with dark, longitudinal stripes. Two head tentacles. Cannot be confused with other species.
Habitat: On rocky bottoms, under stones, usually at 10-20 m depth, but can be found also in the intertidal.

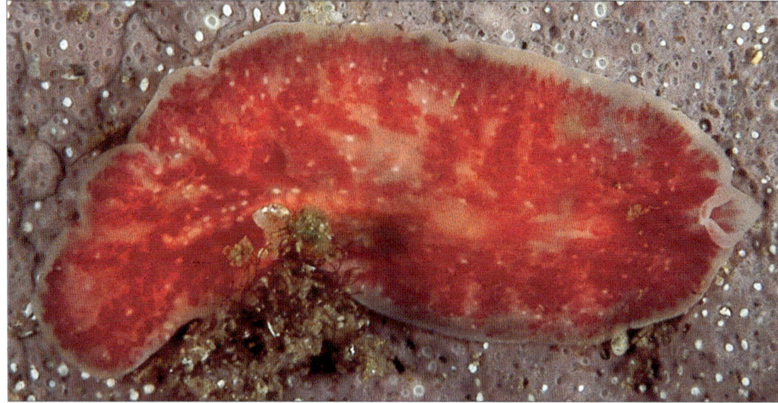

Oligocladus sanguinolentus *photographed in shallow water in south-western Norway*

LEARNING ABILITIES

These animals can learn, in spite of being very simple. Turbellarians, if exposed to light followed by an electrical charge, will develop 'conditioned reflexes', and afterwards react just to the light. If a specimen after that is cut in two, both regenerated specimens will 'remember' the experience.

Prostheceraeus vittatus *has characteristic longitudinal brownish-black stripes. Very common.*

RIBBON WORMS

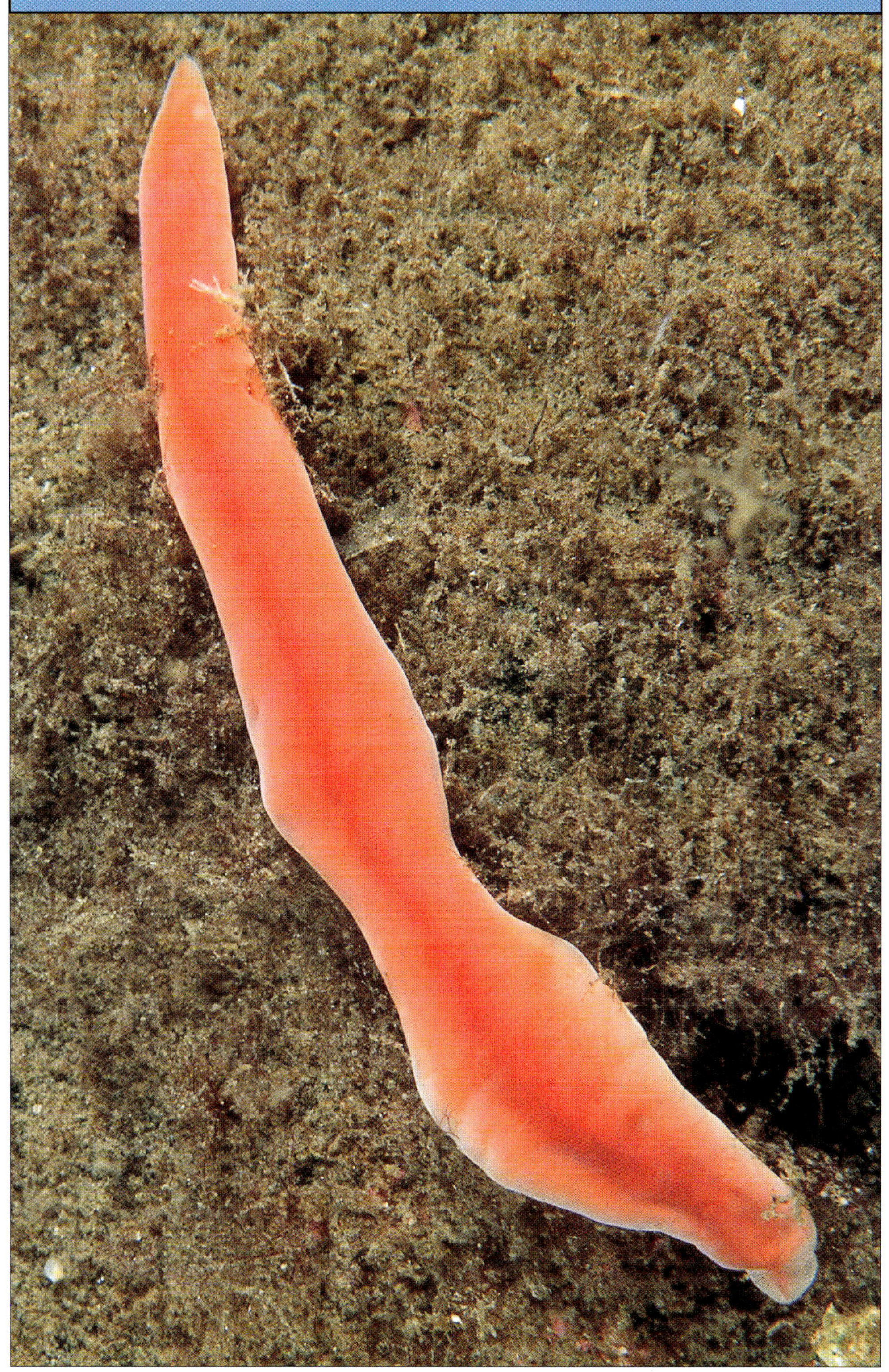

Phylum Nemertea

Nemertea, sometimes called ribbon worms (Gr. myth. Nemertes: one of the nereids - sea-nymph, daughter to Nereus and Doris), includes about 900 species. They are non-segmented worms, usually very thin with a highly extensible body. They vary in length from less than 1 mm to about 60 m in extended condition. Many are slightly flattened, and there is a more or less distinct head region, usually with eyes. They differ from the flatworms in having a ***two-way alimentary system*** *with an anus, a* ***closed circulation system****, and an eversible proboscis with a unique construction. The body is covered by* ***cilia*** *and* ***gland cells*** *that produce mucus. The majority are benthic and free-living, but there are a few pelagic representatives, and some that live in symbiosis with bivalves and other invertebrates. They are common in sand and mud in the intertidal.*

The large majority of nemerteans are carnivores, either feeding on small invertebrates which the catch, or on dead animals. They have separates sexes, and asexual reproduction by means of fragmentation occurs but is less common. They are split into the two classes Anopla and Enopla, each further subdivided into two orders. Species identification is often based on internal features, although many can readily be identified by their shape, and especially, by their pigmentation patterns.

Many ribbon-worms can stretch to several metres in length.

ACCUMULATORS OF POISONOUS SUBSTANCES

Most nemerteans are carnivores. They do not produce any tubes and many move around completely unprotected and visible on the bottom. However, they are protected by their ability to accumulate poisonous substances. Some also produce a nerve poison that immobilises their prey.

Tubulanus annulatus

Tubulanus annulatus *is recognisable by its three longitudinal white lines interrupted by cross-stripes.*

Distribution: Widely distributed on the northern hemisphere. In Europe from the Mediterranean to western Norway.
Description: Distinguished from most other nemerteans by the presence of three longitudinal white lines, one dorsal and two ventral, which are interrupted by rings. Body colour brown to reddish. Length, when extended, up to 1 m, although usually 10 or so cm. A characteristic feature of the genus is a rounded head that is wider than the body. *Tubulanus annulatus* can be confounded with two other members of the genus, ***T. superbus***, which has a third ventral longitudinal line, and ***T. nothus,*** which has a white head with two black spots.
Habitat: Occurs in the intertidal down to 30 - 40 m, both on soft and hard bottoms, and often in empty polychaete tubes where the original inhabitant has been consumed.

Tubulanus superbus *is distinguished from* **T. annulatus** *by having an additional longitudinal stripe, i.e. it has four in total. This specimen photographed near Stavanger in Norway.*

Cerebratulus marginatus

This specimen probably is **Cerebratulus marginatus.**

Distribution: Widely distributed on the northern hemisphere, both in the North Pacific and the North Atlantic.

Description: Members of this genus are more or less flattened, with almost fin-like lateral edges, and many are good swimmers. They are wider than most other nemerteans, and *C. marginatus* often is about 10 cm long and 25 mm wide, although it can stretch up to a meter. The anterior end is wider than the posterior end. A lateral slit is present alongside the head. The colour is greyish brown, greyish green, weak blue, or brown, with lighter sides. A large number of eyes, although barely visible. Can be confused with ***Cerebratulus fuscus***, which also has large number of eyes; other members of the genus lack eyes.

Habitat: Rarely shallower than 20 m, down to 150 m. Soft bottoms.

The head end is conspicuously wider than the rear end in species within this genus. They are further characterised by a flattened body with fin-like margins.

Lineus longissimus

UP TO 60 M LONG!

The extension capabilities of *Lineus longissimus* are beyond competition, and up to 60 m length of worm may appear from what previously was only a small, entangled ball. When seen extended it appears mostly like sewing-thread spread out on the bottom. If disturbed this sewing-thread will constrict and then shrink to 'only' a few meters length.

Distribution: Widely distributed in North-east Atlantic, all British, Isles, west coast of Norway to Trondheimsfjord.
Description: When observed in nature, it appears as a very thin, brown to near black, thread among the algae. If disturbed it will withdraw into a small knot, but can extend to a length of up to 60 m, although more commonly it measures 5 - 15 m in length with a width of 5 mm. There is a row of 10 - 20 eyes each side of the head.
Habitat: Both on hard and soft bottoms, shallow water and intertidal.

Detail of the head end of this up to 60 m long ribbon worm.

Lineus bilineatus

This species is very characteristic with its two narrow stripes along the mid-line.

Distribution: All British Isles, North Sea, west-coast of Norway, Iceland and the Mediterranean.

Description: As indicated by the species epithet, it is characterised by two dorsal parallel, longitudinal white or light yellow stripes. At the head they dissociate and form a bright triangle. Eyes absent. Length up to 70 cm, width to 6 mm. Colour reddish brown, purple, or chocolate brown. Unlikely to be confused with other species.

Habitat: Intertidal and in shallow water, on many different kinds of bottoms, but often among red calcareous algae (*Corallina*), under stones, and among mussels.

We have often recorded this ribbon worm on gravel and stony bottoms.

Lineus ruber

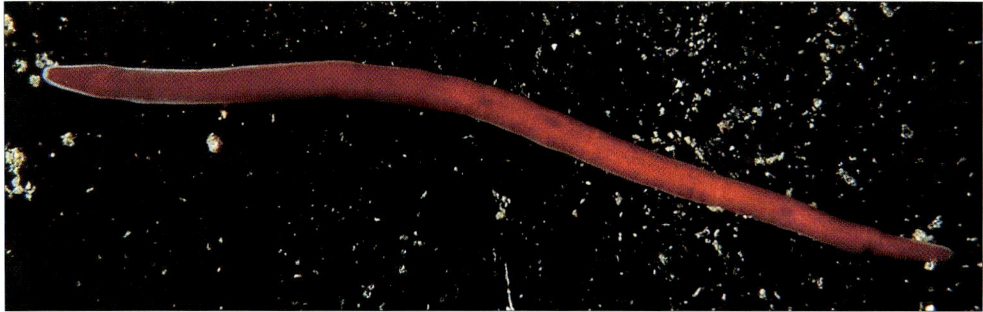

Lineus ruber *is relatively abundant in shallow water.*

Distribution: Widely distributed across the northern hemisphere.
Description: Rarely over 8 cm long and 2 - 3 mm wide. With 2 - 8 eyes in irregular row each side of the head. Colour reddish brown to orange, ventral side brighter. Similar to ***L. viridis***, which differs in being green.
Habitat: Common under stones on mud in the intertidal, but also on rocks among mussels.

Nipponnemertes pulchra

Distribution: The species is widespread in the northern hemisphere, with a distribution extending from the east coast of North America to the Atlantic coasts of France and Scandinavia.
Description: The species epithet refers to the external appearance of this animal–it

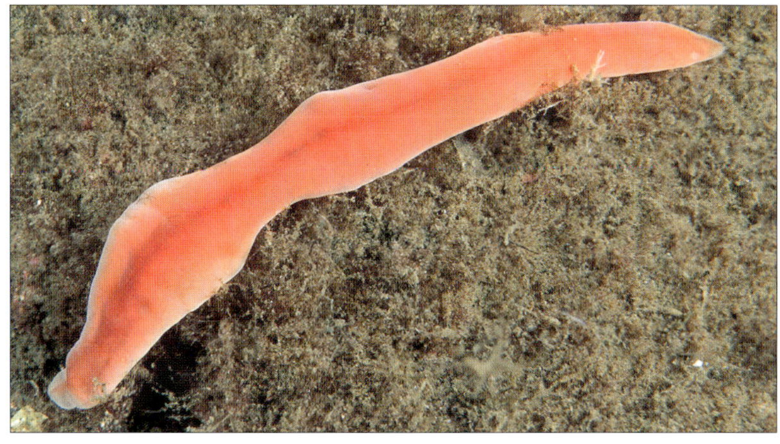

The species epithet pulchra *means beautiful!*

We have most often seen this ribbon worm in fjords during wintertime, from 10 m depth and below.

means simply "beautiful". The colour is sharp red; reddish brown dorsally, but considerably brighter ventrally. Characteristic features include a median keel and two lateral furrows on the head. Length up to 9 mm, body stout.

In contrast to the previously mentioned nemerteans, *N. pulchra* belongs to Enopla.

Habitat: Mud bottoms, from about 5 down to several hundreds of meters.

Phylum Priapulida

This phylum consists of entirely marine, benthic, cylindrical, 'worm-like' animals. They are non-parasitic and occur in all seas, down to several thousands of meters. The body is divided in two or three regions: an anterior and fully eversible proboscis; the main trunk, which is nonsegmented but often carries transverse and /or longitudinal furrows; and, in some taxa, a posterior appendix. Priapuloids dig through sediment in search of food. Larger specimens (up to 30 cm long) feed on other invertebrates that are not protected by external shells, such as polychaetes. It is a small group with only 16 species known worldwide. Apart from the species described below, we may also encounter **Halicryptus spinulosus** *and* **Tubiluchus arcticus** *in our waters.*

Priapulus caudatus

This specimen was hidden under a stone at 25 m depth in Lurefjorden, southwest Norway.

Distribution: Widely distributed in the northern hemisphere. Arctic, southwest British Isles, North Sea. All Norwegian coast.
Description: Up to 20 cm long. Body in three distinct regions, with a posterior, branching tentacle-like appendage. The proboscis constitutes about 20-30% of total body length, and is provided with 25 distinct longitudinal furrows.
Habitat: Wide range of depths, 6 - 7500 m.
Biology: Sexes are separate, and the animals molt their cuticle (outermost dermal layer) before reaching maturity.

ANNELIDA

Phylum Annelida

The annelids (Fr. annelide < Fr. annele- ler = organised in rings, Lat. annulus = ring, a term introduced by Lamarck 1802) include over 15 000 described species. They are highly variable in size, range in length from less than 1 mm to three metres, and are characterised by a number of repeated and, more or less, similar segments (serial homology). They have a mouth, a gut and an anus, usually a closed circulatory system with blood, a well-developed nervous system, and some taxa are provided with gills. There are two major groups, ***Polychaeta*** *and* ***Clitellata****, of which the latter includes the oligochaetes, such as earthworms, and* ***Hirudinea****, leeches.*

Class Polychaeta

The vast majority of polychaetes (Gr. Polys = many + Lat. chaeta = bristle)–or bristle worms as they are sometimes called–are marine, and include about 10 000 described species. There are a few limnic, and even fewer terrestrial ones. Polychaetes have bristles occurring in bundles, usually inserted on muscular appendages, **parapodia**, whereas clitellates have either few chaetae or no chaetae at all, and no parapodia. The head region is often provided with various **appendages**, which can have sensory or feeding functions, referred to as antennae and palps. Many species have one or several pairs of **eyes** on the head. In many groups the anterior part of the gut is eversible and forms a **proboscis**, sometimes also equipped with teeth or jaws. There are both separate sexes and hermaphrodites, and the larval development can be more or less direct, or include a planktonic trochophore larva. A few taxa are pelagic also as adults, although the majority are benthic and live in tubes, digging or staying in the sediment, or moving around on the sea floor. The classification of polychaetes is currently in a state of flux, as new knowledge about their evolution is gained. An older classification recognised two groups, mainly based on their way of living, **Errantia**, and **Sedentaria**. Although now replaced, these names still are encountered in the literature and are retained here for practical purposes. As is usual, we also present most of them family by family below. Species identifications often demand detailed knowledge of the group. Many of those included here can be identified from high-resolution photographs, although accurate determinations usually require closer studies.

Here the head segment carries palps, eyes and antennae. Further back, the segments have parapodia ('feet') with bristles.

Errantia

Errant polychaetes usually do not construct tubes, and tend to move around actively on the bottom. Some are pelagic. The body is not divided into distinct regions, and most segments have well-developed parapodia and bristles. Most are predators or scavengers.

Superfamily Aphroditoidea - Scale worms

Characterised by having the back partly or fully covered by tiled scales (elytra). Most species are flattened and short, and many live as commensals with tube-living or burrowing animals. They are often difficult to identify to species.

Aphrodita aculeata - Sea mouse

The sea mouse is highly characteristic, covered in fine bristles, which give it a furry appearance. The bristles located along the sides have a greenish metallic sheen.

Distribution: Widely distributed in European waters, from the Mediterranean to northern Norway. All around the British Isles.
Description: Highly characteristic. Large and stout, up to 200 mm long and 35 mm wide. Elliptical in outline, with a convex dorsal and flat ventral side. A fine felt of chaetae covers the back, whereas the lateral chaetae include stouter ones that are strongly irridescent. The colour of the body is yellowish brown. A similar species, ***Laetmonice filicornis***, has a flat, rather than convex, dorsal side, reaches a length of 90 mm, and is dorsally bluish to violet in colour.
Habitat: Moving through the sediment in muddy sand, from 12 down to several hundred metres depth.

Lepidonotus squamatus

This is a common scale-worm in our waters, often found under stones.

Distribution: North sea, English Channel and most North Atlantic coasts
Description: Back covered by tiled elytra. Colour variable, often yellowish with darker spots on the oval elytra. Up to 50 mm long. Similar to ***L. clava***, which also has 12 pairs of elytra, but differs from the latter in that the elytra cover also the mid-dorsal part of the animal. Coils if disturbed.
Habitat: Common intertidally under stones, but present also in deeper waters, often associated with tube-living worms.

Eunoe nodosa

Distribution: Common throughout large parts of northern Europe. Circumpolar.
Description: Dorsum (back) covered by 15 pairs of scales (the photographed individual has lost some). Scales greyish and/or brown, often with violet tones and a pale or dark patch in the centre. Antennae reddish, with brown bands and white tips. Dorsal cirri (lateral antenna-like projections on each segment) with similar pigmentation as the antennae and considerably longer than the bristles. Relatively large, up to 90 mm in length and 30 mm wide.
Habitat: Common both on coarse sandy or shellsand substrates and muddy bottoms. Down to several hundred metres depth.

This most likely is Eunoe nodosa. *Note the projections at the posterior scale margins. Here from Egersund, southwest Norway.*

Alentia gelatinosa

The scales are typically soft and translucent, forming a jelly-like covering.

Distribution: Northeast Atlantic north to northern Norway. Around most of the British Isles.

Description: Large scale worm, up to 90 mm long, with 18 pairs of elytra, which are transparent, dirty white or brownish. Back often with transverse brown and white bands; ventral side orange. The number and shape of the elytra are characteristic.

Habitat: Shallow soft bottoms, rarely intertidally.

Some individuals are almost colourless.

Harmothoe propinqua

Distribution: Northwest Atlantic, in European waters from the Mediterranean up to Norway, possibly to the northernmost parts.

Description: Back partly covered by 15 pairs of oval elytra, leaving both the middorsal part and the posterior ten or so segments naked. Length up to 30-40 mm, width 8-10 mm. Colour highly variable, elytra with brown, grey or reddish spots, usually with a brighter central spot. Difficult to separate from many other species of *Harmothoe*.

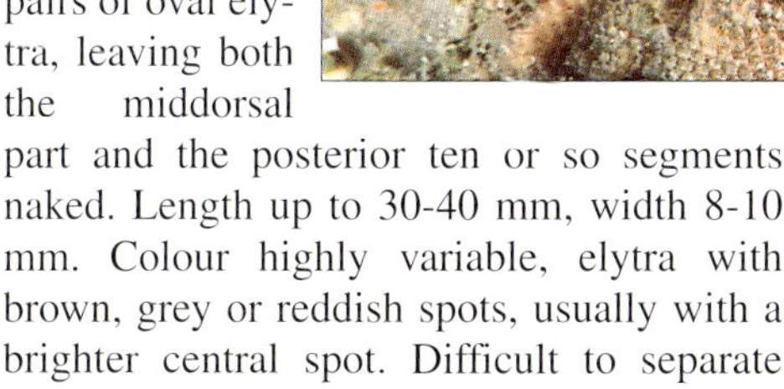

This scale worm belongs to the genus **Harmothoe**, *possibly* **H. propinqua**. *Certain identification is not possible from photographs alone.*

Habitat: Intertidally among stones, but also in deeper waters.

Harmothoe *sp.*

Sthenelais boa

Distribution: In European waters from the Mediterranean to northern Norway.
Description: A long and slim scale worm, up to 200 mm long. Back completely covered by elytra, overlapping both mid-dorsally and posteriorly. Elytra bordered by fine papillae. Colour light grey to golden brown, sometimes with reddish brown transverse stripes. There are several species of *Sthenelais* in the area, and accurate identifications demand closer examination. However, none of the others exceed 120 mm in length
Habitat: Shallow waters on various kinds of bottoms, under stones, in gravel, or in eelgrass (*Zostera marina*).

The body typically is golden-brown and elongate. Individuals longer than 12 cm probably are **Sthenelais boa**, *but smaller individuals may belong to one of several species.*

Family Nereididae

Characteristic free-living polychaetes often referred to as "rag worms". The body is long, often with more than a hundred similar-looking segments, and with well-developed ***parapodia****. Considering their size, they can be fearsome animals, armed with a pair of chitinous jaws that are powerful enough to harm a curious human finger. The eversible proboscis usually has small teeth (****paragnaths****) on the surface, which are highly useful for identifications of the beasts (applying a bit of pressure on the segments after the head usually forces the specimens to extrude the proboscis). Other diagnostic features are the details of the parapodia and the bristles. The head is provided with two pairs of eyes, a pairs of small antennae, and a pair of large* ***palps****. The first rings following the head also carry long cirri. Most nereidids are active predators. We include three of the most common species in the area. A fourth species,* **Neanthes fucata,** *lives inside the shells of hermit crabs.*

Anterior end of **Neanthes virens.** *The head carries palps and antennae, immediately followed by the cirri of the first segments.*

SWARMING

In association with reproduction, large numbers of nereidids can be found swimming in the water. They develop into so-called **epitokes**, and go through a number of morphological changes, which, among other things, facilitate swimming, such as the development of paddle-like bristles.

Nereis pelagica

Nereis pelagica *often is found on the shore, but take care - it bites!*

Distribution: In European waters from the Mediterranean to the Arctic. Also in the south Pacific. Most coasts of the British Isles.

Description: Apart from the features previously mentioned, which are characteristic for the family, this species has very long palps. The colour is metallic golden brown, also with a greenish hue, and with a distinct red longitudinal blood vessel. Length up to 210 mm. It is difficult to separate from several other nereidids, and examination under a microscope generally is required.

Habitat: In shallow water among stones and algae, often in the holdfasts of *Laminaria*.

Members of the family Nereididae often are found under stones on rocky substrates. Here with another polychaete **Eupolymnia nebulosa**.

Hediste diversicolor (= Nereis diversicolor)

Here the parapodia ('feet') are relatively widely-spaced.

Distribution: In European waters from the Mediterranean to northern Norway. Also in the Baltic. Most British coasts.

Description: Colour variable as indicated by the Latin specific epithet: greenish, yellowish, or orange, but with a highly distinct middorsal blood vessel. Length up to 120 mm. Usually requires examination in microscope of parapodia and chaetae for identification.

Habitat: Lives in galleries in sand, mud or mud, often in brackish water. Intertidal and in shallow waters.

Utility: Commonly used as bait by anglers.

Neanthes virens (= Nereis virens)

Distribution: European Atlantic coasts up to mid-Norway. Common along the south and west British coasts.

Description: Often called "king rag" since it can reach a length approaching 1 m! Can have several hundreds of segments. Slightly more flattened and wider than other nereidids. The dorsal parts of the parapodia have characteristic leaf-like extensions, and the species is a good swimmer. The colour is dark green with a bluish, metallic hue. Dorsal blood vessel not distinct. Easily

A large specimen, here from southwest Norway.

It is a spectacular sight when **Nereis virens** *swims in serpentine movements across the sea floor. The parapodia have rythmic movements.*

confused with ***N. grandis,*** which has a yellowish green dorsal side, lacking blue colour.
Habitat: Lives in galleries in soft bottoms. Swims with sinusoidal movements. Intertidal and in shallow waters.
Utility: Very popular as bait. Caught commercially, and cultured in the UK.

Family Phyllodocidae

Long and thin worms, often with hundreds of segments. The dorsal part of the parapodia is large and flattened, and the animals are for this reason sometimes called "paddle worms". They are separated from nereidids by having a single pair of eyes, rather than two, and by lacking jaws. Many produce large amounts of mucus when disturbed. We here include a picture of one specimen, likely belonging to **Phyllodoce groenlandica**, *distributed from the Artic and all around the British Isles. However, there are a number of species in the area, and many are difficult to identify.*

Eggs from **Eulalia viridis**. *We have seen such egg clusters, with the female worm coiled within it. Main picture probably* **Phyllodoce groenlandica.**

Family Hesionidae

Rather small animals, usually with less than 50 segments. They are active predators and can move and swim fast. The first segment carry 6 - 8 pairs of cirri, which are usually much longer than on following segments. Several species live in shallow water among stones, mussels and algae, often in Laminaria *holdfasts.*

Ophiodromus flexuosus

Distribution: In European waters from the Mediterranean to northern Norway. All British coasts.

Description: Up to 70 mm long. Highly characteristic pigmentation, with brown colour and regularly occurring transverse white-bluish bands across the back.

Habitat: Usually on shallow mud bottoms, but occurs down to 500 m depth. In the Mediterranean it lives commensally with the sea star *Astropecten aurantiacus.*

The highly characteristic green or blue stripes make this species easily recognised.

Ophiodromus flexuosus *also is found at polluted or oxygen-depleted sites.*

Family Eunicidae

Large cylindrical worms with many segments. Occurs in sand and gravel or mixed bottoms and in **Lophelia** *reefs.*

Eunice norvegica

*Here on a coral reef (***Lophelia pertusa***) in the Trondheimsfjord, Norway, at 52 m depth.*

Distribution: In the northeast Atlantic from Morocco and the Mediterranean in the south to northern Norway and southern Iceland in the north.
Description: Large, up to 410 mm long and approx. 10 mm wide. Colour whitish or yellow, with a bluish metallic hue. They produce a thin parchment-like tube with two or more openings, slightly larger and longer than the worm itself, and within which it can turn. In *Lophelia* reefs the tubes are attached to corals. There are several species of *Eunice* within the region, and they can be difficult to separate.
Habitat: Often lives associated with *Lophelia pertusa*. It is mostly found at depths below 100 m, and down to 1500 m. It occurs both among dead and live corals.

Sedentaria

Usually with less well-developed parapodia than errant polychaetes, and lacking a large eversible and armed proboscis. Body usually divided into regions, such that all segments and bristles do not look similar along the body; often divided into a thorax (anterior part) and abdomen (posterior part).

Family Chaetopteridae

Species within this family live in tubes, which are either buried in sediment and with only the openings emerging, or attached to rocky substrata. The body is divided into highly distinct regions, which are specialised for different purposes–a feature that gives the animals a highly characteristic appearance. Chaetopterids are all filter feeders.

Chaetopterus norvegicus (= *C. variopedatus*)

Distribution: Species of *Chaetopterus* currently are confused. The genus has a cosmopolitan distribution, and the closely related *C. variopedatus* has been mixed up with *C. norvegicus*, such that the distribution at present is difficult to assess.

Description: Median region provided with five pairs of enlarged parapodia with wing-like extensions, serving to create water currents through the tube. Extremely slim 'waists' connect these segments to each other. The animals are very delicate and fragment easily; however, the regeneration capacity is high. The colour is greenish yellow to yellowish white, with the gut forming a darker line through the animal. Mature males are whitish, mature females reddish yellow. The closely related *C. sarsi*, which also occurs in the area, can be distinguished by lacking the slim 'waists', and by having the tube buried in soft bottoms (preferably gravel).

Habitat: From the sublittoral down to 150 m depth. Tube usually attached to stones or other hard substrates.

Chaetopterus norvegicus *lives in a parchment-like tube. Here the tube is under a stone. Rocky bottom, 20 m depth.*

The body is divided into three very distinct regions.

Family Pectinariidae

Pectinarids build highly characteristic cone-shaped tubes of sand grains.

Detail of the tube opening. Note the golden bristles.

Polychaetes in this family construct cone-shaped tubes of carefully selected sand grains or pieces of shells. The tubes have openings at each end. They occur in sand in shallow water, often in areas exposed to waves or currents. The tubes are orientated upside down in the sediment, with the larger anterior end directed downwards. A part of the posterior end usually emerges from the sediment. The actual animals are short and stout with a limited number of segments. The body consists of three regions. The head is provided with stout, golden anteriorly-directed bristles that are used for digging. The mouth is provided with a large number of tentacles that are used for feeding, and on the first segments there are two pairs of pectinate gills. The median part consists of 12-13 segments, and is followed by a much smaller posterior part. Pectinariids live from deposited organic particles. There are five species of Pectinaria *in the region, of which* P. koreni *is the most common. It can be recognised by having a very weakly curved tube, up to approx. 80 mm long. The inhabitant reaches a length of about 60 mm.*

In our waters are five species within **Pectinaria**. *This is the tube of one of them.*

Family Flabelligeridae

Rather short animals, up to 60 mm long and approx. 50 segments. The long, anteriorly-directed chaetae are characteristic, and surround the mouth region, which often is completely retracted, but carries green gills and feeding tentacles. The green colour is due to the fact that the group has green blood. They are usually found under stones or in sediment.

Pherusa plumosa

A highly characteristic polychaete often found under stones on hard bottoms.

Distribution: Widely distributed along the European coasts.
Description: Short and stout, up to 60 mm long. Posterior segments gradually smaller than anterior ones. The anteriorly-directed bristles are located on the first three rings, forming a cage surrounding gills and tentacles. Four brown eyes. Colour of younger specimens dark orange to reddish yellow, adults greenish, brownish or grey. Another species, ***P. falcata***, which also occurs in the area, only reaches a length of 20 mm, lacks eyes, and occurs below 100 m depth.
Habitat: Under stones, or in burrows in sediment. From a few m depth down to 1400 metres.

Family Terebellidae

Anterior part of body (thorax) much stouter than posterior part (abdomen). Thorax often with 17 rings, abdomen with varying number. The large tentacular crown, with its highly contractile tentacles, is characteristic. Most species have gills on segments 1 - 3. Terebellids build a tube that is attached by one side to a surface, or emerging from the sediment, with the very long tentacles radiating from the opening.

Thelepus cincinnatus

Terebellid polychaetes can be difficult to identify with certainty, and usually a photograph is not enough. This most likely is **Thelepus cincinnatus.**

Distribution: Circum-Arctic. Along European coasts up to the Arctic. Mediterranean.
Description: Length up to 200 mm, with approx. 100 segments. Colour variable, brownish, pink, or orange-yellow, with slightly lighter ventral side. Two of the anterior-most segments have groups of filamentous, red gills. The tentacles, large in number, are orange or light pink, sometimes with red spots. The tube is thin, opaque, with small, attached particles.
Habitat: Common among kelp holdfasts or on old shells. Recorded down to 4000 metres depth.

Eupolymnia nebulosa

These polychaetes often are visible by their tentacles extending across the sediment.

The white spots are typical for **Eupolymnia nebulosa**. *Note the red gills.*

Distribution: In Europe from the Mediterranean to northern Norway, but also occurs in other oceans. Most of the British Isles.
Description: Up to 150 mm long and approx. 8 mm wide. Body soft, delicate. The colour is pink, orange or brownish, covered by white spots. Approx. 100 segments. Three pairs of branching gills, often with white spots. Tentacles pink or greyish white with white rings. May be confused with several other species from the region.
Habitat: Thin, sand-incrusted tubes that are attached to stones or dead shells. From shallow waters down to approx. 500 meters depth.

Family Cirratulidae

Usually constructing galleries in soft bottoms, by divers often detected by the long red gill filaments emerging from a under a stone or a crevice. Both tentacles and gills are easily lost during attempts to pull the animal from its burrow. Body shape usually long and slender, with poorly developed parapodia and bristles emerging more or less directly from the body wall.

Many segments typically bear single gill filaments.

Cirratulus cirratus

We have observed aggregations of this species around fish processing plants. Here on a sediment/organism covered rocky substrate.

Distribution: Large parts of northwest Europe, up to northern Norway.

Description: Long and thin with up to 130 segments, length up to 120 mm. Head with 4 - 8 conspicuous eyes in a transverse row. First segment appears to be divided into three rings, of which the posterior-most is provided with two groups of 2 - 8 long and slender tentacles, red to yellowish in colour. Long, thin, red to yellow gills occur all along the body. Colour of body is orange, pink or brownish red.

Habitat: We have observed this species in heavily polluted areas, under fish farms or around the quays of fishery industries.

Family Serpulidae

Serpulids are characterised by having white, calcareous tubes. They have 30 - 40 tentacle-like appendages called **radioles** *emerging from the tube opening, and most have one of these transformed into an* **operculum**, *serving as a lid for the opening. Many can be identified from their tubes and opercula. Most species have tubes that are attached to the substratum, often occurring in colonies.*

Filograna implexa

Filograna implexa *forms large colonies, and can even build reefs. The worm itself is only 4 mm long. Note the hydroid* **Tubularia larynx** *near the centre of this colony.*

Distribution: Cosmopolitan in all oceans.
Description: *Filograna implexa* may constitute a species complex. Occurs in large colonies. Each specimen is only up to 4 mm long, with about 30 segments, and is pink in colour. There are eight radioles. Colonies consist of a dense network of fine, delicate, and almost transparent tubes, each with a diameter of about 0.5 mm. Each specimen has two opercula.
Habitat: Usually subtidally on bottoms exposed to currents, in depressions between stones. In addition, among dead branches of ***Lophelia pertusa,*** soft corals such as ***Paragorgia arborea*** and ***Paramuricea placomus***, and on shipwrecks, where they can be abundant on exposed sites such as the masts.

Ditrupa arietina

Distribution: Widely distributed in the northeast Atlantic.
Description: The tube of this species resembles to those of tusk shells (*Antalis* spp.), but is of a more even width, more strongly curved, and has the opening with narrowing outer sides. Observation of the radioles and body makes the affinity obvious. Tubes up to 40 mm long, animal up to 20 mm. Tentacular crown with circular red bands.
Habitat: On soft bottoms, tubes not attached, from 10 - 1100 metres depth.

This species cannot be confused with any other polychaete, but it does resemble the tooth shell, which is a scaphopod mollusc.

At some localities, huge aggregations of individuals can be seen.

Hydroides norvegica

Hydroides norvegica *is found on hard substrates and commonly fouls ship hulls and buoys.*

Distribution: Cosmopolitan, in European waters from the Mediterranean to Norway.
Description: Characteristic calcareous tubes, having an elevated and coiling anterior end. The animal is up to 30 mm long, and the tube has a diameter of up to 3 mm. The radioles carry red and white stripes. The older parts of the tube are brownish, whereas the younger parts are shining white.
Habitat: Hard substrata, up to the intertidal zone. Common fouling organism on ships, buoys, and jetties.

Serpula vermicularis

Distribution: Cosmopolitan. Western parts of the British Isles, Norwegian coast north to Finnmark.
Description: The tubes are characteristically pink to light red, with conspicuous ledges, but also with longitudinal ridges. The tube has a diameter up to approx. 5 mm, and the worm is up to 70 mm long, with about 200 segments and 70 radioles. The colour of the worm varies from light yellow to brick red.

The tube of this species gradually widens towards the head end, which is flared and trumpet-like.

The attractive radiolar crown is used in feeding. To the right in the picture is the red operculum ('lid').

Habitat: Lives on hard substrata, on stones and various shellfish. Occurs up to the intertidal zone. The animals often grow several together, but also with other serpulids such as ***Pomatoceros triqueter*** and ***Hydroides norvegica***.

Characteristic 'ring' on the tube - see left of picture.

Pomatoceros triqueter

The tube is narrower at the posterior end and often ends in a spiral.

Distribution: Atlantic, in Europe from the Mediterranean to northern Norway. Very common.

Description: Tube up to 12 cm long, irregularly winding (especially the posterior end), and usually attached to the substratum all along its

Certain hard-bottom areas support thousands of serpulid polychaetes.

length. The tube has a characteristic keel, in contrast to the very similar, more southerly species ***P. lamarckii***, which has three keels. The tentacles are feather-like. The animals are highly polymorphic in colour, the most common ones being red to reddish-brown with yellow to white transverse bands. The collar is bluish.

Habitat: Attaching to virtually all hard substrata, including stones, crustacean and mollusc shells, kelp holdfasts, ships, buoys, jetties, etc. Animals attached to crustaceans can actually serve as indicators for the time since the last moulting.

The colour of the radiolar crown often is red, with yellow or white perpendicular bands. Other individuals can have orange or bluish radioles.

Placostegus tridentatus

At the entrance to the tube, only the operculum ('lid') of the worm can be seen. The radiolar crown is retracted within the tube.

Distribution: Widely distributed in the northeast Atlantic. All Norwegian coast.
Description: Characteristic tubes with the outermost and youngest part white and partly transparent, remaining parts are brownish white. They have a median serrated keel and two lateral ones. The tube opening is provided with three conspicuous pointed teeth (explaining the name "tridentatus"). The tentacular crown is dark red with two brown transverse bands. Up to 300 mm long.
Habitat: On hard substrata, usually below 25 metres depth, recorded down to 2500 metres.

Left: **Placostegus tridentatus,** *together with a colony of the bryozoan* **Reteporella beaniana.**

Subfamily Spirorbinae

Highly characteristic serpulids that have tubes looking like a post-horn. There are many species, all of which are small. The tube is a spiral, usually lying flat against the substratum. The direction of the spiral is used in identification, and those that coil in a clockwise fashion are called ***sinistral****, whereas the anti-clockwise ones are* ***dextral****. Many species are associated with highly specific substrata, such as a single species of algae.*

Spirorbis spirorbis (= *S. borealis*)

Spirorbis spirorbis *can completely cover the surface of certain brown algae.*

Distribution: In the northeast Atlantic from the English Channel to northern Norway.

Description: This small serpulid builds white, smooth and clockwise coiling tubes looking like post-horns. The diameter of the whole tube is 3 - 4 mm. It is easily confused with a number of other species of the same genus. Of these, ***S. corallina*** always grows on the red alga ***Corallina officinalis***, and ***S. tridentatus*** (previously regarded as the same species as *S. spirorbis*) has rounded keels on the tube.

Habitat: Usually on brown algae, especially on ***Fucus serratus*** and ***Laminaria*** spp. From the intertidal down to 280 metres.

Circeis spirillum

Distribution: Widely distributed in the north Atlantic from the Azores to northern Norway.
Description: The tube, in contrast to the previously mentioned ***Spirorbis*** species, coils anti-clockwise. The orange worm is visible through the more or less transparent and porcelain-like tube.
Habitat: Usually on hydroids and bryozoans, sometimes on red algae, from 5 - 60 m depth.

On close inspection, the radiolar crowns can be seen.

Family Sabellidae - Feather-duster worms

This family contains some of the most strikingly beautiful polychaetes, many of which attain a considerable size. They are characterised by their large and often colourful, feather-like radioles, forming a symmetrical crown. The radioles are ciliated, creating water currents through the crown, collecting suspended particles for food and transporting them to the central mouth. The radioles also serve as gills for uptake of oxygen and parts of the crown are used in tube-building. Some sabellids have eyespots on the radioles, and will quickly withdraw into their tube if crossed by a shadow. Great care has to be taken in photographing these animals, such that they are not shadowed, or exposed to sudden movement in the water. The group includes a series of animals, which are not easily identified.

The tube of these attractive 'feather-duster' worms is around 30 mm long. Note the tubes 'close' when the occupant withdraws. Here from near the Arctic Circle at 10 m depth.

Sabella pavonina (= S. penicillus) - Peacock worm

Distribution: Widely distributed in the Atlantic. North to northern Norway. Mediterranean. Common throughout the British Isles.
Description: An elastic tube of protein that is covered with fine mud particles; the tube is up to 500 mm long and 8 mm wide, and the body length is up to 250 mm. Without detailed examination, it may easily be confused with other sabellid polychaetes. There are about 100 long radioles arranged in two groups, one sometimes forming a spiral. The animal is orange to weak violet, whereas the tentacles are banded in different colours, including violet, brown and red.
Habitat: On muddy bottoms, but also among stones and under cliffs. Intertidally down to 750 metres depth.

The systematics of **Sabella** *recently has been much revised. Certain identification requires close examination of specimens.*

Myxicola infundibulum

The radiolar crown projects a few centimetres from the sediment like a parasol. Note the almost transparent web joining the individual radioles. Here from 20 m depth.

Distribution: From the Mediterranean to the Arctic, although several different species may actually be involved.
Description: Most part of the mucus tube submerged within the sediment, with only a small part emerging. Animal up to 200 mm long, and each segment is divided into two rings. There are approx. 60 radioles, and a thin web surrounds the whole crown.
Habitat: On bottoms with mud or shell sand, from about 15 m down to 500 m depth. Often on sites exposed by currents.

Several colour variations exist.

Family Arenicolidae

This group of polychaetes lives within the sediment, and most people have probably noticed the casts from Arenicola marina. *The animals feed on sand and digest the particles and the small organisms within this sand. They process large amounts of sediment each day, and the undigested remains form coiling casts on the surface. The worms lack eyes, and have a cylindrical shape. On the median segments there are a series of characteristic bush-shaped gills. The worms are large and the adults have a length of 120-350 mm.*

Arenicola marina - Lug worm

Distribution: In Europe from the Mediterranean to the Arctic.
Description: One of our largest polychaetes; up to 350 mm in length. Body cylindrical. Colour usually reddish to pink, but can also be yellowish, red with bluish tints, green or olive. The head is devoid of appendages, and the chaetae are inconspicuous. Median segments with 13 pairs of bush-shaped, red gills.
Habitat: Lives in sand and muddy sand where it forms u-shaped galleries; intertidally down to 20 m depth.
Utility: Commonly used as bait by fishermen.

Right: *Characteristic casts, or 'faecal' mounds.*

Below: *This individual was swimming at 8 m depth, probably in connection with reproduction. Usually it lives buried in the sediment.*

Class Clitellata

Includes leeches (below) and oligochaetes (such as earth worms). Most oligochaetes are small species, less than 1 mm long, and they have an important role in the decomposition of organic matter, both in sea, lakes and on land.

Sub-class Hirudinea - Leeches

Generally lacking chaetae. With the exception of two species, the body consists of 34 segments, each of which is divided into several rings. A conspicuous feature is the suckers, usually two in number. Many have two or a larger number of eyes. Most leeches suck blood from vertebrates, or are predators on crustaceans, other worms, or insect larvae, and some live as scavengers. Leeches are hermaphrodites, and undergo a direct development in cocoons, without pelagic larvae. Leeches living in the sea mostly belong to a family of fish parasites, and live from the blood of their hosts. These tend to be highly host-specific, with one species of leech parasitising one species of fish. We here include two such species. Some flat heads and bull heads may actually have large amounts of such leeches attached to the head, especially on or around the eyes.

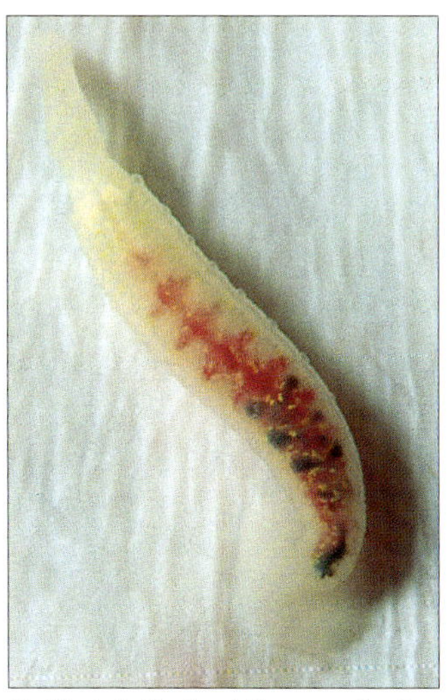

Calliobdella lophii, *attached to the belly of an anglerfish.*

Calliobdella lophii

Distribution: In European waters from the Mediterranean to Norway.
Description: Two large suckers, posterior one three to four times as large as anterior. Body cylindrical, up to 60 mm long. Eyes lacking. In species of *Calliobdella*, the posterior part has four to six rings per segment.
Habitat: Ectoparasite on anglerfish (*Lophius piscatorius*).

The posterior sucker is conspicuously larger than the anterior one.

Calliobdella nodulifera

Distribution: Known from Scotland, Norway and Iceland in European waters.
Description: Resembling the previous, ***C. lophii***, but is smaller, up to 30 mm long, it has the posterior sucker about twice the size of the anterior, and it lacks eyes.
Habitat: Ectoparasite on a number of fish species, especially of codfishes, but we have also observed it a number of times on the shorthorn sculpin (***Myoxocephalus scorpius***).

We often see large amounts of leeches on flatheads and bullheads

Sometimes leeches can be seen moving like caterpillars. They often attach to the eyes of fish.

Phylum Echiura - Spoon worms

This is a small phylum (Gr. echis = viper + Gr. oura = tail) of exclusively marine non-segmented worms. There are less then 10 species in the area. They have separate sexes, often with a microscopic male that lives as a 'parasite' within the female, although for some species the male is currently unknown. The females are often pear-shaped with a long, contractile proboscis. The anus is terminal. The animals usually live partly or completely buried in sediment, feeding on deposited matter.
We include one species that occurs in deeper waters, often in high abundances, on soft bottoms.

Bonellia viridis

Distribution: Widely distributed in the northeast Atlantic.
Description: Very long and flexible proboscis that is distally split and has a longitudinal furrow leading to the mouth. Body pear-shaped, colour bluish-green. Length up to 100 mm. *Bonellia viridis* most likely constitutes a species complex.
Habitat: Below 30 m depth, usually below 60 m. Soft or mixed bottoms, usually with mud. When the animals are caught in the spotlight from a diver, they slowly withdraw the proboscis within the sediment. They may locally be very abundant.

POISONOUS

The outer layer of the skin contains high concentrations of the green pigment bonellin, which actually resembles chlorophyll. It is weakly poisonous, and can paralyse small animals that come in contact with the skin. Furthermore, a settling larva of *B. viridis* will develop into a male if it comes into contact with the skin of a female, if not it develops into a female. The males live as internal parasites within the females.

This echiuran is bright green in colour. It is likely that what is currently recognised as the species **Boniella viridis** *may actually comprise several species.*

Most of the body of Boniella viridis *is buried within the sediment, such that we only see its flexible proboscis (mouthpart).*

Phylum Sipuncula - Peanut worms

Sipunculans are marine, non-segmented worms. Their body is divided into a long and thin retractable anterior part, the introvert, and an inflated trunk. The mouth is situated distally on the introvert, and is surrounded by ciliated papillae. Sipunculans feed on detritus from the upper layer on soft bottoms. There are more then 20 species in the area.

This sipunculid most likely is **Golfingia vulgaris**. *Photographed at 15 m depth in Lurefjorden, southwest Norway.*

ARTHROPODS

Phylum Arthropoda

Arthropods (Gr. arthron = joint + Gr. podos = foot) include insects, crustaceans, spiders and sea spiders, and represents more than 80% of all known species. The number of described species approaches one million. This number is rapidly increasing, not least due to ongoing studies of the huge numbers of insects in the tropical rain forests, and some biologists estimate that the true number of arthropod species exceeds 30 million!

Arthropods are characterised by a segmented body and an exoskeleton. Each segment is provided with a pair of segmented extremities, which during the evolution have been changed in various ways, and have become reduced, enlarged, modified into mouthparts, into reproductive organs, into gills, or into swimming appendages. They have an outer, chitinous skeleton, or cuticle, which in the crustaceans also may be calcified. The joints have a much thinner cuticle, which allows for movement of the parts.

Growth of arthropods is complicated by the presence of the hard exoskeleton, and can only be accomplished by moulting. The old skeleton is dissolved from within as the new one is forming. The animal pulls out of the old one, which splits along predetermined lines, and the new one hardens after some time. It is only during this short time, often a few days, that actual growth can take place. The nervous system is generally well-developed, as is the blood vascular system, especially in marine species. In the sea the subphylum **Crustacea** is the dominant arthropod group. Sea spiders (superclass **Pycnogonida**), in spite of their common name, are not closely related to spiders (class Arachnida), which, together with scorpions, mites and ticks, belong to **Chelicerata**. They are, however, referred to the same subphylum, **Cheliceriformes**. A third subphylum, **Uniramia**, includes the insects (superclass **Insecta**). It should be noted that arthropod relationships are still uncertain, and other classifications exist. For example, some authors refer to pycnogonids and crustaceans as distinctive phyla. There are few marine spiders and insects, and they are therefore not included here.

Here **Caligus elongatus** *has attached itself to a male lumpsucker fish. This copepod lives on the gills and around the mouth of fish such as cod, seith and lumpsuckers.*

Despite its relatively wide known distribution, the sea spider **Pycnogonum littorale** *is rarely observed by divers. Note the strong 'snout' (proboscis) between the first leg pair.*

Subphylum Cheliceriformes

Includes scorpions, spiders and sea spiders

Superclass Pycnogonida - Sea spiders

*The first sight of a pycnogonid is bound to astonish the observer. They seem to consist of little more than legs. The illustrated specimen (**Nymphon** sp.) carries egg sacs on the legs, probably in various stages of development, as seen from the different colours. Note the shorter head appendages chelicerae and palps, seen to the left. Their length and segments provide useful features for identification.*

Sea spiders (Gr. pantos = all + Gr. podos = foot) are exclusively marine. They are divided into eight families, and encompass 1200 described species. Their systematic position remains uncertain, and they have during the last century been referred to virtually all main groups of arthropods, as well as some non-arthropods.

"Lice and giants"

Sea spiders vary greatly in size, the span of legs when extended ranging from about 5 cm in *Pycnogonum littorale* to over 60 cm in *Colossendeis gigas*. The latter is found only at great depths, as are the majority of pycnogonids, although some species extend to the foreshore.

A typical pycnogonid has two pairs of simple eyes, at least four pairs of jointed legs, and a segmented trunk that is extremely reduced, and is easily recognised by the overall size dominance of legs over the trunk. The first segment is fused to the **cephalon** (head), which carries a tube-shaped **proboscis** and the first pair of extremities. Many species have thin ventrally displaced palps and dorsal, stouter **chelicerae**, both situated on the cephalon, in parallel with the proboscis. The proboscis is used to suck body fluids from various invertebrate prey, notably hydroids and sea anemones. The smallest pycnogonids do not exceed 2 mm, whereas the largest can have a leg span of up to 70 cm. It is the males that carry the fertilised eggs until they hatch.

Pycnogonids occur from the intertidal down to about 7000 metres depth. Some live as parasites, but most are free-living. There are more than 100 species in the area, and most require detailed examination for identification, often with a dissecting microscope.

The genus ***Nymphon*** is the largest and most common in our waters, but the distributions of the different species are often unknown due to uncertainties in identification. The trunk of

Nymphon is little more than a meeting point of the legs, and internal organs, including the guts, are displaced into the legs. Species of *Nymphon* tend to be large, and most are characterised by having longer chelicerae than palps. In ***N. gracile***, one of our most common species, the two distal-most segments of the palps are of equal length. Another common species, ***N. brevirostre***, has palps that are shorter than the chelicerae. The latter also differ from *N. gracile* in having a short and thick proboscis, and in being reddish. Both species have a body length of about 8 mm, and a leg length of about 15 mm.

In mortal combat! This sea spider is caught in the tentacles of a polychaete worm.

Pycnogonum littorale

Distribution: From southwest Greenland and the US east coast to the White Sea and the Mediterranean. Common all around the British Isles.

Description: Length of females 15 - 18 mm, males slightly smaller. Leg length up to approx. 30 mm. Legs very short and stout, colour yellowish white to brown. In the North Atlantic it can only be confused with a deep-water species: ***P. crassirostre***. However, middorsally on the trunk, *P. littorale* has a pointed tubercle on each joint, whereas these are rounded in *P. crassirostre*. Also, the latter is only up to 10 mm long, and rarely appears shallower than 100 metres.

Habitat: Predator, mainly on sea anemones and hydroids, rarely on polychaetes. Avoids light, and occurs in light-sheltered places from the intertidal and down to 1000 metres depth. Relatively common in shallow waters and on rocky shores, where it often sits under stones on the pedal discs of sea anemones.

Pycnogonum littorale *uses its strong proboscis to suck the body tissues out of sea anemones.*

Biology: Sexes are separate (as in other sea spiders), and the male carries the developing eggs until they hatch. The larvae are parasites on the hydroid ***Clava***.

Subphylum Crustacea

Like most other crustaceans, the northern stone crab **Lithodes maja** *is heavily armoured.*

Crustaceans (Lat. crustaceus = covered by shell) are among the most successful animals in the sea, and dominate marine life in a way comparable to insects on land. There are more than 30 000 species described, but estimates indicate that four times that number are awaiting description and naming. The majority are marine. Even though most crustaceans are small, the American lobster can reach a length of 60 cm and a weight of 20 kg. The largest reported crustacean in terms of length is a the Japanese giant crab, which can have a leg span (distance between tips of the extended, anteriormost pair of legs) of up to 3.6 m.

Crustaceans include crabs, lobsters, shrimps, and anomurans. These all belong to the order **Decapoda**, which together with the orders **Amphipoda** and **Isopoda** constitute the class Malacostraca. The subclass **Cirripedia** (including the barnacles) and the class **Copepoda** are referred to as the superclass **Maxillipoda**.

Crustaceans are segmented animals with two pairs of antennae. Due to its high morphological diversity, few other generalisations can actually be made about this group. The body can be subdivided into **cephalon**, **thorax**, and **abdomen**. Commonly the cephalon and thorax are fused and form a cephalothorax. The paired appendages of the segments can have highly diverse functions. They form antennae (**antenna/antennula**), mouthparts (**maxillule, maxilla, maxilliped**), thoracic legs (**pereopods**) used for walking or swimming, and abdominal legs (**pleopods**), also used for swimming and/or holding on to and caring for developing eggs. The first pair of pleopods often is modified into reproductive organs.

Crustaceans display a high diversity of life styles. In the sea they can be split into those that live their whole their life in the free water (**holoplankton**), and those that spend only the early stages in the free water, but then settle and develop into adults (**meroplankton**). Others instead live as parasites and attach to a host.

Superclass Maxillipoda

This superclass (Lat. maxilla = jaw + Gr. podos = foot) includes crustaceans with five head segments, six thoracic segments, usually four abdominal segments, and one posterior segment. Fusion of some of these segments occurs within some taxa. Maxillipoda represented here include members of the class Copepoda and the subclass ***Cirripedia*** *(in class* ***Thecostraca****). Members of the other classes of Maxillipoda rarely exceed 0.5 mm. One of these, the* ***ostracods****, or mussel shrimps, includes over 2000 species. Ostracods are common among algae and as plankton in all oceans, and are characterised by having the whole body enclosed within two more or less transparent shells.*

Class Copepoda

Copepods (Gr. kope = oar, handle, Gr. ***podos*** *= foot), in terms of biomass and numbers of specimens, are the dominant group of animals on earth. There are currently over 9000 known species. A number are parasites on other animals, although the majority are free-living as plankton, on mud bottoms, or among algae and stones. Some planktonic forms can be comparatively large, but a common size is 0.5 - 5 mm. Some parasites, such as* **Lernaeocera branchialis** *(see below) can reach a length of several cms.*

A copepod within the order Harpacticoida. Enlarged 100 times.

Copepods include, among others, the orders **Calanoida, Harpacticoida**, and **Cyclopoida**. Most calanoids live as holoplankton, and constitute a major food source for larger marine animals. The long, straight antennae (>17 segments) are characteristic, as is the separation of the anterior (prosome) from the posterior part (urosome), which occurs behind the last segment provided with appendages. One of the most common species in our waters is the relatively large ***Calanus finmarchicus***. Cyclopoids usually have shorter antennae with fewer segments, and the separation between prosome and urosome appears before the last segment carrying appendages. There are both parasitic and free-living cyclopoids. The species-rich genus ***Oithona*** is more abundant than *Calanus* in terms of numbers, but the animals are small and the total biomass much less. Harpacticoids lack a clear delineation between prosome and urosome.

HUGE ABUNDANCES

Copepods occur in huge numbers as zooplankton in all oceans, and are one of the most important groups in marine food webs.

Lernaeocera branchialis

Close-up of the parasitic copepod **Lernaeocera branchialis.**

Distribution: This parasite occurs mainly on codfish in the North Atlantic, with a southern limit in Spain.
Description: Females strongly modified and are unlikely to be confused with any other copepods. The body is s-shaped, up to 30 mm long, red, and the eggs are carried in long coiled bands. The males are 'dwarfs', and resemble other copepods.
Habitat: Parasites on the gills of codfish.
Biology: The body of the females is highly modified for parasitic life, and suck blood from the gills of codfish. The males are mainly planktonic and much smaller than the females. It is notable that the species has two different stages. In the first one both males and females live as parasites on a series of different flatfish, and mating take place during this stage. It is followed by a short planktonic stage in the females, where after she attaches to the second host, usually a codfish.
Economic impact: This species has a serious impact on fish populations. Studies indicate 28% mortality in some haddock (***Melanogrammus aeglefinus***) populations due to the parasite, with more severe effects on smaller than on larger specimens. It increases mortality rates in coastal populations of cod, whereas Norwegian-Arctic populations become infected only when moving into more shallow water for breeding, and are therefore also much less affected.

Lepeophteirus salmonis - Salmon lice

*A salmon (***Salmo salar***) infested by salmon lice.*

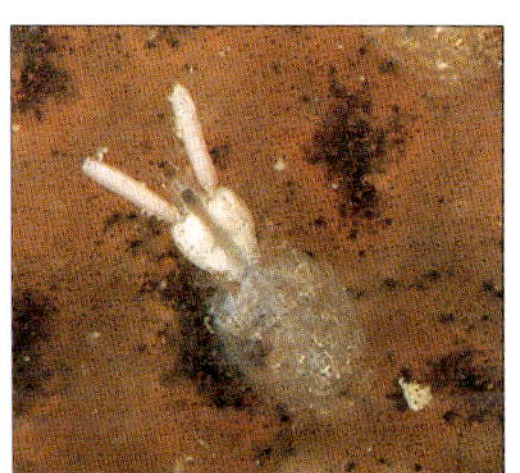

The sea louse above is attached to a lumpsucker fish - **Cyclopterus lumpus.**

These parasitic sea lice are feeding on a saithe - **Pollachius virens.**

Distribution: Co-occurs with salmon in most parts of its marine distribution.

Description: The parasite is a highly modified copepod that attains a length of 10 - 30 mm. The females have long, posteriorly attached egg sacs.

Habitat: External parasite on salmonid fish.

Biology: The first three larval stages are planktonic, and the host is infested during the fourth stage. The life cycle is simple with a single host. Similar to other crustaceans, growth is accomplished by moulting, and this occurs nine times before reaching maturity. It lives on blood, body fluids and dermal tissue of the host. Reproduction takes place on the host. Heavily infected fish have conspicuous and deep wounds, and often die from bacterial and fungal infections. The parasitism only takes place in the sea, as the parasites will rapidly die when the host enters fresh water.

Economic impact: Salmon lice are an important problem in fish culture. Infected fish lose weight and may die from secondary infection by bacteria and fungus. Lice infections can be countered by chemical means, or in a highly efficient way with the help of wrasses (cleaner-fish). The wrasses eat the parasites, and there are no pollution effects.

Class Thecostraca

*This is a newly revised group, and includes taxa such as the subclasses **Cirripedia** (barnacles) and **Rhizocephala**. Members of the latter group are highly modified and live as parasites on other crustaceans.*

Subclass Cirripedia - Barnacles

*Barnacles, or cirripedes (Lat. cirrus = (hair) lock + Lat. pedis = foot), include about 1000 species. The morphological diversity within the group makes a general description difficult. Some species are highly modified parasites. The following, however, is valid for most members. Adult cirripedes are sessile animals that use the first pair of antennae to attach themselves, and are therefore literally standing on their heads. The body is incompletely segmented, and is surrounded by (usually) calcareous plates. The most familiar ones, goose barnacles and acorn barnacles, belong to the superorder **Thoracica**, and we here present some of these. The superorder includes about 700 species. Most members are hermaphrodites.*

Barnacles filter particles from the water.

Chirona hammeri *is a deep-water species that can reach up to 5 cm in height. Photographed at Magerøy, northern Norway at 22 m depth.*

Lepas anatifera

Distribution: Mediterranean, Atlantic, North Sea, rarely in the western parts of the Baltic. Stranded specimens have been reported as far north as Spitsbergen and Greenland.
Description: A common length is 150 mm, although specimens are reported up to 800 mm. It consists of a long stalk and a capitulum or 'head', enclosed by five calcareous plates. The plates are usually white with a bluish tone, but can also be yellowish white, grey or rust red with brown spots. The margin of the capitulum is usually bright orange. The brown stalk makes up about 2/3 of the total length of the animal. The species can be confused with other, but more rarely occurring members of *Lepas*. One non-drifting species, ***Scalpellum scalpellum***, lives attached on rocky bottoms at approx. 50 - 100 metres depth. It has a very short stalk and 13 - 14 calcareous plates.
Habitat: Lives attached to floating objects or ships, and can therefore occasionally appear also outside its 'normal' distribution range. Animals that are found in the north parts in Europe tend to have their origin in more southerly waters.
Biology: Hermaphrodite, as most cirripedes. Pelagic larvae. Capture suspended objects and plankton with the extended appendages for feeding. Similar to other *Lepas*, it does not reproduce in colder waters.

DID YOU KNOW...

The name goose barnacle has its origin in a previous belief that they were the juvenile form of geese. This was partly due to the fact that the breeding places of many geese species were then unknown (such as the Brent goose and Barnacle goose, that breeds on Arctic islands).

These goose barnacles were attached to a metal tin. Goose barnacles can reach all of 80 cm in length, but the usual length is 15 cm.

Note the 'stalk' and relatively strong, stout thoracic legs used to filter food from the water.

Semibalanus balanoides

Distribution: North Pacific and North Atlantic. In Europe from the Iberian Peninsula to northern Norway, with occasional records from Greenland and Spitsbergen.
Description: The animal is constructed much like a box with a lid. The 'box' consists of six plates, and the 'lid' of a further two pairs. Size, shape and position of the plates are important characteristics, and species identification from a high-quality photograph is often feasible. The basal 'floor' plate of the box is not calcified and consists only of a thin membrane. Length or diameter is usually about 10 - 15 mm, and the animals are mostly wider than high. Highly variable.
Habitat: Usually in the upper intertidal, such that the animals are covered by water only at high tide, although they can occur down to 10 - 20 metres depth. Usually on exposed shores with strong currents.

The position, shape and relative sizes of the calcareous plates in barnacles is a certain identification character.

Biology: As in other intertidal species, ***S. balanoides*** is extremely resistant to environmental fluctuations. At low tide they are exposed to desiccation, and they also have to withstand severe temperature fluctuations. Some specimens are situated so high up on the shore that they are covered by water only during spring tide or by extreme waves. The food consists of phytoplankton, small zooplankton, and detritus (dead parts of plants and animals), and is collected with the six pairs of branched 'feet'. They are hermaphrodites, and reproduction is accomplished with a long penis that reaches over to the neighbour barnacle. A single specimen can produce up to 20 000 larvae each spring, although only a tiny fraction of these will survive and settle. At settlement the larvae show preferences for areas where adults already are present, attracted by certain proteins (arthropodines) that occur in the adult plates. The obvious advantage for the larvae is that they will settle in a suitable environment.

Balanus balanus

Distribution: Northern hemisphere, in Europe from the English Channel to Franz Josef Land.

Description: Larger and stouter than the preceding species, with a diameter of 20-30 mm, sometimes up to 50 mm. As high as wide. The lid plates together form a characteristic, pointed 'beak', which also usually is violet. The shape is highly variable, but the plates tend to be strongly folded. A calcareous bottom plate becomes visible if the animal is detached. The colour varies from white to yellowish brown. Based on these characters *B. balanus* should not be confused with other species. ***Chirona hammeri*** (p. 204) is encountered at greater depths; 40-300 metres, and may reach a height of 90 mm and a diameter of 70 mm.

Habitat: Attached to stones or shells, from the low-water mark down to several hundred metres depth.

Biology: Occurs slightly deeper than ***Semibalanus balanoides***, and does not appear intertidally. The biology of the two species is otherwise fairly similar.

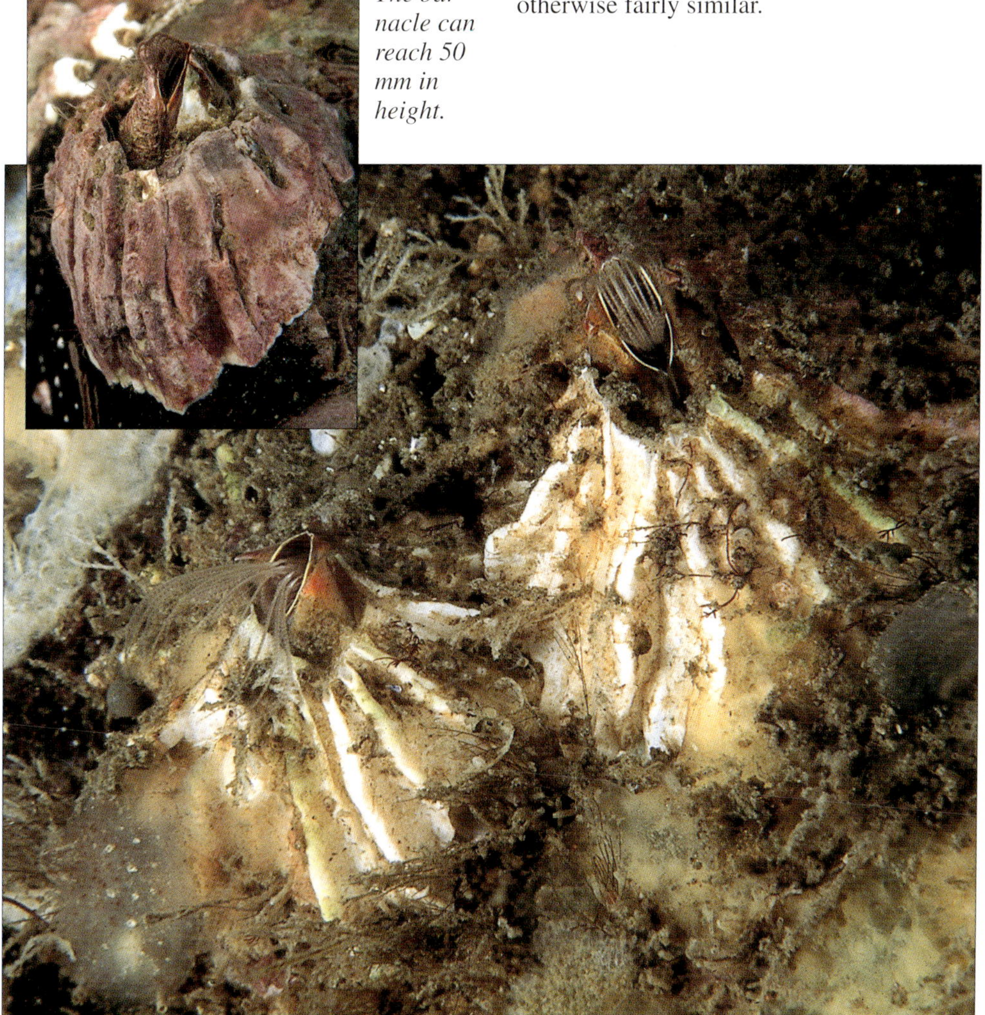

The barnacle can reach 50 mm in height.

The barnacle **Balanus balanus**, *with the thoracic legs extended to catch food. The plates are markedly folded in this species.*

Balanus crenatus

Distribution: In European waters from the Mediterranean to the Arctic.

Description: Externally, the plates forming the lid do not fit entirely together, such that there appears to be a small opening. The side plates are smooth and usually bright white, and diverge slightly from each other in the upper parts. The base plate is calcified. Rarely exceeding 20 mm in diameter, with a height up to 45 mm.

Habitat: Found on boat hulls, jetties, buoys, stones, and other hard substrata. Also on blue-mussels (***Mytilus edulis***) and horse-mussels (***Modiolus modiolus***). Usually in shallow waters, but can occasionally be found below 100 metres.

Balanus crenatus *has characteristic smooth and usually shining white calcareous plates.*

Superclass Malacostraca

Even though malacostracans (Gr. mala-kos = soft + Gr. ostrakon = shell) include the large and most familiar crustacean groups, the majority of them are actually small in size. They have in common a cephalon with five segments, a thorax with eight segments, and an abdomen with six (rarely seven) segments. Malacostracans in our waters include the class **Phyllocarida**, *with the single order* **Leptostraca** *and less than 20 species, and the class* **Eumalacostraca**, *which is highly diverse. Here we only include members of the most common groups. Malacostracans encompass over 20 000 species.*

Superorder Peracarida

Includes, among others, mysids, amphipods, and isopods

A mysid shrimp, **Praunus flexuosus**, *sits on the green alga* **Codium fragile**.

Order Mysida

*Mysids (Gr. mysis = sleep, close mouth), sometimes referred to as opossum shrimps, look like small shrimps. Most are slender, 10 - 30 mm long. The eyes are situated on stalks, and the exoskeleton is soft and not calcified. The outer branch of the second antennae is modified into a plate. Mysids have eight pairs of biramous thoracic legs, in contrast to shrimps, which have five pairs. They can be distinguished from krill (***Euphansiacea***) by the lack of gills (except for the* *suborder* ***Lophogastrida****). The illustrated species can be confused with several congeneric ones. It is common to encounter mysids in shallow water among algae and in kelp forests, although they are well-camouflaged and difficult to spot. Many mysids also occur in deeper water, where they often spend the day on the bottom, but in the free water during the night - some migrate 600 metres vertically each night. Shallow water species tend to migrate into deeper and warmer waters during winter. Many shallow forms also have a well-developed ability to shift colour, depending on background and light conditions. The life span is commonly 1-2 years, and the females tend to breed 1-2 times. Mysids feed on plankton and detritus.*

Hemimysis lamornae

Distribution: Black Sea, Mediterranean, and north to northern Norway.
Description: Large wide eyes, antennal plates lancet-shaped, not longer than the stem of the first pair of antennae (in contrast to ***H. abyssicola***, which also occurs deeper than 200 metres). The colour is light red to orange. Length up to approx. 13 mm.
Habitat: On rocky bottoms from the intertidal down to about 100 metres of depth. Avoids light, and only emerges from the cavities where it dwells at full darkness in the night.

Members of this species are usually found in caves or other places with low light conditions.

Praunus flexuosus

Distribution: Large parts of the northeast Atlantic, including Iceland and the Baltic.
Description: Up to 25 mm long. Well-developed ability to shift colour. Can be identified from the long antennal plates, although it can be confused with several other species within the genus.

This individual appears to have 5 or 6 pairs of legs. However, on closer inspection, a larger number of legs are visible tucked along the shell margin (cephalothorax). This separates mysids from shrimps, which, in common with other decapods, have 5 pairs of walking legs.

Habitat: Often on different wrecks, kelps and eel-grass, well camouflaged.
Biology: Males of the genus *Praunus* are attracted by odours produced by the moulting females. Reproduction has to take place within 12 hours following her moulting.

DID YOU KNOW THAT...

In *Praunus flexuosus* the pigments are agglomerated in concentrated spots during the night, making the animals virtually transparent, whereas in the daytime they are dispersed and give the animal a darker colouration. The pigment spots are often species-specific.

Order Amphipoda

Amphipods (Gr. amphi = around, double + Gr. podos = foot) comprise over 6000 species. Most are laterally flattened, one exception being the caprellids, or skeleton shrimps (see below). They have legs specialised for grasping, walking and swimming, and sometimes even for jumping. The largest ones attain a length of 140 mm. Most species live associated with the bottom, often to hydroids and algae, others are pelagic, and some are parasites. Identification is demanding, and often requires a microscope. There are three suborders in our waters: ***Gammaroidea, Hyperiidea,*** *and* ***Caprellidea****, of which the first is the most species-rich. Amphipod relationships are highly uncertain, and there are ongoing revisions of the groups.*

Amphipods have a characteristic body form. Most are flattened from the sides, such that a marked keel is formed along the back.

Suborder Gammaroidea

Most of the 6000 species belong within this group. A number of species live in fresh water. Identification of the large number of species in our waters is complicated and requires special literature. The use of powerful macro-lenses has permitted us to obtain detailed photos, which in many cases allow species determinations. Photographic documentation can therefore be a useful tool for this group.

Stenopleustes latipes

This species has characteristic colouration.

Distribution: Widely distributed in the northeast Atlantic, from Greenland to the Azores.

Description: Stout animal. A backward-turned tooth is present on the seventh thoracic segment, and on each of the three first abdominal segments. Second pair of legs much larger than first pair. Very large lateral plates on the first four pairs of legs. Large, kidney-shaped, dark red eyes. The colour is greyish white with conspicuous brown-orange transverse bands.

Habitat: We have only encountered this

species associated with the hydroid ***Eudendrium rameum*** (p. 76). Previous literature gives a depth range of 60 - 1400 metres, although we have observed it at 30 metres depth in the Trondheimsfjord.

Detail of the hydroid **Eudendrium rameum**, *densely populated with the characteristic amphipod* **Stenopleustes latipes.**

Iphimedia obesa

Close-up views show that many of our amphipods are colourful and beautiful creatures. Note the effective gripping tools.

Distribution: Widely distributed in the north-east Atlantic, from the Mediterranean to northern Norway.

Description: *Iphimedia* species have a strongly laterally compressed and curved back. Some segments have the posterior borders provided with paired teeth. *Iphimedia obesa* attains a length of 12 mm, more than twice the size of the otherwise similar ***I. minuta***. The pigmentation is characteristic: pale yellow with 2 - 3 red-brown transverse stripes on each segment, and with additional golden stripes inserted between many of these. The eyes are red-brown to violet.

Habitat: Common on sand bottoms at 10-300 metres depth.

Paramphithoe hystriix

This beautiful arctic amphipod was encountered at Spitsbergen Island. South to mid-Norway. Here on the soft coral **Gersemia rubiformis.**

Gammarellus homari

Distribution: Northern distribution, from Spitsbergen and Greenland to the Arctic parts of North America. In European waters south to France.

Description: Amphipods of this genus can be identified by their marked dorsal keel, formed by dorsal prolongations or teeth. The up to 10 mm long ***G. angulosus*** has relatively small dorsal teeth, which are not pointing back-

Gammarellus homari, *photographed on a hydroid.*

On guard. Many amphipods are effective predators and will readily pounce on small prey whenever the opportunity arises.

wards. The illustrated ***G. homari*** attains a length of 20 mm (in the northernmost areas up to 44 mm), and has large dorsal teeth that are pointing backwards. The eyes are large and red, often with darker areas (commonly described as dark brown in older literature). The colour is variable, often reddish brown with lighter spots, but also olive green to yellow with brown spots.

Habitat: Shallow water, often among algae and stones. We have often observed this species on brown algae in kelp forests. Another common place is between the egg strings of the nudibranch ***Archidoris pseudoargus*** (p. 334). The depth distribution largely follows that of the algae, 0 - 20 metres.

This amphipod has typical dorsal teeth.

Gammarus locusta

Distribution: North Atlantic down to the Canary Islands, Baltic. One of the most common amphipods in our waters.
Description: Black, kidney-shaped eyes. Males length up to 20 mm, larger than females. Arctic specimens, or specimens from deeper waters, can attain a length of up to 50 mm. Colour brownish-green, with a red spot on abdominal segment 1 - 3. Easily confused with other species of *Gammarus*.
Habitat: Very common among algae and stones, from the intertidal down to 100 metres. Also in brackish waters and estuaries. Can easily be found under stones or among algae at low water.
Biology: One can often observe couples that are joined together, and this is part of the reproduction. The male attaches to the female long before the actual copulation, which can take place only when the female moults. Already fertilised females can be identified by the presence of a ventral egg sac. The pair formation takes place immediately following the hatching of the previous brood. The male helps the female through the moulting, and directly afterwards transfers the sperm to her ventral side, whereafter he abandons her. The female emits the eggs into the egg sac where the fertilisation takes place. The mating has to be carefully timed as the fertilisation can only take place within a few hours following the moulting. The duration of the pair formation is temperature dependent, and lasts 9 - 25 days. It takes only about a month (also temperature dependent) for the young to become reproductive. The females can reproduce up to seven times before they die.

DID YOU KNOW...

Gammarus locusta reproduce all year round. Try and lift some stones on your next visit to a shore - there is a good chance that you will find pairs, since they spend about 90% of their time in this condition.

Suborder Hyperiidea

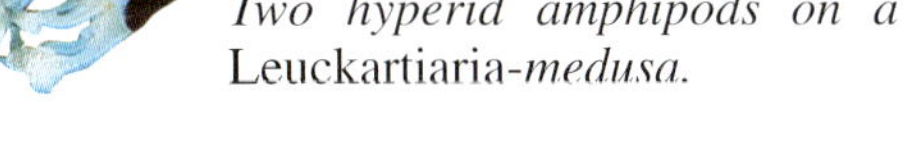

Two hyperid amphipods on a Leuckartiaria-*medusa.*

This group (Gr. hyperos = morter pestle, or from Gr. mythology Hyperia = a source, also a village on Sicily) are characterised by being short and thick, and by the huge eyes that cover most part of the head. In contrast to the majority other amphipods, they are pelagic (living in the free water masses). Several species live on the outside, or in the gastric cavity, of various jelly fishes. Most occur in deeper waters. The most common one in our waters is **Hyperia galba** *(illustrated), which lives commensally in various jellyfish. It is characterised by the very large eyes and a relatively large head. Males have short, and females long antennae.*

The large picture shows **Hyperia galba** *and the smaller picture shows another hyperid amphipod on a jellyfish.*

Suborder Caprellidea

*Caprellids (Gr. caper = goat), or skeleton shrimps as they are sometimes called, is a group of amphipods with a long and tiny body-shape, and reduced number of appendages. The first segment is fused to the head, they have small talons on the first pair of legs, and larger ones on the second. Most species lack appendages on segments 3 and 4. They live on hydroids, algae, bryozoans and sponges, to which they attach with the posterior most pair of legs. Some sit on sea stars. Caprellids follows the water movements, and occur in areas with currents or wave action. They are predators on smaller invertebrates. Identifications are not simple, and many specimens exhibit intermediary conditions between related species. The whale lice (***Paracyamus boopis***), which are ectoparasites on humpback whales, are members of the same sub-order.*

Caprella linearis

Skeleton shrimps have strong claws.

Distribution: In all northern seas.
Description: Relatively large species, males up to 32 mm long, females up to 14 mm. The colour is red or greenish, with light red eyes. Characteristic features are the dorsal 'thorns' on segment 5, and one on segment 6. Segment 3 and 4 lack appendages, but have keel-shaped gills. This species can be confused with ***C. septentrionalis***, that also is common in the area, but which usually has darker red eyes and pale green to olive body with black spots. A third species, ***Phtisica marina,*** with similar size and distribution, can be separated from the previous two by the presence of legs on segment 3 and 4, and club-shaped gills on segments 2, 3 and 4.
Habitat: Holds on to the substratum with the claws of the posterior most legs. Tends to appear in waters with strong current or wave action, and can be found in large numbers on the breadcrumb sponge (***Halichondria panicea***). Also on algae and hydroids. Depth distribution from low intertidal to about 65 metres.
Biology: The females have brooding pouches on segment 3 and 4. Similar to most other caprellids, they are 'sit and wait' predators.

Skeleton shrimps attach themselves to the substrate by means of three pairs of claws on the rear-most body segments.

The combination of the slender body shape and the ability to shift colour makes them very difficult to detect.

Order Isopoda

*Body dorso-ventrally flattened, with uniramous appendages (Gr. isos = similar + Gr. podos = foot). Most species are marine, some are limnic, and a few live on land. A number of species live as parasites on fish and other crustaceans, and have lost the characteristic body shape of other isopods. The gribble (***Limnoria lignorum***) bores in wood, and can cause extensive damage on jetties and boats. There are several hundred species i northern Europe, many living as parasites.*

Idotea balthica

GOOD FRIDAY

Good Friday is actually an excellent time for intertidal studies, since the tides are very good and permit collecting of animals that otherwise are covered by water.

The genus **Idotea** *contains many species. A distinguishing character is the shape of the tail segment (telson). As shown on the small picture, the tail plate of* I. balthica *has two small points at the corners.*

Distribution: Widely distributed in the northeast Atlantic.
Description: Dorso-ventrally flattened. Males 30 - 40 mm long, females much smaller. The colour is variable, often green or brown with white spots. Females tend to be darker than males. The shape of telson (the median posterior-most plate) is useful for separating different species of *Idotea*.
Habitat: Lives intertidally and in shallow water among algae; very common.
Biology: Feeds on epiphytic algae. The brooding sac in females is often infested by another parasitic isopod.

Idotea neglecta

The tail plate in this species has one central spine and straight or convex sides.

Distribution: Northeast Atlantic from France to northern Norway.
Description: Similar to the previous species, but can be separated by the shape of telson, which is more pointed and carries only one spine in *I. neglecta*. Body rather wide. Length up to 30 mm in males, 16 mm in females. Often with uniform colouration, sometimes with white spots. Similar also to ***I. granulosa***, although the latter have a slimmer body and more pointed telson with slightly concave sides.
Habitat: Intertidally and to a few metres depth among algae.

Idotea granulosa

I. granulosa *is the only member of this genus with slightly concave sides on the tail plate.*

Distribution: From northern France to northern Norway.
Description: Similar to the two previous species, but is slimmer and smaller. The telson is pointed and forms a rather sharp angle, and the sides are slightly concave, in contrast to all other *Idotea*, in which the sides are convex. Males up to 20 mm long, females about 13 mm. Often of uniform colouration in brown, green or red, sometimes with longitudinal white spots.
Habitat: This is a common isopod among various brown algae (especially knobbed wrack and serrated wrack) in shallow water.

Idotea pelagica

Distribution: Northern parts of the northeast Atlantic, from France to Norway.
Description: Easily separated from the previous species of *Idotea*, in that telson is rounded and has a poorly delineated tip. Up to 10 mm long. Short antennae. Colour usually greyish or brownish.

Habitat: down to a few metres depth. Usually with sponges, algae, mussels and barnacles on exposed rocky bottoms. Easiest found among lumps of blue mussels or with the red algae ***Corallina***.

Idotea pelagica *is both smaller and rounder in shape than the other members of the genus. The antennae are very short, and the tail plate is rounded, with a very small point.*

Astacilla longicornis

Distribution: From the coasts of Portugal to northern Norway.
Description: Easily identified from its very slender and long shape. There are a number of species within the genus in our waters, and we here only include the most common one. Females are larger than males, up to 30 mm long, males about 10 mm. The segment at the middle of the body represent half the length of the whole animal. This segment lacks appendages, which are present only on anterior and posterior segments. The second pair of antennae has two very long segments and an accessory flagellum.
Habitat: Occurs from about 10 metres down to 100 metres. It is common on the gorgonians ***Paramuricea placomus*** and ***Primnoa resedaeformis,*** plus other soft corals.

This characteristic isopod **Astacilla longicornis** *is distinguished from other, smaller members of the same genus, by its smooth carapace. Here on the gorgonian fan* **Paramuricea placomus.**

BROOD CARING

Astacilla has brood caring, and newly hatched larvae attach to the antennae of the female.

Munnopsis typica

Distribution: Widely distributed in North Atlantic and Arctic waters.
Description: Belongs to a very typical family of isopods. ***Munnopsis typica*** is, among other things, characterised by reaching a length of 15 - 18 mm. The thorax legs number 3 and 4 are more than twice the length of the body, and the antennae are also extremely long. Eyes are lacking. The colour is reddish brown. The three posterior most pereopods are modified into swimming appendages.
Habitat: Deep-water species on muddy bottoms, 80 - 1200 metres depth, although more shallow in Arctic waters.

Syscenus infelix

This is a 'giant' of a parasitic isopod. Here its host is **Trisopterus minutus**.

Distribution: Widely distributed in the North Atlantic.
Description: Tail distinctly less wide than anterior part of body. Telson wide, slightly pointed. Large, up to 44 mm.
Habitat: Adults live as parasites on codfishes (***Gadidae***), whereas earlier stages are free-living and pelagic. The illustrated specimen is attached to a young poor cod, approx. 20 metres depth.
Biology: One of many isopods that lives as parasites on fish. Adults suck the body fluids from their host, whereas younger stages feed on zooplankton.

Superorder Eucarida

Includes malacostracans with complex eyes. The group (Gr. eu- = true + Gr. caridis = shrimp) is characterised by having all thoracic segments fused into one part, the carapace. In our waters we mainly encounter members from two orders, krill and decapods.

Krill are typical inhabitants of deep water, but can move into shallower waters during darkness.

Order Euphausiacida - Krill

Krill (Gr. eu- = true + Gr. phausis = illumination), in cold waters, occur in huge numbers and is one of the major food sources in these areas. In the Barents Sea it makes up 40-45% of the weight of zooplankton, and densities often surpasses 1000 specimens per m^3 (614 g wet weight per m^3). Many euphausiaceans are luminescent. They have light-producing organs, usually ten, one pair at the bases of the eye-stalks, one pair at the bases of the 2nd and 7th thorax leg, and one on each of the 4 foremost abdominal segments. They resemble shrimps, but have a straight, rather than angled, body outline. Furthermore, parts of the gills are visible and unprotected by the carapace. The krill fauna in coastal waters tend to be different from open waters. All species are pelagic.

Krill has a row of luminiscent organs along the body.

DID YOU KNOW...

Krill is important as food for whalebone whales, and likely also for young stages of many fish species. Following the over-exploitation of the capelin in the second part of the 1980s, krill became an important food source also for cod.

Order Decapoda

This group (Gr. Deka = ten + Gr. Podos = foot) includes the most familiar crustaceans, such as shrimps, lobsters, crabs and hermit crabs. Most are marine, although there are also some that live in fresh water or on land. As given by the name, they have five pairs of legs that function as walking legs and talons. The eyes are mobile and situated on stalks. They display a huge variation in size, from a few millimetres long crabs with a weight less than 1 g, to the American lobster that can weigh up to 20 kg. All mentioned infraorders are part of the suborder ***Pleocyematina****.*

Many decapods have a well-developed anterior spine, called the rostrum. Here the squat lobster **Galathea strigosa.**

Suborder Dendrobranchiatina

Most of our decapods belong to the suborder Pleocyematina. The 'shrimp' **Sergestes arcticus**, *however, is referred to sub-order Dendrobranchiatina. As given by the name, it has a northern distribution, and occurs from Skagerrak to northern Norway and Arctic. Species of Dendrobranchiatina are characterised by not carrying the eggs with the pleopods, and in that the larvae hatch as nauplia.* **Sergestes arcticus** *is pelagic deep-water species, normally occurring at depths of 400 - 700 metres, but migrates upwards during the night. It is about 5 cm long and semi-transparent with red pigment spots; it is luminiscent, as often is the case with deep-water forms.*

Sergestes arcticus *is a deep-water species. However, at nights, it often accompanies zooplankton on their vertical migrations towards the surface. Here pictured in the Trondheimsfjord, Norway, at only 5 m depth.*

Suborder Pleocyematina

Infraorder Caridea - Shrimps

Shrimps (Lat. Caridis = shrimp) are swimming decapods. The body is laterally compressed, and forms a more or less distinct angle at the third abdominal segment. The abdomen is longer than the carapace. Most are more or less transparent, and the shell is thin. The posterior segments have swimming legs that are used for forward movement. The have a well developed tail fan that functions as a powerful paddle for rapid movement backwards. There are a large number of species in the area, many of which live in deeper waters. We here include a sample of the most common ones from shallow water.

Divers most commonly encounter this shrimp, **Pandalus montagui.**

Palaemon elegans

Palaemon elegans *is commonly found low on the shore, preferably on sandy bottoms or in inter-tidal pools. The legs have characteristic yellow rings; a feature shared with the related* Palaemon adspersus. *The latter has red pigment on the rostrum.*

Distribution: Widely distributed in the North Atlantic, Mediterranean, and western Baltic.
Description: Up to 60 mm long. More or less transparent with yellowish brown or red bands and spots. Legs often with alternating yellow and blue rings. Rostrum rather long,

Palaemon elegans *is an exceptionally curious creature, and will often swim right up to the camera lens.*

DID YOU KNOW...

Whereas caridean shrimps are most frequently seen wandering on the surface of rocks in rock pools, some burrow in sediment, some hide among fronds of algae, and others swim in the midwater of the oceans.

slightly bent upwards, with seven-ten dorsal, and three ventral teeth. The two innermost dorsal ones are situated on the actual carapace. Lower part of rostrum without red pigment spots. Can be confused with the co-occurring ***P. adspersus***, which reaches 70 mm, has six-seven dorsal teeth on rostrum of which a single one is on the carapace, and that has red pigment spots on the lower part of the rostrum.

Habitat: In shallow water among algae, eelgrass, stones and gravel. Migrates to deeper and warmer waters during winter.

Biology: *Palaemon elegans* has an amazing capacity to adjust its colour according to background, and during the night the pigment is concentrated such that the animals become virtually transparent. Females migrate from shallow water for laying eggs, but soon return. They attain an age of about four years, whereas the much smaller males probably rarely are older than two years. They feed mainly on dead plant and animal remains. A friend of our who took a bath in the sea one warm summers day confirmed this, when his feet were invaded by *Palaemon* - they were feasting on the dead epidermal cells of his feet.

Pandalus montagui

Pandalus montagui *pictured showing its characteristic turquoise-coloured eggs.*

Distribution: North-east Atlantic from the British Isles to northern Norway.

Description: Congeneric with the commercially important ***P. borealis***, the northern shrimp. *Pandalus montagui* can be identified by the following features: rostrum nearly 1.5 times as long as carapace, slightly bent upwards, with ten-twelve dorsal teeth, of which four sit on the carapace, and five-six ventral teeth; teeth are absent from the distal half. *Pandalus* can be separated from *Palaemon* in that, among other features, the second pair of walking legs lack distinct talons (note: the second pair is easiest identified by counting from the back, starting with the fifth pair, since the long mouth parts otherwise may be confused with legs). *Pandalus montagui* is semi-transparent with marked reddish-brown bands and spots, and dorsally often with a large number of small, yellow spots. The pigmentation often differs during the night, and the animals become more transparent. The eggs are turquoise, and the length is up to 160 mm, rostrum included. Following examination of these features, it should not be confused with any other species in the area. Another species, ***P. propinquus***, occurs in deeper water, is uniformly light red, and has the rostrum strongly bent upwards and with three teeth continuing on the carapace.

Pandalus montagui, *in contrast to its more deep-water relative* **P. borealis,** *has attractive pigmentation.*

Habitat: Many divers encounter and note this species. It is common in shallow water, from a few down to several hundreds of metres, most common at 5 - 15 metres, with adults occurring deeper. They live on various kinds of bottoms, but are more common on hard bottoms, often among algae.
Commercial value: Fished commercially in England and Scotland.

Pandalus borealis - Northern shrimp

Distribution: Northerly distribution, from Skagerrak to Barents Sea, Spitsbergen and Greenland. Also found off Northumberland.
Description: Live specimens light red to rose (darker red when cooked). Similar to ***P. montagui***, but lacks the red stripes, has the dorsal side of rostrum with 12 - 16 teeth, which almost reach the distal tip. It is a protandric hermaphrodite, meaning that the animals at some point shift sex from males to females. Males are therefore up to 120 mm long, whereas females are up to 185 mm.
Habitat: Lives on or above muddy bottoms. Such habitats exist as vast planes on the sea floors, or more patchy in coastal areas. The animals occur in deeper waters, below 70 metres, where temperatures are low. Cold and deep water may under some conditions be transported up into shallow waters, and ***P.***

SEX CHANGES

Deep-water shrimps have a special life cycle. Those who have enjoyed ***P. borealis*** probably also have noted that only the largest specimens carry external eggs. The smaller ones are much thinner and devoid of eggs. This is due to the circumstance that deep-water shrimps start out as males, and shift sex to become females after attaining a certain size.

borealis may then occur at lesser depths and be observed by divers. Although mainly bottom living, the shrimps may migrate into the free water column. Following bright days they may move upwards as the light intensity decreases, to return to the bottom as it gets dark. This also affects fishing conditions, which are best in early mornings and around midnight, and least good in the afternoons. This relates to the light conditions, and varies therefore with time of year and latitude. The fishing is also less good during cloudy days, when the shrimps may move upwards. They also perform horizontal migrations, and occur in shallower waters in the first months of the year.

Biology: Larger specimens carry eggs that appear as yellowish-orange (cooked) lump in the carapace. The eggs are spawned later in the autumn, and attached to the pleopods. Spawning times vary with area, and are earlier (August) in the northern parts. During mating, the smaller male transfers a spermatophore (sperm package) to the ventral side of the female. The eggs are spawned through her oviducts, the eggs are fertilised, and attached ventrally to her abdomen. In January-February the larvae have developed eyes, which can be seen as small black spots in the eggs. The brooding time varies a lot between areas. In southern Norway it is about five months, and the larvae start to hatch by the beginning of March, whereas in Spitsbergen the brooding time is nine months. The newly hatched larvae live as plankton for three months before they go down to the bottom. The following observations have been made in the Oslofjord: in the autumn, about one year after hatching, the shrimps are around 90 mm long. At this size they are mature males that will breed. They start to change sex the following winter, and the whole of this generation will have completed the sex change by the next summer. At two years of age, and at an average length of 120 mm, the shrimps will breed as females. In Spitsbergen this cycle instead takes three years, and five years before they carry external eggs. The food consists of dead plant and animal matter (detritus).

The red coloured material in the head and anterior thoracic region is roe not yet spawned.

Commercial value: The fishing of ***P. borealis*** is extensive and commercially important in northern areas. They are fished with small trawlers in coastal areas, whereas more industrial fishing is carried out in the Barents Sea and off Greenland, mainly at depths of 100 - 700 metres. The total catch in Norway in 2002 was 68 000 tons, with a landing value of approx. 95 million Euro.

Crangon crangon - Common shrimp/sand shrimp

Crangon crangon *often is found partially buried in sand or fine gravel, as pictured here.*

The common shrimp is compressed from the top, a feature distinguishing it from most other shrimps.

Distribution: North Atlantic and North Pacific. White Sea to Morocco. Mediterranean. All British Isles.

Description: Length up to 90 mm, but usually 50 - 70 mm. Greyish with dark spots. More dorso-ventrally flattened than most shrimps. Rostrum very short, half the length of the eyes. Carapace with one dorsal and, on each side, one lateral spine. Can be confused with ***C. allmanni***, which lives in slightly deeper water (10 - 250 metres), and has distinct lateral keels dorsally on the posterior most segment.

Habitat: Usually not deeper than 10 metres, but has been recorded down to 150 metres. Occurs also in brackish water.

Biology: During the day it is usually buried in sand, with only the eyes emerging. Feeds on algae, polychaetes and smaller crustaceans, but can also catch fish larvae and other shrimps, including conspecific ones. They hunt during nighttime, where smell plays an important role. ***Crangon crangon*** breeds twice a year, first time in mid-April. The larvae are pelagic for about five weeks before settling. A specimen of 75 mm length is minimally three years old.

Commercial value: Eaten in several countries, although the taste largely depends on the food of the shrimp, and in some areas they are not considered edible.

Hippolyte varians

This shrimp is small and extremely variable in colour. A characteristic feature is the rostrum (anterior spine), which is relatively short, with a single spine located just in front of the eyes.

Distribution: Large parts of the northeast Atlantic, Mediterranean to northern Norway. All British coasts.

Description: Sometimes referred to as 'the chameleon prawn'. Highly variable in colour, and can be red, green or brown, depending on habitat. It is notable that they turn blue to bluish green during the night, independently of habitat: when it is dark they need no longer invest in camouflage, and return to their basic colour. Small, up to 32 mm long. Rostrum straight, with a single dorsal tooth situated above the eyes. Belongs to the family Hippolytidae, characterised by having well-developed talons on the first pair of walking legs, and by this pair of legs also being considerably stouter than the following pair. The five following species also belong to the same family.

Habitat: In shallow water among algae and eelgrass. Down to 150 metres depth

Eualus pusiolus

Distribution: In Europe from Bay of Biscay to Iceland and northern Norway. All British Isles, less abundant in the northern parts.

Description: Small shrimp, up to 28 mm long. Colour variable, often with distinct patches of green, red and/or brown pigment. A spine is present on the carapace, under the eye, and another one in the antero-ventral corner. Carapace rather large. Rostrum short, evenly pointed, slightly longer than eye stalks, with 2 - 5 teeth on the dorsal side, and none on the ventral. The rostral tip is not bifid, as is the case in the close relative ***E. occultus***, which also occurs in the area.

This individual is pictured on the dead man's finger coral **Alcyonium digitatum**.

Habitat: In shallow water, down to approx. 150 metres, usually on hard bottom.

Eualus gaimardii

The head and thoracic region of this individual is filled with roe, and it is soon ready to spawn. Here at 55 m depth, between living branches of the stone coral **Lophelia pertusa.**

Distribution: Northern distribution, from the northern parts of the British Isles to northern Norway and Spitsbergen.

Description: Rather similar to the commercially fished ***Pandalus borealis***, although the rostrum is much shorter, the carapace is stouter, and the shell in the genus *Eualus* is virtually transparent. *Eualus gaimardii* can be distinguished from other members of the family Hippolytidae by the shape and length of the rostrum. It is about half the length of the carapace, has 5 - 7 dorsal teeth, of which two sit on the carapace, and 3 - 5 ventral teeth. It is the largest member of the family, up to 100 mm in length. It is more or less transparent, with reddish brown pigmentation. The internal organs are visible through the shell.

This individual has released its eggs and is carrying them in brood pouches under the abdomen.

Habitat: On sand, hard and soft bottoms, from the intertidal down to approx. 300 metres depth.

Thoralus cranchii

Distribution: In Europe from the Mediterranean to mid-Norway. All British Isles.
Description: Similar to the previous species, ***Eualus pusiolus***. Up to 22 mm long, usually dark brown in colour with lighter spots. Rostrum short, with three dorsal and no ventral teeth. Rostral tip bifid, with the inferior prong shorter than the superior one. Rostrum slightly longer than eye stalks, and medially wider than in previous genus, including ***Eualus occultus***. ***Thoralus cranchii*** is further distinguished from *Eualus* by the presence of six rings on the next last segment (carpus) in the second pair of walking legs, whereas *Eualus* has 7 - 8 rings on this segment. Accurate identification requires examination of these features.
Habitat: On various types of bottoms, usually above 30 metres depth, but occurs down to 70 metres.

Thoralus cranchii *is only 22 mm long; even smaller than* **Eulalus pusiolus.**

Lebbeus polaris

This species has a northern distribution. This photograph taken at Spitsbergen at 10 m depth.

Distribution: Arctic distribution, in the northeast Atlantic from the Hebrides and further north. Spitsbergen, Greenland, North Pacific.
Description: Colour variable, with red and yellow pigment spots. Often brightly coloured, and pigmented rings on the legs. Distal-most part of the

Lebbus polaris *is beautifully patterned, with pigmented rings around the legs.*

talons are brownish black. Rostrum straight, or weakly bent downwards. Up to 9 cm long, usually 6 - 7 cm.

Habitat: Surface down to 1000 metres depth, more shallow in northern areas.

Spirontocaris liljeborgi

Distribution: Northern distribution in Europe, from the northern parts of the British Isles to northern Norway, Greenland and Spitsbergen.

Description: With one exception, ***Spirontocaris liljeborgi*** can be distinguished from all other shrimps in the area by the shape of the rostrum, which is very high and pronounced. Carapace has spines above the eyes. The colour is usually light red, often with reddish brown spots. The teeth of the part of the rostrum behind the eyes are of similar size, and their posterior side is smooth, rather than finely serrated as in ***S. spinus***. In the latter species the dorsal rostral teeth reaches almost the posterior border of the carapace, whereas in ***S. liljeborgi***

S. liljeborgi: *another attractive shrimp, often seen together with the sea anemone* **Bolocera tuediae**.

Spirontocaris liljeborgi *has a very characteristic rostrum (anterior spiny structure), and therefore can only be confused with one other species within the same genus in our waters;* S. spinus.

they only reach the median part. Furthermore, the rostrum of ***S. spinus*** is anteriorly more cut-off, with 2 - 3 anteriorly directed spines of similar length. In ***S. liljeborgi*** the distal part is extended into a dorsal tooth. ***Spirontocaris spinus*** also has a large, distinct, posteriorly directed spine on the third abdominal segment, whereas this is much smaller in ***S. liljeborgi***. Up to 75 mm long, slightly larger than ***S. spinus***, which is up to 60 mm long.

Habitat: ***S. liljeborgi*** occurs both on hard and soft bottom, normally at depths of 30 - 300 metres, although it can occur from 20 - 1000 metres.

SEEKING PROTECTION

Spirontocaris liljeborgi often occurs together with the sea anemone ***Bolocera tuediae.*** The shrimps may appear in several dozens in varying sizes, standing in ring below the sea anemone, which provides protection against predators by means of the powerful nematocysts.

Infraorder Astacidea

Astacids (Gr. astacos = kind of crustacean) include crayfishes and lobsters, with over 700 species. The shell is cylindrical, with a straight abdomen and a well developed tail fan. The first three pairs of thorax legs have talons, and the legs are well developed for walking. Swimming abilities are usually restricted, and most species walk on the bottom. The most familiar ones in our waters are the common lobster and the Norway lobster.

Nephrops norvegicus - Norway lobster

The characteristic orange colour, and the long, slender but still robust claws make the Norway lobster almost impossible to confuse with any other species. The favoured building material is clay.

Distribution: In the northeast Atlantic from Morocco and the Mediterranean in the south to Iceland and mid-Norway in the north. Common all around the British Isles.

Description: Length up to 250 mm, carapace up to 100 mm. Males much larger than females. Colour pale orange (does not change when cooked). Body and claws slender. Eyes kidney-shaped, thereby the generic name Nephrops, meaning kidney-eye. Rostral spine long and thin, with up to five additional lateral spines each side. Can hardly be confused with any other species in our waters.

Habitat: May occur as shallow as 20 metres depth, but usually at 40 - 800 metres. Preference for soft bottoms with mud or muddy sand, where it constructs galleries and finds shelter during daytime. May be very abundant locally.

Detail of the head, showing the large, kidney-shaped eyes.

Biology: Reproduction in the Norway lobster and the common lobster are fairly similar. Release of eggs and fertilisation takes place during summer, although the larvae do not hatch until next spring or summer, 8 - 9 months later. Adult females moult only every second year.

Commercial value: Shrimp trawling may yield important side-catches of Norway lobster. An increasingly popular method in later years is to catch them with pots instead of trawling.

DID YOU KNOW...

The Norway lobster is caught by Danish and Swedish fishermen in the Kattegatt and Skagerrak, and considerable amounts are caught in Scotland and around Iceland. A special trawl is used, provided with a chain that cuts into the mud and forces the lobsters out of their galleries. However, this kind of trawling has been severely criticised, because it causes extensive damage to the soft bottoms.

Poised and prepared either to attack or retreat.

Carrion-feeders of the sea

Crustaceans have a highly developed sense of smell. As carrion appears, olfactory substances are spread through the water. These are registered by the crustaceans' smell organ, which can triggered by only a few molecules. The direction is localised by means of tracking the increasing density of the molecules. Even though decaying animals are on the menu of shallow water crustaceans, this often represents merely an additional food source. In deeper water, however, carrion can be the dominant food source for many crustaceans, especially some large amphipods and isopods. On the picture we can see a fish head attracting the attention of the anomurans **Munida** *and* **Lithodes maja**, *and of* **Cancer pagurus**, *the edible crab.*

Homarus gammarus - Common lobster

The common lobster has asymmetrical claws; one large claw for crushing and the other thinner and more scissor-like.

Distribution: Common from the English Channel, British Isles to northern Norway, sporadically south to Morocco, in the Mediterranean and the Black Sea.

Description: Old males can attain a size of 50 cm, measured from the tip of the rostrum to the telson and a weight of 5 kg, although specimens up to 10 kg have been recorded (Shetland). In live specimens shell greenish black with bluish tubercles and joints. Occasionally entirely blue. Talons and body stout, males with more powerful talons than females. Female abdomen wider than in males. However, the best clue for sex determination is that the first pair of abdominal legs in the males are transformed into reproductive organs. The talons are different, with one being stouter and used to crush prey, and the other slimmer and used as a weapon. No risk for confusion with other species.

Habitat: The common lobster is associated with rocky bottoms, and lives among heaps of stones and in clefts, with good water circulation. It occurs in shallow waters and rarely goes beyond 40 metres depth, and in summer it may appear intertidally. The depth distribution is related to the fact that they need relatively high temperatures for their reproduction, which takes place in shallow water and, preferably, in temperatures above 15°C. Smaller specimens will tend to hide under stones or in cavities in day-time, whereas larger ones often appear unprotected. There is little information on the younger stages. A close relative, the American lobster (***Homarus americanus***), occurs in the northwest Atlantic from Labrador to Cape Hatteras. This species spends its first years buried in coarse gravel, and this may be the case also for our species, although actual observations are lacking. Lobsters tend to stay in the same spot for many years, interrupted only by short

excursions in search of food or mates, usually no more than a few hundred of metres away. From times before the populations were heavily decimated by fishing, there are tales about mass wanderings of lobsters, even swimming at the surface.
Biology: The lobster is primarily nocturnally active, feeding on mussels and snails that it breaks open with the crushing claw, but also feeds on polychaetes and sea urchins. It will also feed on whatever carrion turns up, and has even been observed feeding on the Lion's mane jellyfish. It does not feed at temperatures below 5°C. Timing of reproduction varies from year to year depending on water temperatures. After spawning the eggs, the females will carry external hard roe until next summer. The time between hatching and settling of the larvae is 2 - 4 weeks, and during this period the larvae have moulted five times. Based on observations made on lobsters in aquaria, the newly hatched larvae are strongly attracted to light (positive phototaxis), and will therefore aim for the surface, indicating that they have a pelagic phase. This is further indicated by the presence of lobster larvae in plankton samples, and we have observed these a number of times *in situ*. The settling larvae are about 18 mm long. Between this size and up to approx. 120 mm, it will rarely occur in any catches or samples. It is estimated that it takes 8 - 10 years, depending on temperature, for them to reach a carapace length of 95 mm. The minimum landing size allowed for capture in British waters is 87 mm carapace length measured from rear of eye socket to rear of carapace. The females reach maturity at a total length of 250 mm; males can be mature already at a total length of 180 mm. A lobster female of 250 mm produces about 8 000 eggs, whereas one of 375 mm length produces about 32 000 eggs, and has thus a much greater reproductive potential.
Commercial value: Lobster fishing have been practised for several centuries. The northermost stable populations are found in northern Norway (Tysfjord). The lobster is one of Britain and Ireland most valuable crustaceans. In Ireland an average of 600 tonnes/year are landed and exported to markets worldwide. Sale values in a good year can reach over 9 million Euro. They are caught in baited pots in shallow water. In Scandinavia they are protected during parts of the year, varying depending on area. Catches have declined steadily since the 1960s, both in Scandinavia and in British waters. For the peak year 1932 in Norway 1300 tonnes were caught, corresponding to over 2 300 000 specimens. During the last decennium only about 30 - 40 tonnes/year have been registered. On Shetland the increasing fishing pressure in the 1960s resulted in over-exploitations of local stocks. After peaking at around 60 tonnes per year, the catches subsequently collapsed. Despite a substantial decrease in fishing effort since then, stocks have never recovered and catches currently remain around seven to eight tonnes/year (approx. corresponding to 5000 specimens). However, 2003 proved to be a good year, in that catches in southern Norway showed a threefold increase, in addition to large numbers of specimens below the minimum size.

FISHERY IN UK

Generally the lobster season in the UK runs from March to October.
Full-time fishers will set 250-1200 creels or pots, while part-time fishers will use 20-100 pots. The minimum landing size in the UK is 87 mm, and average daily catch rates vary from 10-15 lobsters per 100 pots in poor fisheries, to 40-80 pots in better fisheries. Catches in the UK have varied between 750 and 1550 tonnes since 1945.

EL NIÑO

Year class strength of lobsters is thought to be strongly affected by sea surface temperature variations. The peak in landings in UK in 1984 has recently been attributed to the 1982 El Niño Southern Oscillation event.

This lobster has rather unusual colouration. In our experience, lobsters living in caves and on muddy sediments generally are coloured in shades of blue.

Right: *A diver displays two colour variants of lobster. This photograph was taken at the Shetland Islands, where it is still (2003) allowed for divers to take lobsters.*

SPAWNING

The lobster spawns in late summer or autumn, directly following the moulting of the female. The male places the sticky mass of sperm near the opening of the female oviducts, and the eggs are fertilised when they are discharged through the oviducts. Subsequently they are attached to the abdominal swimming legs of the female.

Infraorder Thalassinidea

The systematic position of this group (Gr. thalassa = the sea) is uncertain. At first glance they appear to be a cross between a shrimp and a lobster. In contrast to shrimps, however, they have stout talons on the first pair of walking legs, but lack the talons of the third pair of walking legs, which are present in lobsters. They further differ from anomurans in that the fifth pair of walking legs is not reduced.

Calocaris macandreae

Calocaris macandreae *is pale pink in colour, and has blind white eyes. The first pair of legs carries stout claws.*

Distribution: European waters from the Mediterranean to northern Norway.
Description: The generic name *Calocaris* is based on the Greek kalos = beautiful, and the Latin caris = shrimp, and at first sight it certainly looks like a shrimp, or maybe a crayfish. It is distinguished from astacids in lacking talons on the third pair of walking legs, from shrimps in having well-developed talons on the first pair, and from anomurans in having well-developed walking legs on segment 5. The shell is light red to pale pink, and the violet internal organs are visible through the shell. Rostrum pointed, triangular, with marked keels. The claws are well developed on the first pair of legs, one slightly larger than the other. Eyes very small, reduced (white, without pigment). Length up to 50 mm.
Habitat: Lives on soft bottoms, but may also appear on mixed bottoms. May occur subtidally in shallow waters, but is common below 50 metres depth.
Biology: Like many shrimps it is hermaphroditic.

Upogebia deltaura

This species is seldom seen by divers. Here pictured at Halden, southeast Norway, at 20 m depth.

Distribution: In European waters from the Black Sea, the Mediterranean, and the Atlantic coasts of Spain, north to western Norway.

Description: Species within this genus are highly characteristic. The abdomen is stout, elongated, and distinctly tapered towards the carapace; the tail is wide and fan-shaped. The first association is to landliving preying mantis. The carapace is usually yellowish red whereas the dorsal side of the abdomen is more light red. Body length commonly to 80 mm, but specimens up to 150 mm have been recorded. The first pair of legs has stout talons, where the outer fixed 'finger' is almost as long as the motile part (dactylus). Furthermore, it lacks distinct thorns on the talons, at the insertion of the dactylus, which separates it from the congeneric and co-occurring ***U. stellata***. The latter is also considerably smaller, with a length up to 50 mm. Both species can be separated from other members of the infraorder Thalassinidea by the lack of talons on the second pair of legs.

Habitat: On mud bottoms from a few down to about 40 metres depth. It lives in burrows made by other animals.

Upogebia stellata *is common on all British coasts. Here at 35 m depth.*

Infraorder Anomura

*Anomures (Gr. anomos = uneven, asymmetrical + Gr. oura = abdomen) include squat lobsters (***Galatheidae***), king crabs (***Lithodidae***), hermit crabs (***Paguridae***) and porcelain crabs (Porcellanidae). The crayfish-like anomurans (squat lobsters and hermit crabs) can be separated from astacids by the lack of talons on the third pair of legs, whereas crab-like ones (king crabs and porcelain crabs) can be separated from 'true' crabs (***Brachyura***) by the reduced fifth pair of legs.*

The squat lobster **Galathea strigosa** *is very commonly observed by divers at night-time.*

Galathea nexa, *in common with other squat lobsters, lives hidden during the daytime. However, in contrast to* **G. strigosa,** *which is often seen in rocky crevices,* G. nexa *is commonly found under stones. This lobster often has a pale longitudinal stripe along the carapace.*

Family Galatheidae - Squat lobsters

Galatheids are crayfish-like animals with very long talons. The abdomen is wide, with a large tail. The posterior part of the abdomen is usually bent forward under the body, but can by powerful strokes be used for rapid swimming backwards. In European waters there are 12 species of *Galathea*, and in the northernmost part 6 species, all occurring in shallow water. For two of these, the adult stages can be identified from their colour patterns, but often the determinations require detailed studies of appendages, notably the third pair of maxillipeds (mouth appendages).

Squat lobsters, as their name suggests, somewhat resemble lobsters, but the abdomen is firmly tucked under the thorax. Here a young **Galathea strigosa.**

Galathea intermedia, *with a length of up to only 18 mm, is the smallest of the squat lobsters.*

Galathea strigosa

Galathea strigosa *is our most common squat lobster and is the only one with blue 'marbled' pigmentation.*
Left: *Juvenile* **Galathea strigosa** *on a starfish*

DELICIOUS BY-CATCH

Galathea strigosa is often obtained as by-catch in crab and lobster pots, and they constitute a veritable treat!

Distribution: Very common in the northeast Atlantic, from the Canary Islands to northern Norway; also the Red Sea. All British Isles.
Description: Highly characteristic colouration, reddish brown with blue stripes. Our largest galatheid, attaining a length over 100 mm. Rostrum with three distinct teeth each side. First pair of legs with hairy and spiny talons. The row of spines on the outer side of the talons, and the blue stripes, are characteristic.

Habitat: May occur in high densities on hard bottoms, from a few down to 600 metres depth. Hidden in crevices and under stones during day-time, and it is only during night-time, when they come out in search for food, that divers may appreciate the actual densities of this species.

Galathea squamifera

Distribution: From the Mediterranean to western Norway. Common all around the British Isles.

Description: This is the next largest species of the genus, with a total length up to 65 mm and carapace length up to 32 mm. Similar to ***G. strigosa***, but lacks the blue colour and has a brownish black shell. The rostrum has four teeth on each side, of which the innermost are smaller than the others. Sub-adult specimens are often red, and can easily be confused with other species of *Galathea*, especially ***G. dispersa***. Adult specimens can be separated from all others by the brownish black colouration.

Habitat: On hard bottoms dominated by coarse sand, from the intertidal down to some 20 - 30 metres. Can often be found by overturning smaller stones.

The dark colouration distinguishes this species from all other squat lobsters.

Left: *This individual has lost a claw and is in the process of regenerating a new one.*

Galathea nexa

Distribution: From the Canary Islands and the Mediterranean to northern Norway. The northern part of the British Isles, including all Irish.

Description: Similar to other *Galathea*, but can be identified from the following features. Rostrum short and wide, with long and anteriorly directed teeth. The carapace gives a rectangular impression. The first pair of legs (those provided with talons) is rather large and hairy. They are provided with two or more spines on the subdistal segment (the one before the talons), and several spines on the preceding one. ***Galathea nexa*** is medium-sized, attaining a total length of 40 mm, of which the carapace is up to 20 mm. Females are smaller. The colour is reddish to reddish green, often with darker and lighter spots. It was previously considered to be the same species as ***G. dispersa***, which, however, has a more slender and pointed rostrum, a hairy median tooth on the rostrum, and less hairy first pair of legs.

Habitat: Commonly on stones, sand, or shell gravel, usually below 15 and down to a few hundred metres.

This squat lobster has characteristic hairy legs and a short, stout rostrum.

Galathea dispersa

Galathea dispersa *is rarely seen by divers. Often confused with the previous species.*

Distribution: From the Canary Islands and the Mediterranean to Iceland and northern Norway. Common all around the British Isles.

Description: Very similar to the preceding species, ***G. nexa***, and the two have often been considered conspecific. They can be separated based on shape of rostrum and talons. Rostrum in ***G. dispersa*** is thinner, with the median tooth more protruding and distinctly hairy. Furthermore, ***G. dispersa*** lacks the hair present on the first pair of legs in ***G. nexa***. Small frayed scales cover the talons. Length up to 45 mm. The colour is usually red, sometimes with white spots. Blue colouration never present.

Habitat: From 10 - 500 metres depth, on various kinds of bottoms.

Galathea intermedia

Distribution: In northeast Atlantic from Dakar and the Mediterranean to northern Norway. Common all British coasts.

Description: A small species of *Galathea*, length up to 18 mm, carapace up to 10 mm. The talon fingers are almost half the length of the whole talon, width of talons near uniform. The first pair of legs (those with

The tiny **G. intermedia** *has long thin pincers on the first pair of walking legs.*

The colouration patterns are often striking, and the long, thin legs are partially transparent.

talons) are up to three times as long as the carapace, and may remind of species of *Munida*. Rostrum has four small teeth each side, with a distinctly protruding median one, reminding of a spear. Two spines can be observed if the innermost segment of the first pair of antennae is examined with a magnifier - this in contrast to other species, which have three spines. Colour usually salmon red.
Habitat: Common and abundant species on shallow bottoms with rock, coarse sand or gravel, usually at 8 - 25 metres depth, but has been recorded down to 660 metres.

Munida rugosa

Distribution: From the Mediterranean to northern Norway. Fairly common, all British Isles.
Description: The genus *Munida* is easily separated from *Galathea* from the shape of the rostrum, which has a long thin median tooth or spine, and one very pointed lateral spine each side. The first pair of legs is very long with slender talons. Colour usually red, but can also be reddish yellow. There are two additional species occurring in deeper waters (usually below 80 metres), ***M. tenuimana*** and ***M. sarsi***. The former is characterised by the lack of scales on the thoracic sternum (the ventral part of the

In defence posture outside its hiding place.

Species of **Munida** *have a very characteristic rostrum. A single spear-like spine is flanked on either side by smaller spines (one broken off).*

carapax), and occurs on muddy bottoms at 200 - 1300 metres depth. Both ***M. sarsi*** and *M. rugosa* have small scales on these segments, fringed with fine hairs. The latter two can be distinguished from each other in that *M. rugosa* has scales that are elongated and longitudinally orientated, and that the fourth segment counted from the distal part of the third maxilliped has a distinct spine in the outer, anterior corner, whereas *M. sarsi* has short scales in transverse rows, and lack the spine on the third maxilliped. *Munida rugosa* also has smaller eyes and a more full red colour. A member of *Munida* encountered in shallow waters is likely to be *M. rugosa.*

Habitat: Usually appears below 20 metres, and down to several hundreds metes of depth. Usually hidden in their permanent refuges, under a stone or in a crevice. Also observed in association with the reef-building coral ***Lophelia pertusa.***

Munida rugosa *is known as the 'long-fingered crayfish' in Norwegian.*

Munida sarsi

Distribution: From The Faeroe Islands and Northern Norway to Spain.

Description: Similar to the previous species, *M. rugosa*, as well as to the mentioned ***M. tenuimana***. The latter species is only found deeper than 200 metres, whereas *M. sarsi* and *M. rugosa* can be separated by the features listed for *M. rugosa.* Earlier descriptions report red colour for both species, but *in situ* observations indicate that *M. sarsi* is less red and more light orange than *M. rugosa*. Furthermore, *M. sarsi* is less hairy on the innermost segment on the first pair of walking legs, and the eyes are much larger. Recorded length up to approx. 75 mm.

Habitat: A deep-water species that rarely appear above 70 metres. Observed on a *Lophelia* reef at 52 metres.

Munida sarsi *is paler in colour and much less red than its close relative* **Munida rugosa**.

DID YOU KNOW...

"*In situ*" means 'in place'.

Munidopsis serricornis

Munidopsis serricornis *is one of the few species that live amongst living branches of the stone coral* **Lophelia pertusa.**

This species has a characteristic triangular rostrum.

Distribution: Common in the northeast Atlantic from northern Norway south to the Azores and the Canary Islands.

Description: Easily distinguished from our other squat lobsters by the shape of the rostrum and the characteristic yellowish brown colour. The rostrum is rectangular, with the median keel anteriorly prolonged into a spine, and with a small lateral tooth each side (this species was previously called "tridentatus" = with three teeth). The first pairs of legs are short and stouter than in *Munida*, with a row of spines and tubercles. The length of the carapace is up to about 20 mm.

Habitat: Rocky bottoms, often associated with *Lophelia*-reefs. Probably common in deeper waters on vertical cliffs. Commonly associated with deep-water corals at 50 - 180 metres depth. Off the west coasts of the British Isles common at depths down to 2000 metres.

Family Porcellanidae - Porcelain crabs

Crab-like anomurans with the fifth pair of legs reduced. Carapace smooth, without spines; rostrum lacking. The abdomen is small and bent forwards under the carapace.

Pisidia longicornis - Long clawed porcelain crab

Distribution: From the Canary Islands and the Mediterranean to middle Norway. Common all around the British Islands.
Description: Small crab-like anomuran with a carapace length up to 10 mm. Colour is brownish red to dark red, lighter ventrally. One talon is notably larger than the other. Three frontal teeth between the eyes. Five pair of legs, of which the last pairs is very small. May be confused with ***Porcellana platycheles***, which has hairy appendages.
Habitat: Common in shallow water under smaller stones or in gravel. Found down to 100 metres.

The long clawed porcelain crab is often found under small stones, usually on gravely sediments.

NOT 'TRUE'

Porcelain crabs and king crabs are not 'true' crabs (Brachyura), in spite of their English names. They are distinguished from brachyurans in having only four pair of visible walking legs (including the ones carrying the talons). Brachyurans have five pairs of well-developed legs.

We have observed a number of individuals that lack pigmentation on large parts of the carapace. Note the characteristic asymmetry between the claws.

Family Lithodidae - King crabs

The northern stone crab, **Lithodes maja**, *has graceful movements underwater. Larger individuals do not hide from divers. Most are orange in colour, but some individuals are more reddish.*

This family includes large, crab-like anomurans with the fifth pair of walking legs reduced. A rostrum is present, and the carapace is spiny. The abdomen is small, asymmetrical, and bent under the carapace. Can be regarded as a derived form of hermit crabs that have left their houses.

Here a reddish-coloured northern stone crab prepares to investigate a sea anemone.

Lithodes maja - Northern stone crab

Distribution: From the English Channel and the northern part of the British Isles. Iceland, Greenland, Spitsbergen and the Scandinavian coasts.
Description: Shell light reddish brown, densely covered with rather long spines. Legs long, fifth pair reduced and covered by the abdomen. Talon small, right one larger then left. Abdomen asymmetrical. Shell width up to 134 mm. Large northern stone crabs may be confused with small ***Paralithodes camtschatica***, which occurs in northern Norway.

The northern stone crab is often seen beneath the protective tentacles of the sea anemone **Bolocera tuediae**.

Habitat: On soft and hard bottoms between 10 - 800 metres depth, shallow only in the more northern areas. Both in coastal and open waters.
Commercial value: The meat is highly appreciated, and experimental fishing for evaluation of commercial potential is carried out in Greenland and Newfoundland. Only the meat in the legs and talons is used, and the taste is sweeter than the edible crab (***Cancer pagurus***).

This juvenile (young) individual was observed on a coral reef in the Trondheimsfjord in mid-Norway.

Close-up showing the eyes well protected by spines.

Paralithodes camtschatica - Red king crab

The red king crab is a considerably larger 'edition' of the northern stone crab. Note one of these individuals shows a more greenish colour variation.

A GIANT

Paralitodes camtschatica can have a leg span (distance between tips of extended anterior most pair of legs) of up to 1.8 m, and weigh over 7 kg. The largest 'crab' in the world, the Japanese giant crab, is also an anomuran, and can have a leg span up to 3.6 metres.

Distribution: In Europe originally from the Kola Peninsula (1961-69), but has spread to northern Norway during the 90s. The southernmost record is currently Vesterålen. Further distribution includes the North Pacific and Bering Sea, from Vancouver Island to Japan.

Description: Known under several names in English, including Alaska king crab, red king crab and Kamchatka crab. Adults of this huge anomuran cannot be confused with any other species in our waters. They have a leg span up to 1.8 metres, and weigh up to 7 kg. Like the northern stone crab, it has the fifth pair of walking legs reduced and inserted under the carapace. The carapace is covered by spines/tubercles. Brownish dorsal side, yellowish white ventral side.

Habitat: Usually in rather shallow, coastal waters, but displays annual migrations. Animals in Varangerfjord migrate down to 150 metres depth in spring and summer, and migrate up to more intermediate depths in November. In the spring the ripe animals appear in shallow water for reproduction.

Biology: Like other crustaceans, *Paralithodes* has to moult in order to increase in size. Moulting frequencies decrease as the animals grow. Adult males in northern Norway usually do not moult every year, whereas the females moult in spring, soon after the hatching of the eggs. Mating takes place directly after the moulting, followed by the fertilisation of the eggs, which then are attached to the abdomen of the female. The females carry the eggs until next spring. The larvae are pelagic and settle after about eight weeks in shallow water. Observations from Alaska and Varangerfjord show that sub-adults can form huge colonies of up to 100 000 specimens. They feed largely on mussels, snails, polychaetes and echinoderms. Being an intrusive species, there is some concern about its relatively rapid increase in density and distribution. In later years it has moved further west and southwards along the Norwegian coast. It seems to have few enemies, and it is not unlikely that it will spread even further south. This may have effects on other species, and yield negative impacts on local marine ecosystems.

Commercial value: This species is considered as a gastronomic treat. It was artificially introduced for commercial fishing in the Murmansk Fjord at the Kola Peninsula, and has spread west to Finnmark during the 90s. This has lead to important catches, most part of which is exported to Japan. The price during the 90s was between Euro 3,6-5,1/kg, depending on quality. In the 2003 the value had increased dramatically - the Norwegian quota was 190 000 crabs to a total initial value of approx. Euro 11 million. However, the immigrant is not entirely well received by Norwegian fishermen, since it destroys fishing-nets and attacks both bait and catch on fish lines.

INTRODUCED IN RUSSIA

Paralithodes camtschatica is sometimes referred to as the Russian crab, because it was introduced to the Kola Peninsula in the early 60s. It is also called Kamchatka crab after the place of its natural distribution (compare with the Latin name).

These giants are most likely preparing to mate.

Family Paguridae - Hermit crabs

The abdomen of these anomurans, which is soft and lacks a distinct shell and segmentation, is usually hidden in empty snail shells, although some species also use sponges or sea anemones for protection. The fifth pair of walking legs is reduced, and for hermit crabs belonging to Paguridae, the right claw is much larger than the left.

HERMIT

Hermit comes from the Greek word *eremos*, meaning solitary, referring to the solitary life of hermit crabs. This, however, is usually not the case. As seen from the descriptions below, they commonly have symbiotic relationships with one or several other organisms. The common species *Pagurus bernhardus* is often associated with three other species–one polychaete, one hydroid, and one isopod.

Without protection from a snail shell house or a sea anemone, a hermit crab is doomed.

Pagurus prideaux

Above: *Protective nematocysts are discharged from the sea anemone.*

Right: *Roe held at the posterior of the body.*

Distribution: From Cap Verde and the Mediterranean to Norway. Common, especially on the west coasts of the British Isles.
Description: Adult specimens usually found with the sea anemone ***Adamsi palliata,*** and are thereby easily identified. Carapace length up to about 14 mm. Colour brownish red with lighter spots, talons salmon pink.
Habitat: In shallow waters from a few metres, usually not deeper than 40 metres. Occurs deeper in winter. Most common on bottoms with mud, sand or gravel.
Biology: Hermit crabs, as is well known, have to change snail-shell as they grow. Most ***P. prideaux*** attach sea anemones–***Adamsia palliata***–as protection from the stinging nematocysts of the anemone. If the couple becomes threatened, the sea anemone will discharge stinging threads (acontia) from the body wall. Initially the crab will search for a small anemone that it holds against the shell until it attaches. The sea anemone probably reacts to chemical signals emitted by the crab, and lets go of the original attachment; other species of sea anemones are much less cooperative. In due time the anemone will cover the whole shell, always with the mouth opening directed towards the opening of the shell, where it obtains food spill from the host. Therefore, the cohabitation is mutually beneficial. Furthermore, the anemone actually enlarges the house for the crab, which therefore will have to shift less frequently. And when it does shift, it brings the anemone to the new shell. Male and female ***P. prideaux*** will form pairs in March, and moulting and mating presumably takes place soon thereafter. The males hold on to the female appendages, and will even pull the females around if disturbed.

DID YOU KNOW...

Symbiosis in biology stands for interspecific relationships, and these can be of several kinds. It is called mutualism if it is to the advantage of both species, commensalism it is for the benefit of one but neutral to the other, and parasitism if it is for the benefit for one but disadvantageous for the other.

Pagurus bernhardus - Common hermit crab

Our most common hermit crab has strong claws equipped with many small spines and 'warts'. The hydroid **Hydractinia echinata** *often grows on the snail shell.*

Like all other crustaceans, hermit crabs must change its hard outer layer in order to grow. Here an individual has just cast off its old armour.

Distribution: From Portugal in the south to Iceland and northern Norway. Very common all around the British Isles.

Description: This is the largest and most common hermit crab along northern European coasts. The talons are covered with densely distributed fine warts, and lack hairs. The colour is variable, many are brownish green. In British waters 18 species are recorded, most with a deeper distribution.

Habitat: Occurs intertidally and down to several hundred metres depth. Smaller specimens will use any kind of snail shells, whereas larger ones often will use the common whelk (***Buccinum undatum***), and the largest ones the red whelk (***Neptunea antiqua***). Intertidal specimens tend to be small, although large specimens can occur in very shallow water where there are strong currents. They are most common on bottoms with shells and sand or gravel.

Biology: Hermit crabs have to change snail shells as they grow. By preference they will take empty shells, but can also attack damaged or otherwise weakened live snails. The new house is carefully examined before leaving the old. Commonly, the shells inhabited by the crabs harbour other organisms, such as the hydroids ***Hydroides echinata*** and ***H. carnea***. The relationship between hydroids and crabs is mutualistic, in that the hydroid obtains more food through the movements of the crab, and the crab gets protection from the poisonous nematocysts of the hydroid. The hydroids can also grow on the inside of the shell, and in those cases the crab is without the parasitic crustacean, which otherwise is common. In some specimens the shell is overgrown by the sponge ***Suberites ficus***, serving to enlarge the shell for the crab that will have to shift shell less frequently. Several anemones may occur on the shell: ***Calliactis parasitica*** and ***Hormatia digitata*** (the latter on shells with, or without, hermit crabs), and further north also ***Hormathia nodosa***. Another co-occurring animal is the polychaete ***Nereis fucata***, using the shell for protection and obtaining food remains from the crab, and keeping the soft abdomen of the host free from parasites. A common parasite is the bopyrid isopod ***Athelges paguri***. Hermit crabs are the cleaners of the sea floor, and virtually anything within reasonable size ranges will be eaten, including other hermit crabs.

Pagurus pubescens

Distribution: Widely distributed in all northern seas; in the north-east Atlantic from the northern parts of the British Isles to Iceland and northern Norway.

Description: Similar to ***P. bernhardus,*** but can be distinguished from all our hermit crabs on the distinct keel on the left talon. It is also hairier than ***P. bernhardus***. Carapace length up to 23 mm. Right talon distinctly larger than the left. On the inner side of the left talon there is an incision where the right talon fits in, serving to

Pagurus pubescens *is more hairy than* **P. bernhardus**. *Here photographed on the hydroid* **Eudendrium rameum**.

Pagurus pubescens, *to the left photographed at Tromsø, northern Norway, and to the right on a* **Lophelia** *reef in Trondheimsfjorden, mid-Norway.*

provide a tight-fitting lid to the shell. Colour speckled dark red to reddish brown.
Habitat: Occurs deeper than ***P. bernhardus,*** usually below 30 metres and at least down to 250 metres. Both soft and hard bottoms.

Anapagurus chiroacanthus

Distribution: Europe from the Mediterranean to mid Norway. From British waters it is recorded in the northern North Sea.
Description: This small hermit crab can be identified from the densely hairy talons and legs. Carapace up to 6 mm long. Front end of carapace rounded, with two marked spines on each of the anterior corners. Long eyestalks of even width, slightly longer than the hardened part of the carapace. The eyestalks reach beyond the second segment of the basis of the smaller pair of antennae. Several hermit crabs have hairy talons, but in our waters the hairy legs are unique for this species.
Habitat: Usually on rather shall soft bottoms, from 10 - 140 metres.

This hermit crab has very hairy legs, making it easy to confuse with the deeper-living **Pagurus cuanensis**.

Infraorder Brachyura - Crabs

Crabs (Gr. brachys = short + Gr. oura = tail) have a characteristic shape. The width usually exceeds the length, and the abdomen is bent in and 'attached' ventrally under the thorax. The first pair of legs is provided with talons, whereas these are always absent from the second and third pairs. In contrast to king crabs and porcelain crabs, which belong to the ***anomurans****, the fifth pair of legs is well developed.*

Upper left: **Inachus dorsettensis**. ***Upper right:*** **Liocarcinus pusillus**.
Above: *The edible crab,* **Cancer pagurus**, *photographed at Lysefjorden, southwest Norway. Most likely a female carrying roe.*

Ebalia tumefacta

Distribution: Recorded from West Africa to western Norway. Very common around the British Isles.
Description: Characteristic shape of carapace, as wide as long or slightly wider. Carapace width up to about 13 mm, length 11-12 mm. Carapace with two inflated parts on each side, above the gills. Colour variable, but often reddish white or flesh-coloured. Easily confused with two congeneric species, ***Ebalia tuberosa*** and ***E. cranchii***, that also occur in the area. The former lives deeper and has a cross-shaped rise on the dorsal side of the carapace, and also has an uneven granulated sculpture. The latter is very similar to ***E. tumefacta***, although the carapace is less wide, and the shell rhomboid in outline with cut corners. It also has elevated areas over the gills, although these are not inflated.
Habitat: Bottoms with stones, gravel or sand, about 5 - 100 metres depth.

Ebalia tumefacta *has a shell wider than its length.*

Above: **Ebalia chranchii,** *in contrast to the above, has a shell width equal to its length.*

Ebalia tumefacta *shows a large variation in colour. Compare this individual to the one at the top of this page. At some localities, we have observed several tens of individuals.*

Cancer pagurus - Edible crab

Distribution: The edible crab has its main distribution from Morocco and the Mediterranean to northern Norway. Very common all around the British Isles.

Description: Easily identified from the colour of the carapace, which dorsally is brown-red and ventrally yellowish white, and the black tips of the talons (may be pale just after moulting). Small crabs (carapace width less than 15 mm) often have white shells. The carapace is much wider than long (over 3/2), and the crabs can reach a width of 300 mm and weigh 5 kg. Males are considerably larger than females. In mature specimens the males have larger talons than the females.

Habitat: In Norwegian waters the edible crab occurs on various kinds of bottoms, both hard and soft. In more exposed areas there are high densities on hard bottoms, dominated by males. In fjords there is instead a dominance of females, which seek sheltered areas for breeding when they become reproductive. Most crabs in the kelp forests of the more exposed areas are not mature, and the smaller ones use holdfasts of the cuvie, ***Laminaria hyperborea***, for shelter. Crabs in British waters show a preference for rocky bottoms, but may also be abundant on mixed sediments or muddy sand.

Biology: The edible crab is omnivorous, although larger specimens have a preference for soft bottom bivalves and polychaetes. It is well adapted for catching shell-protected prey,

MOULTING

Similar to other crustaceans, the edible crab needs to moult in order to grow. At each moult the carapace width increases with about 25%. There are often reports from divers of large number of dead crabs in the autumns, but these are usually just moulted shells. Note that also the gills are moulted - they can be observed inside the empty shell in the lower part of the picture. The upper picture shows the shell sizes before and after moulting.

Edible crab with roe.

COASTAL FISHERY

Crab fishing is confined mainly to coastal waters within 6 miles of the shore.

Young edible crabs with a shell width less than 10 mm are usually white. However, already at this stage, the tips of the claws are black.

DID YOU KNOW...

Not all crabs are 'full', and should be properly examined to avoid cooking 'empty' ones. Crabs need several months after moulting to put on weight, for this reason those with old shells tend to be best. Such shells appear worn and are very hard, usually with various attached epifauna, including calcareous tube worms, occasionally also with some small blue-mussels and other bivalves. A crab with a clean and neat shell will tend to have a much less rewarding content. Generally speaking, they are best in late summer and in the autumn, prior to the moulting. But even following these rules, some meagre ones may be found. Some areas are known for poor crabs, often in exposed areas where there are large number of small specimens and intense food competition. The highest quality usually occurs in more sheltered areas. In areas with large number of small crabs, the smallest specimens tend to be best.

which are crushed with the powerful talons. Dissecting the gut of a crab reveals three teeth, used for further crushing of the food. Crabs living in kelp forests prey on a variety of animals, including crustaceans such as amphipods, isopods and squat lobsters, blue mussels and horse mussels, sea urchins, and a series of different snails. Dead fish are also on the menu. The edible crab has a complex life cycle. The eggs hatch in summer, and the larvae live as plankton for about eight weeks, and moult seven times. At settling they are about 2.5 mm wide, and after a year they have grown to a width of about 15 mm. During the first year they moult a number of times, but after that usually not more than twice a year (temperature dependent). After two years they have a width of about 30 mm. They are not mature until reaching seven-eight years of age and a width of about 120 mm. Mature specimens moult during the autumn, the females about one month before the males.

Mating takes place immediately after the moulting of the female, when the shell is still soft. Pair formation has been observed by divers, often about a week before mating, and the pairs remain together also a few days after. The actual fertilisation occurs a long time after the mating, normally not until the following autumn, 14 - 15 months later. During this period the female carries the sperm in a spermatheca ('sperm chamber'), and uses this time to find a suitable home and to gain weight. Following fertilisation, the eggs are attached to the abdomen, where they will remain for eight months before hatching. The female does not feed during this period, and lives partly buried in sand at 10 - 20 metres depth. Sperm from one mating event can be stored for several years and be used for two, possibly three, fertilisations. Tagging experiments have shown that on the east coast of Britain mature female crabs can move distances of between 180 and 200 miles in twelve to eighteen months. These migrations are associated with the offshore movement of females for spawning. Male crabs rarely moved far from the point of release.

Commercial value: The edible crab is fished all over its area of distribution. Largest catches are taken in the English Channel by French and English fishermen. In British waters about 80 percent of the total catch is landed on the east coasts of England and Scotland. However, there are important fisheries in Devon and Cornwall, along the English Channel and on Shetland. The main season extends from March to September, with peak catches in May and June. Landings are affected by seasonal movements of crabs into deep water in winter and back inshore in spring, and by the moulting season. Both the white meat, the brown liver, and the eggs are used for food.

Males have very well developed claws.

Carcinus maenas - Common shore crab

Left: *The common shore crab is our most usual crab in shallow waters.*

Below: *Common shore crabs usually form pairs already several days before they actually mate. The males are usually larger than females.*

Distribution: Distributed in all oceans. In the northeast Atlantic up to northern Norway.
Description: Dorsally greenish or brownish with darker marmorations, ventrally light yellow. Carapace width up to 80 mm, length 60 mm. Provided with five wide, pointed teeth each side of the eyes, and three more rounded teeth between the eyes.
Habitat: Very common in shallow water, also in brackish water. Common under stones in the intertidal, both on soft and hard bottoms. More rarely in

DID YOU KNOW...

Similar to many other crabs, common shore crabs mate belly to belly. The process starts with the female emitting pheromones that inform males that she is reproductive. A male will sense the pheromones and form a pair with the female, and this may occur several days before she is ready to mate. The actual mating can only take place after her moulting, a process where the male helps the female. The pair formation also often continues a few days afterwards. This increases the chances of survival of their common offspring, since the female during this period is very vulnerable. Males moult up to a month after the females.

deeper waters, down to 200 metres.

Biology: The common shore crab eats virtually anything. Females with eggs can be found all year round, and the time of settlement of the larvae varies. Mating procedure are largely similar to those described above for the edible crab. This is a fast-moving animal, and can appear aggressive towards divers.

Many colour variations exist, but green is the most common. This specimen has some orange pigmentation.

TREAT

The common shore crab is considered a delicacy in the south, but is rarely eaten in northern Europe. However, it is known that they were caught with baited pots in shallow waters in Norway already in the 18th century.

Liocarcinus depurator - Harbour crab

The harbour crab has an aggressive behaviour and moves quickly, both when swimming and walking. Photograph from Høgsfjorden, southwest Norway.

In swimming crabs, the last segment in the last pair of walking legs is modified into a characteristic paddle-shaped structure used for swimming.

Distribution: Distributed from West Africa to Norway. Very common all around the British Isles.

Description: Carapace width over 50 mm. Colour weak reddish brown to lighter brown with darker spots. The upper side of the carapace is provided with small hairy tubercles. The outermost segment of the fifth pair of legs are paddle-shaped and violet.

Habitat: Common on bottoms with sand and gravel, and can swim rapidly if disturbed. Often observed buried in coarse sand. Aggressive. Occurs at 1-450 metres depth.

Individuals are often observed buried in coarse sand between 10 and 30 m depth.

Liocarcinus pusillus

Distribution: From northwest Africa to northern Norway. Common, all around the British Isles.
Description: A small species, as indicated by the Latin name. Colour varying from brown to reddish brown and yellow. Between the eyes there are three rounded teeth, with the median one protruding. Some specimens with a dark longitudinal stripe on the back.
Habitat: From the intertidal down to about 100 metres. Usually on bottoms with fine gravel, into which they hide rapidly whenever necessary.

PADDLES

Swimming crabs are characterised by having a paddle-shaped distal-most segment (dactylus) on the fifth pair of legs . This is an adaptation for swimming, and the legs are used as propellers. This is used mainly for rapid escape.

This crab has three characteristic 'teeth' between the eyes. Many individuals have a dark stripe along the middle of the carapace.

Liocarcinus navigator (= L. arcuatus)

Distribution: From the Mediterranean to mid-Norway. Common on the south and west coasts of the British Isles.
Description: A small and characteristic swimming crab. It can easily be separated from others by the smooth line between the eyes, without teeth, and by being conspicuously hairy. Carapace length up to 30 mm, width 33 mm. Laterally to and behind the eyes there are five pointed teeth each side. The last segment on the fifth pair of legs is flat and paddle-shaped although less extremely so than in ***L. depurator***. Colour dark brown.
Habitat: Usually on sand bottoms at 5 - 10 metres, down to about 100 metres.

Liocarcinus navigator *is distinguished by the straight, hairy edge between the eyes.*

Necora puber - Velvet swimming crab

Distribution: Northwest Africa, the Mediterranean and the Black Sea, north to the Faeroes and northern Norway. Very common in British waters.
Description: A large swimming crab with a carapace width exceeding 80 mm. The last segment on the fifth pair of legs is paddle-

Velvet swimming crab.

The velvet swimming crab is the largest of our swimming crabs. It has characteristic red eyes.

shaped. The dorsal side of the carapace is covered by small hairs that give a dirty brown impression. Naked (non-hairy) areas on the legs are blue, sometimes red. The eyes are characteristically red. Size and colour distinguish adults of this species from other swimming crabs. Between the eyes there are 7-10 small teeth, the median ones being largest.

Habitat: Among stones in shallow waters, but have been observed down to 70 metres. The fastest and most aggressive one of our crabs, and will 'attack' a diver if threatened.

Pilumnus hirtellus - Hairy crab

Distribution: From North Africa to western Norway. Very common all around the British Isles.

Description: The reason for its English name should be obvious from the photograph. It is a small crab with a carapace width less than 20 mm. The colour usually is brownish red. There is a considerable size difference between the two talons.

Habitat: Rocky bottoms, often intertidally, down to 70 metres.

This crab has a very hairy appearance and this distinguishes it from all other crabs in our waters.

Xantho pilipes

In common with the much larger edible crab, **Xantho pilipes** *has black claw tips.*

DID YOU KNOW...

The sex of a crab can be determined by examining the abdominal segments. In males these are narrower than in females. When lifting the tail to see its ventral side, the female genital pores become visible, alternatively the male reproductive organs. These differences can be less obvious in small crabs.

Distribution: West Africa to Norway, Mediterranean. Very common on the west coasts of the British Isles.
Description: Carapace width 30 - 40 mm, wider than long. Colour often brownish red or pale brown, marbled. Tips of claws black. Smooth carapace, without teeth between the eyes. Difficult to detect because they are often buried in sand, but may be fairly common across the area of distribution.
Habitat: Rocky bottoms, or bottoms with stones and sand, from a few metres depth down to about 40 metres. We have observed it a number of times, always in connection with reproduction, probably because they then leave their hiding places.

We usually see individuals of this species in pairs. The colour varies, but the shell is usually distinctly marbled.

Hyas araneus - Great spider crab

Distribution: Large parts of the North Atlantic, from Spitsbergen, Greenland and North America, and down to the English Channel. Very common.

Description: Pear-shaped carapace. Dorsal side reddish brown or yellowish brown, ventral side dirty white. Carapace often overgrown by a variety of organisms. Carapace width up to 90 mm, length 115 mm. Males slightly larger than females. Can be confused with ***H. coarctatus***, which has a more marked incision behind the eyes, and a more flattened and usually red shell.

Mating in the great spider crab only lasts a few hours, so it is relatively rare to catch them in the act. The female has the widest abdomen and is lying below the male.

Expert at using other organisms for camouflage.

Habitat: All bottom types, 1-350 metres depth, both in exposed and sheltered areas.

Biology: *Hyas araneus* uses other organisms to camouflage itself, including bryozoans, algae and hydroids, all part of its natural environment. Newly moulted specimens are easily identified by the lack of attached organisms.

Hyas coarctatus

Upper left: *On this newly moulted individual of* **Hyas coarctatus**, *the 'hooks' on the carapace, used for attaching small loose objects, are clearly visible.*
Above: *The carapace has a very characteristic shape.*

This individual with roe was observed in Høgsfjorden, southwest Norway in September.

Distribution: Similar distribution to the preceding species, large parts of the North Atlantic and down to the English Channel. Also Greenland and North America. Very common.
Description: Triangular shell, although less pronounced than in previous species. Dorsally reddish brown, ventrally dirty white. Gives often a more reddish impression than ***H. araneus***. Carapace width up to 50 mm, length 65 mm. The two rostral teeth are slightly longer, and a little further apart than in *H. araneus*, and the first pair of legs (those with the talons) is distinctly longer. Can otherwise be distinguished from *H. araneus* in the marked incisions posterior to the eyes.
Habitat: All kinds of bottoms, soft, hard and sandy, at 1-500 metres depth. Often a bit deeper than *H. araneus*.
Biology: Similar to *H. araneus* both in appearance and biology.

Inachus dorsettensis - Scorpion spider crab

The scorpion spider crab has red-violet patches and is usually covered by sponges. Photographed at Egersund, southwest Norway.

Distribution: Large parts of the east Atlantic, from South Africa to northern Norway. Mediterranean. Common. All around the British Isles.

Description: Very long legs. Carapace grey or light brown, with red-violet spots on the back, often covered by sponges (see picture). Carapace length 30 mm, width 27 mm, roun-

Scorpion spider crabs are commonly seen in harbours and other polluted places. Often found amongst algae, as shown on the picture to the right.

Inachus phalangium *is common in south-west of Norway and is more brown-yellowish in colour.*

ded triangular. Rostrum is short and bifid, with diverging processes. One spine behind each eye, plus a horizontal row with four spines. Talons of females about as long as carapace; almost twice as long in males. Another less common species, **I. phalangium** (illustrated above), is similar, but lacks the row of horizontal spines behind the eyes. *Inachus dorsettensis* should easily be separated from ***Macropodia rostrata*** based on the mentioned features.

Habitat: All bottom types at 5 - 30 metres, but recorded down to 110 metres.

Macropodia rostrata - Long-legged spider crab

Distribution: From South Africa to northern Norway. Mediterranean. All around the British Isles.

Description: Greyish brown, yellowish brown or reddish brown, often with algae. Legs very long, 2 - 3 times the length of carapace. Lacks spine behind the eye. The first segment in the

Macropodia rostrata *has a spider-like appearance.*

long, bushy antennae are longer than the rostral process. Carapace length in a large male is 22 mm, width 15 mm. Appears much thinner than the scorpion spider crab.

Habitat: Usually on hard bottoms above 20 metres, but recorded down to 86 metres. Often on brown algae.

Members of this species have very long legs and antennae.

MOLLUSCA

Calliostoma formosum, *photographed at Høgsfjorden, southwestern Norway.*

Order Mollusca

This group (lat. molluscus = "soft"), includes some of the most well-known marine invertebrates; ***snails, clams, squids*** *and* ***octopi****. Although the inner structures and physiology are relatively similar, there is enormous variation in exterior body forms. The largest molluscs are the up to 20 m long eight-armed squids, whereas some snails never grow larger than 1 mm in length.*

A characteristic shared by all molluscs is **bilateral symmetry** (symmetry about one axis) and a lack of segmentation. In most molluscs, the internal area is surrounded by a usually folded mantle, such as to form a **cavity** between the mantle and the body. The mantle can secrete an exterior **calcareous shell**, such as in snails and bivalves, but this is absent in, for example, sea slugs. Within the mantle cavity are usually **gills**, which are used for respiration and sometimes also for filtering food particles. Molluscs have a **heart**, which consists of separate chambers, usually an auricle and ventricle, but the circulatory system generally is partially open (closed in squids and octopi). In many molluscs, the mouth region is equipped with a **rasp-like tongue**, the **radula**, but this is absent in bivalves. There is a well-developed **digestive system**, with a continuous gut that opens into the mantle cavity. Molluscs have a relatively large and well-developed **excretory organ** (equivalent to kidneys). The nervous system is composed of paired ganglia, which in squids is more developed than in any other of the invertebrates. Molluscs can be either hermaphrodites or have separate sexes. Many use external fertilisation, whereas others display mating behaviour. The fertilised eggs usually develop into free-living ciliated **trochophore larvae**, and then further into **veliger larvae**. Veliger larvae have a shell, foot and an outer layer of ciliated cells (**velum**), which assist in vertical movement. In the plankton, veliger larvae resemble miniature snails or bivalves. Some molluscs, such as the sea butterflies, spend all their adult lives in this form, but the majority settle on the bottom after a relatively short time in the plankton.

The lower of these two slipper limpets, **Ansates pellucida**, *has most likely laid its eggs on the upper individual.*

The Mollusca is divided into two sub-orders. The **sub-order Aculifera** (Lat. aculeus = needle + Lat. fero = bearing) comprises four classes of which three, the **Caudofoveata**; **Solenogastres** and **Monoplacophora** are small groups. Most of these live as infauna, buried within the sediment. These are not dealt with here. The fourth class in this sub-order is the **Polyplacophora**, known as chitons, of which many are common on our shores.

The other sub-order; **Conchifera** (Gr. konchion = shell/shellfish + Lat. fero = bearing) includes the already discussed classes **Gastropoda** (snails), **Bivalvia** (bivalves) and **Cephalopoda** (squids and octopi) as well as the **Scaphopoda** (tooth shells).

SPECIES-RICH GROUP

Among the molluscs there are around 100 000 existing known species, and approximately 60 000 fossil species. Because large areas of the world's oceans are insufficiently studied, it is believed that there are at least twice as many species in existence.

Class Polyplacophora - Chitons

Chitons have usually eight shell plates, encircled by a mantle. Here **Lepidonochiton asellus** *(not described).*

*Chitons (Gr. Polys = many + Gr. plakos = plate + Gr. phoreus = bearing) are oval in shape, with a shell consisting of eight **shell plates**, or valves, although atypical individuals can have 5 - 7 plates, surrounded by a mantle/belt/ridge (**perinotum**). Underneath the shell is a strong foot that acts as a sucker to attach the animal to the substratum. The mouth is equipped with a **rasp-like tongue**, the radula, which scrapes off diatoms, green algae and calcareous algae, as well as polyps, bryozoans, foraminiferans and other microfauna. Chitons lack eyes, but they have light-sensitive organs in the shell. Often found on the underside of stones, in cracks in rocks and other shaded places in shallow water.*

LIVING FOSSILS

These molluscs have changed very little in form since the Cambrian period, 600 - 500 million years ago.

Chitons have separate sexes and use external fertilisation. When stimulated by the presence of sexually mature females, the males release sperm directly into the water masses. In the presence of sperm, the females release eggs that are placed either singly or in groups on solid substrates. From these, ciliated larvae (**trochophores**) develop that live in the plankton for a few hours or in some cases up to two weeks. The different forms of chitons can be recognised mainly by the shape of the mantle and shell plates, as well as the number and position of the gills. For identification, it is often necessary to have the animal in hand, under magnification, but good close-up pictures are often sufficient. There are around 40 species known from around Europe.

Lepidochitona cinereus

Reliable species determinations usually require magnification using either a magnifying glass or a microscope. To distinguish this species, **Lepidochitona cinereus***, from its close relative* **L. asellus***, requires microscopic examination of the tiny details of the mantle surface. In* **L. cinereus***, the mantle has a fine granular surface, whereas in* **L. asellus***, the mantle surface is covered with striped rectangular overlapping scales.*

Distribution: Along the European west coast from Gibraltar and north to the Barents Sea. Western Mediterranean. Common around all British Isles.

Description: Relatively wide/oval, around 1.5 times longer than wide. Approx. 25 mm long. The mantle forms a relatively narrow girdle covered in small rounded tubercles, with short, thick, cigar-shaped bristles at the outer margins of the girdle. The shell plates, or valves, have a slight ridge in the middle, such as to form a low keel along the animal. Shell plates with a finely granular surface, becoming coarser towards the outer margins. Growth lines are visible at the outer margins of the plates. The head plate has 8 or 9 indentations at the anterior lower margin (not visible on photographs). 16 - 19 pairs of gills. Colour varies; red, brown, yellow and/or green patches and stripes. Often confused with ***Leptochiton asellus*** (pictured on the previous page).

Habitat: Very common under stones and in shaded places in shallow water. Most abundant closest to the lower tidal limit, but has been found as deep as 275 m.

Biology: Around the British Isles, reproduction occurs in the period between July - September. Sperm and eggs are released freely into the water, by means of the animal lifting the rear part of the mantle from the substrate.

Tonicella marmorea

Distribution: Northern distribution. Northern coasts of the British Isles, Norwegian coast, North Sea, but also recorded as far south as the Canary Islands.
Description: Up to 40 mm; one of the largest chitons found in northern Europe. Relatively broad; length only 1.5 times its width. As its scientific name suggests, often beautifully marbled in appearance, predominantly red and brown in colour. Shell plates smooth and shiny, but show clear growth lines. Posterior margins of the shell plates often with eye-catching small white patches. Mantle broad and thin, with a few microscopic scales and small flat spines at the outer margins. ***Tonicella marmorea*** is distinguished from the notably smaller ***T. rubra*** (next page), because the latter has mantle scales that are visible without magnification. The girdle in *T. marmorea* is leathery to the naked eye, dorsally clothed with widely separated, small, conical, longitudinally ribbed, sharply pointed spicules. *T. rubra* has abanal gills with about 12 ctenidia on each side, *T. marmorea* has merobranchial gills and 19-26 ctenidia on both sides.
Habitat: Lower part of the shore and in shallow water. As with other chitons, firmly clamped to stones and similar hard substrates. Grazes on calcareous red algae.

Tonicella marmorea *is a large chiton, reaching up to 40 mm in length. It is distinguished from* **T. rubra** *by its size, and by the smooth, leather-like appearance of the shell plates. These do have a granular surface, but this is not apparent without magnification.*

Tonicella rubra

This attractively coloured chiton is often seen on the shore.

Spirorbid polychaetes have settled on this individual.

Distribution: Wide geographical distribution. Found along the west coasts of Europe south to Portugal. Mediterranean to the Adriatic Sea. Around all of the British Isles, Iceland, Norway, Svalbard and Greenland.

Description: Up to 20 mm long, relatively broad/oval, with a wide mantle. Mantle with small round tubercles (barely visible without magnification), and short, thin bristles at the outer margins. Shell plates with orange, reddish-brown and/or violet patches or stripes. Girdle often with alternating reddish-brown and pale patches. The eight shell plates show clear parallel growth lines.

Less than 15 pairs of gills, limited to the posterior half of the two longitudinal mantle cavities.
Habitat: Clamped to hard substrates where it grazes on encrusting algae; usually just below the intertidal zone. Found down to 270 m. Common.
Biology: Along the Norwegian coast, spawning has been observed in January.

The mantle usually has characteristic yellow patches at the outer edges.

Callochiton septemvalvis

In fully-grown individuals, the pale overlapping scales along the mantle sides are clearly visible. Although the basic colour still is dark red, the shell plates are more richly patterned than in younger specimens.

Distribution: Found along the entire northeastern Atlantic coast. Mediterranean.
Description: Up to 30 mm in length, and half as wide. Mantle partially covered in small, flattened scales, arranged like roofing tiles. Shell plates smooth, most often green or reddish-brown with white or greenish-white marbling. Occasionally with an addition of purple. Juveniles (young individuals) almost always red, with four pale patches on the

mantle at each corner. Shell plates forming a distinct longitudinal keel. Clear growth rings. Head plate with 15 distinct indentations in its anterior margin. Mantle relatively broad, particularly in younger individuals. After having photographed numerous individuals of this species, we see a gradual development in pigmentation and development of the shell along the sides of the mantle.

Habitat: From the lower part of the shore down to approx. 120 metres depth.

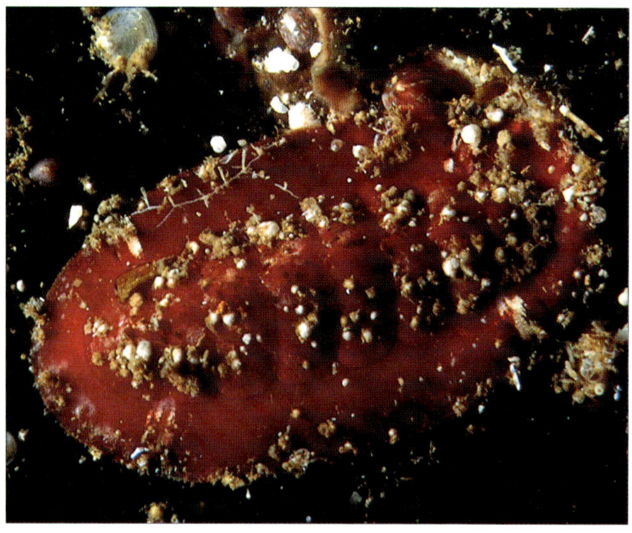

CHANGES APPEARANCE AS IT GROWS

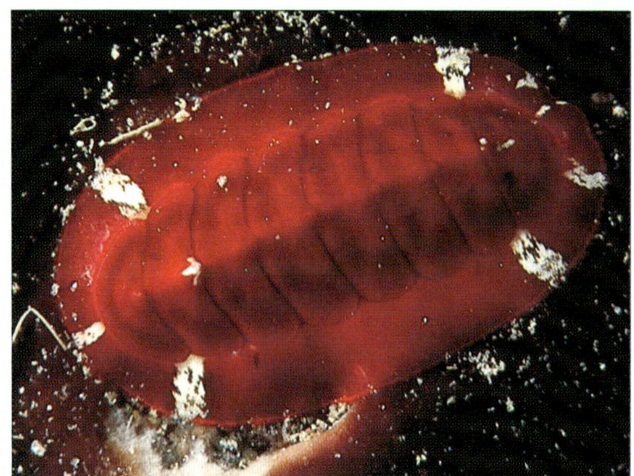

Juveniles of this species are very characteristic. They are dark red in colour and the mantle has four pale patches, one at each corner. On each patch are small scales arranged like roofing tiles and bristles. As the individuals grow, the number of pale patches with their associated scales increases, until they eventually cover most of the side parts of the mantle in adults. However, even in fully-grown adults, a portion of the mantle at the head and rear end retains its juvenile red, smooth, leathery structure.

Upper left: *Juvenile individual.*
Left: *The mantle scales are formed.*
Above: *Fully-grown individual.*

Class Gastropoda - Snails

Above: **Cryptonatica affinis,** *photographed in Finnmark, northern Norway.*

Right: **Nassarius incrassatus***; a small snail that can have up to eight body whorls.*

The Gastropoda (Gr. gaster = stomach + Gr. podos = foot) is the largest group within the molluscs, including in total around 40 000 living species. The main distinguishing features include a flat and broad foot, a radula and a head with tentacles and eyes. The vast majority of snails have a diet of plants, or sessile and/or sedentary animals. The Gastropoda is now divided into two sub-classes; the ***Prosobranchia*** *and* ***Heterobranchia****. The latter includes the former sub-class, now supra-class* ***Opistobranchia*** *and the former sub-class, now supra-class* ***Pulmonata****.*

Sub-class Prosobranchia

Prosobranch snails (Gr. pros = forward, anterior to + Gr. branchia = gills) usually have a coiled or cup-shaped shell. Almost all snail shells found on the beach belong to this group.

Lacuna vincta. *Here the spire comprises three whorls, separated by distinct sutures. The top of the spire is called the apex. This snail is a holostome, lacking a siphonal canal. Head bearing two tentacles.*

The shell is originally cup- or cone-shaped, but in most cases it has become coiled during the course of evolution. As the name suggests, the gills are located at the front of the animal, as is the anus and mantle. On the foot is a horny shield (**operculum**) that covers the opening when the snail retreats into its shell. The exception to this are the limpets, which protect themselves by using their large muscular foot to attach themselves to the substratum by suction. The operculum, or suction, protects the snail from water loss (such as in the tidal zone) as well as predators. Almost all prosobranch molluscs live in water, and most are marine. In contrast to the other two sub-classes, prosobranch molluscs have **separate sexes**. Amongst the marine prosobranchs, development is via pelagic ciliated **trochophore larvae**, which further develop into shelled **veliger larvae**. Species within this sub-class are relatively easy to identify by the exterior shape of the shell, together with the form of the operculum and its growth rings. The following terms are used to describe the exterior of the shell. The individual coils of the shell are known as whorls, which are separated from each other by more or less pronounced grooves, called **sutures**. The largest, ultimate, whorl is called the **body whorl**. The **whorls** above the body whorl are together called the **spire** with the oldest part of the shell, the **apex**, in the uppermost end. The shell surface is sometimes smooth or sculptured with ridges, grooves, ribs, tubercles, or various sorts of **projections** that may run parallel to the sutures or run at right angles to them. The opening of the shell is known as the **aperture**, and is often divided into an outer and inner **lip**. If the aperture is notched or formed with a canal for the protrusion of the siphon the gastropod species is **siphonostomous**. If the aperture is entire, without a canal, notch, or any extension the species is **holostomate**. Some snails with spiral shells, have an indentation or cavity or a circular depression at the axial base of the shell known as the **umbilicus**, which possesses a hollow central axis of the shell. Photographs of species larger than 1-2 cm are often sufficient for identifications. Relationships within this group are continually being revised, and there are many synonymous names. A range of snails that are common, but only a few millimetres in length, are not dealt with here.

Infra-class Docoglossa - Limpets

This group (Gr. docos = ray, spear, lance + Gr. glossa = tongue) includes snails that usually have a cup-shaped shell without coiling. A variety of species occur in shallow water, often right up into the inter-tidal zone. Adult individuals always lack an operculum (horny plate covering the shell opening).

Ansates pellucida (= *Helcion pellucidum*, = *Patina pellucida*)

Distribution: From Portugal to Iceland and Norway. British Isles. Not found around Belgium, the Netherlands, east coast of Denmark or in the Baltic Sea.

Description: The colours of this limpet, together with its limited distribution range, makes it almost impossible to confuse with other species. Thin, somewhat transparent shell, light brown in colour with marked longitudinal blue stripes. Up to 20 mm long, 15 mm wide and 7 mm high. Two sub-species are identifiable based on habitat: ***Ansates pellucida pellucida*** occurs on the blades of kelp (***Laminaria***) or occasionally serrated wrack (***Fucus serratus***), whereas ***Ansates pellucida laevis*** lives within the holdfasts of the cuvie (***Laminaria hyperborea***). The latter form has a thicker, non-transparent shell lacking the characteristic blue stripes and is slightly higher (20 x 8 x 10 mm).

Habitat: As mentioned above, found on kelp blades or within the holdfasts. From 0 - 27 m.

Biology: During grazing, the snail scrapes off the outer layer of the kelp blades, leaving an obvious trace. Also feeds on microscopic diatoms.

This limpet has eaten its way into a kelp stipe.

The characteristic appearance makes for easy recognition.

Tectura testudinalis - Common tortoiseshell limpet (= *Acmaea tessulata*, = *Testudinalia testudinalis*)

Distribution: Northern, circumpolar distribution. From northern parts of the British Isles as far north as northern Norway.

Description: Up to 30 mm long, 24 mm wide and only 10 mm high. Main colour white or pale green, with irregular brown bands giving a tortoiseshell appearance. Smooth, often shiny shell. The characteristic colouring, together with its exterior dimensions are usually enough for identification. The apex lies a little to the front of the centre of the shell. The inside of the shell has a brown patch. Individuals found on the shore do not grow as large as those living in a few metres depth of water. A somewhat smaller species, ***Tectura virginea,*** is similar in appearance, but is paler and has a pink pattern around the edge of the shell.

Habitat: Usually found on small stones in shallow water. From the lower part of the intertidal zone to around 50 m. The maximum depth distribution is limited by the occurrence of the red algae upon which it feeds.

Biology: Grazes on encrusting non-calcareous red algae, diatoms and detritus (dead plants and animal remains). Separate sexes. Reproduction occurs in the spring, when the females lay bands of small red eggs. Can reach around 3 years of age.

The beautiful tortoiseshell-like pigmentation has inspired the name of this limpet. The basic colour is nearly always brown.

Tectura virginea (=*Acmaea virginea*)

Distribution: Common along all western coasts in Europe. South to Cape Verde Islands and the Azores. North to Iceland and northern Norway.

Description: This small limpet resembles its larger relative; ***Tectura testudinalis***. However, ***T. virginea*** is paler and has characteristic pink, relatively broad stripes radiating from the apex down to the shell margins. In many individuals, only the lower part of these stri-

Tectura virginea *always has characteristic pink stripes at the outer edge of the shell.*

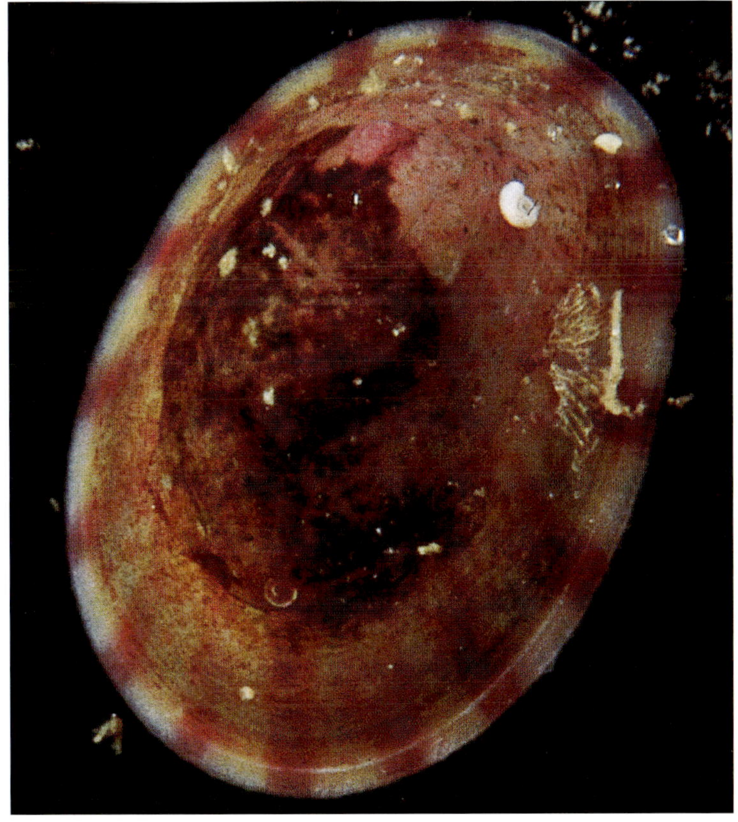

pes are visible, reducing them to pink patches around the edge of the shell. Interior of shell without brown patch. Length up to 15 mm.

Habitat: Hard substrates from the lowest part of the tidal zone down to approx. 100 m. Common in kelp forests.

Biology: Feeds on encrusting red algae, as in the previous species. Probably also feeds on sponges. Lives for around 3 years.

Patella vulgata - Common limpet

The common limpet is very abundant on cliffs and rocks in wave-exposed areas.

Distribution: Widely distributed across large parts of the northeast Atlantic. Mediterranean. Common, but decreases in abundance at northern latitudes such as northern Norway.

Description: This limpet is recognisable by its size and the rough texture of the shell. Up to 60 mm long. Shell height varies with habitat. Individuals living in the upper part of the intertidal zone

This individual lost its grip on the stone it was sitting on, and we found it lying on its side. It is doubtful whether it is able to turn itself the right way around again.

have a rounded pyramidal form whereas younger individuals and those living lower on the shore are flatter. The shell has marked radial ribs extending from the apex to the shell margins. The strong foot is pale, with greyish or khaki tones. Interior of shell greyish-green, with blue tones. Head tentacles transparent. Another limpet, ***Patella ulyssiponensis*** (= ***P. aspera***), can reach up to 30 mm in length, but it is much flatter than its relative ***P. vulgata***. Further, the foot of ***P. ulyssiponensis*** is yellowish-orange in colour.
Habitat: Lives within, or directly below the intertidal zone. Prefers high salinities, but can tolerate salinities down to 25‰. Hard bottom, preferably rock faces or cliffs, or large stones offering a certain room for movement.
Biology: Grazes on algae. By scraping off newly-settled algae, it effectively keeps rocks and cliffs bare of algal growth. Moves within a radius of around one metre, but consistently returns to its more or less fixed place. Movement generally occurs at night, and the shell is generally lifted some mm from the substrate. If disturbed, the animal attaches itself to the substrate by suction and can be extremely difficult to remove. This is one of the few snails that are hermaphrodite. Individuals less than around 4 years are males, before they change sex and become females. External fertilisation, with spawning occurring from October to December. The larval stage lasts for only a few days, after which the around 0.2 mm long juvenile snail settles to the bottom. Individuals have been found to be up to 16 years old.
Commercial value: The species is edible.

Patella ulyssiponensis *has a yellowish to orange foot and the inside of the shell also is orange.*

Infraclass Archaeogastropoda

As the name suggests (Gr. arch- = first, primitive + taxon Gastropoda) this group is an old form, in an evolutionary sense. Some have retained the ancestral uncoiled form. Radula, often with rows of teeth, suitable for a vegetarian diet. Shell with mother-of-pearl sheen. Paired gills. In most taxa, the males lack a penis, and in these, both fertilisation and larval development occur in the water.

Emarginula fissura - Slit limpet

The slit limpet is found between around 20 - 700 m depth. The diet consists of sponges.

Distribution: All western European coasts from the Canary Isles to Northern Norway. Mediterranean.

Description: Small, helmet-shaped, up to 15 mm long and approximately equally high. Colour greyish-white to grey. Apex (tip) markedly bent over. Foot with 10 tentacles on each side, and a single head tentacle on right side. Anterior/ head end of shell with characteristic fissure. Thick, downward-pointing ribs are crossed by perpendicular ridges, such that the shell surface has a lattice-like pattern. Can be confused with ***E. gigantea*** (= ***E. crassa***) which can reach up to 30 mm in length, but this lacks the conspicuous perpendicular ridges, and therefore the latticed pattern is absent. The latter has a thick shell and a very short fissure.

Habitat: Found from the lower part of the shore down to approx. 260 metres depth.

Calliostoma zizyphinum - Painted topshell

Our most beautiful snail? Cannot be confused with any of our other snails.

Distribution: From the Azores in the south to northern Norway. All British Isles.

Description: Very characteristic pyramid-shape. Its pink/violet colouration and size makes it easy to recognise. Up to 30 mm in height. Prominent spiral striations. Width of shell similar to its height. Approx. 10 whorls. Colour can vary from yellow, pale pink, or violet with pink or purple stripes. Seldom only a single colour.

Habitat: On hard bottom from the shore down to approx. 300 m depth. Common in kelp forests, but occasionally found on soft bottom. Tolerates salinities down to around 21‰.

Biology: As typical for members of the 'top-shells' (fam. Trochoidea), eggs are laid in long chains on the bottom, from which creeping juvenile snails hatch.

Gibbula cineraria - Grey topshell

Distribution: Found in Europe from Gibraltar north to northern Norway and Iceland. Common around the British Isles.
Description: Rather unevenly rounded form. Characteristic red-brown stripes radiating somewhat irregularly and curving from the apex in an almost zig-zag form. Up to 16 x 15 mm; 5 - 6 whorls. Basic colour light yellow to ash grey, under the aforementioned stripes or patches. Umbilicus (channel) present; oval in shape or almost covered in adults. Can be confused with ***Gibbula tumida***, but the shell of the latter is less smooth and divided into relatively conspicuous sections.
Habitat: On sheltered shores with hard bottom, on stones and rocks. Can be found from the lower intertidal zone down to approx. 130 metres.
Biology: Reproduction in summer. Hatches pelagic larvae that settle to the bottom after around 9 days. These feed on small algae or dead algal material.

Gibbula cineraria, *in contrast to some of its relatives, has a smooth shell.*

A favourite grazing place is the blades of the cuvie kelp.

Margarites helicinus - Pearly topshell

Distribution: Found in Europe south to the Swedish west coast and northern parts of the British Isles. Entire Norwegian coast.
Description: Small, somewhat compressed shell with 4 - 5 clear whorls. Short spire. Height up to 9 mm, but in colder waters seldom more than 5 mm. Shell opening large and rounded. Colour pale brownish, with a very characteristic greenish mother-of-pearl sheen.
Habitat: Very common on brown algae in shallow water from around the lower part of the tidal zone down to approx. 20 m depth.

The pearly topshell, **Margarites helicinus**, *is only a few millimetres in width and grazes on seaweeds. As the name suggests, the shells have a characteristic sheen (see picture above). On the left is an individual with eggs.*

Sub-class Caenogastropoda

This group (Gr. kainos = new, recent + taxon Gastropoda) comprises the orders Neotaenioglossa (former Mesogastropoda), and Neogastropoda. Most prosobranch snails belong to this group. These are well-developed, distinguished from their older, or more 'primitive', relatives by having only one kidney, a single sexual organ and a single gill. Head equipped with a pair of head tentacles, usually bearing eyes.

Order Neotaenioglossa

Lacuna vincta - Banded chink shell

Striped colour variations occasionally occur.

Distribution: From the Atlantic coast of France north to northern Norway. Very common on many coasts.
Description: Small snail with classic shape. Up

These snails are copulating. Note that individuals of this species are smaller in size than the nail of our little finger.

to 10 x 5 mm. Colour light brown, occasionally with pale reddish-brown bands. Shell thin, partially transparent, with 5 - 6 smooth whorls. Conspicuous v-shaped umbilicus, with a basal opening.
Habitat: Very common on seaweed in shallow water, particularly serrated wrack (***Fucus serratus***). Down to approx. 40 metres. Also on eelgrass (***Zostera marina***). Tolerates salinities down to approx. 20‰.
Biology: Becomes around 1 year old. Reproduction from January to June, during which ring-shaped gelatinous egg masses are laid on the algae. These hatch into free-swimming larvae.

Littorina littorea - Common periwinkle

Distribution: Found in Europe from northern Spain to the White Sea. Common.
Description: Of the 6 different periwinkles within the genus *Littorina* found in our waters, this is the most common. Generally familiar to all who have explored the lower parts of the shore. Relatively large, up to 40 mm tall, with robust shell. Body whorl comprising up to 80 % of the total height of the snail. Shell colour varies, often black or dark grey-brown, occasionally red, orange or white.

PERIWINKLE

Alteration of Old English pinewincle, from Latin pina, a kind of mussel + Old English - wincle (akin to Danish vincle snail shell). Various gastropod molluscs; any of a genus (*Littorina*) of edible littoral marine snails; also any of various similar or related marine snails.

The common periwinkle is one of the commonest snails along our shores.

DID YOU KNOW...

A much-used characteristic for identifying species within this genus is the shape and length of the penis.

Head tentacles with transverse black stripes. Outer lip (extremity of shell opening) meets the body of the shell at an angle, not straight on as in ***Littorina saxatilis*** (= ***L. rudis***). The latter generally is found only in the upper part of the intertidal zone and often has conspicuous, spiral striation. Another species, ***Melarhaphe neritoides***, closely resembles the common periwinkle. However, this only reaches 9 mm in length, and can be distinguished from small common periwinkles by the head tentacles having longitudinal black stripes. All periwinkles lack an umbilicus, the central channel within the shell. Relationships within the genus ***Littorina*** are under debate, and there is uncertainty as to whether different variations should be recognised as species in their own right.

Habitat: Very common on hard bottom on the shore, but also on sandy bottoms. How high up on the shore the animals are found depends on the degree of wave action. The more exposed the shore, the higher up the periwinkles are found. Down to approx. 60 metres.

Biology: Spawns gelatinous egg capsules directly into the water. All periwinkles are herbivorous.

Commercial value: Considered an excellent delicacy in various European countries.

Littorina obtusata - Flat periwinkle

The flattened shell of this periwinkle makes it easily recognised, but be prepared for many different colour variations!

Distribution: Western parts of the Mediterranean to northern Norway. All British Isles.
Description: Typical shore snail. Body whorl comprises more than 90% of the shell's height; apex (tip) almost completely flat. Shell almost completely smooth. Much variation in colour according to diet; usually yellow, but green, orange, brown and black snails, sometimes with stripes, not uncommon. Up to 15 x 17 mm. Head tentacles with two longitudinal stripes. Due to the flattened apex, can only be confused in our waters with ***Littorina fabalis*** (syn. *mariae*), which measures 11 x 12 mm. The shell opening of the latter is often extremely large, extending across 90 % of the length of the shell. However, there do exist individuals of both these species whose shell construction actually most resembles the opposite species. Further, *Littorina fabalis* lives on serrated wrack (***Fucus serratus***), a feature used in identification. For certain identification, it is necessary to examine the penis or carry out genetic analyses.

Habitat: Found on algae, particularly bladder wrack (***Fucus vesiculosus***) and knotted wrack (***Ascophyllum nodosum***), which it grazes on. Generally found in areas of brown algae. Tidal zone or very shallow water.
Biology: Spawns gelatinous egg masses, which hatch into developed, creeping juvenile snails. Can reach at least 3 years of age. *L. fabalis* only lives up to one year.

Turritella communis - Tower shell

As the name suggests, the tower shell has a very long, pointed, turret-like shell. Living individuals usually bury into the sediment with the shell opening barely projecting from the bottom surface.

Distribution: From northern Africa to northern Norway. Mediterranean. Common around the British Isles.

Description: Characteristic tower-shaped snail difficult to confuse with other species in European waters. Approx. 16 - 20 whorls. On each whorl are usually 3 (up to 6) pronounced spiral striations. Grey shell. Length up to 60 mm.

Habitat: Occurs on soft bottoms from a few metres depth to approx. 200 m. Usually lies buried with the shell opening directly under the surface.

Biology: Feeds on most types of phytoplankton and floating dead organic material (detritus). These are filtered from the water across the gills, and transported to the mouth by ciliary action.

Here a hermit crab has acquired a perhaps unnecessarily large house.

Aporrhais pespelecani - Pelican's foot

Extremely modified shell opening.

Distribution: In Europe from the Mediterranean north to northern Norway and Iceland. All British Isles, but less common around the Channel Isles and Scilly Isles.

Description: Very characteristic appearance, making identification easy. Because of the very elongated shell opening, it can only be confused with its relative ***Aporrhais serresianus***. This is not so common, but occurs in the same areas as the pelican's foot snail. *A. serresianus* has even larger, wing-shaped extensions of the outer lip of the shell opening. In both species, there are three outwardly directed extensions of the outer lip of the shell opening. In *A. serresianus*, the extension parallel with the height of the 'house' reaches past the apex (shell tip) whereas it does not reach the apex in *A. pespelecani*. Pelican's foot reaches up to 50 mm in

The lip, which is the outer part of the shell opening, has three extensions.

height, whereas *A. serresianus* only reaches around 40 mm. Creamy in colour, with additions of violet, often with brown patches.

Habitat: Soft bottom, where it burrows. From a few metres depth to approx. 180 m. *A. serresianus* is found as deep as 1000 m.

Crepidula fornicata - Slipper limpet

Distribution: From the British Isles north to northern Norway. Sweden and Denmark. Also around Sicily and in the Adriatic Sea.
Description: Very characteristic body shape, curved, almost without visible whorls. On the underside is a special plate covering just less than half of the shell opening. Up to 25 mm high, and approx. 50 mm long. Robust shell, colour whitish-yellow to reddish-brown, with numerous short brown streaks that are particularly evident on the inside of the shell.
Habitat: From approx. 1 - 10 metres depth, on hard substrates, attached to rocks or other mollusc shells such as oysters (***Ostrea edulis***) and horse mussels (***Modiolus modiolus***).
Biology: The species is known as a protandric hermaphrodite, changing sex as it grows. Often up to 10 or more individuals sit on top of each other, with the largest individuals on the bottom. These are always females. The smallest individuals on top are always males. The individuals in the middle are in the process of becoming females.

The white plate located at the anterior part of the shell opening distinguishes this snail from other species in our waters.

This species is a 'guest' from the northwest Atlantic. The American slipper limpet was introduced in 1887, along with consignments of oysters, with which it competes for food.

Both shape and colour are characteristic. The spire is reduced, with barely two visible whorls. Here three individuals are sitting on top of each other.

Trivia arctica - Arctic cowrie

Distribution: Widespread from the Mediterranean north to northern Norway. All British Isles.
Description: Very characteristic thick, white shell, reminiscent of the more exotic conch shells. Only one whorls visible, covering the juvenile (first-formed) parts of the shell. Perpendicular to the whorl are conspicuous thick striations. When the animal creeps, the shell is completely covered by the more or less transparent brownish mantle. Approx. 10 mm in length. Cannot be confused with other species in Norwegian waters, but four other species within the family Triviidae are found further south in Europe. Around the British Isles, the **spotted cowrie**, ***Trivia monacha*** is much alike, but has pigment spots, and is more common in the south.
Habitat: Hard bottoms, down to approx. 100 metres. Often amongst algae or colonial sea squirts, upon which it feeds. We have found large abundances of this snail in fjords, suggesting that it does not favour exposed conditions.

The snail to the left of this picture has a 'bare' shell, whereas in the individual to the right, the shell is enclosed by the mantle.

Trivia arctica *displaying typical snail-like characteristics. The foot extends considerably beyond the shell opening, and the mantle is folded over the shell. At the head end, the two tentacles and the large respiratory siphon are clearly visible. Our only snail with this characteristic coffee-bean shape of shell.*

Epitonium clathrus - Common wentletrap

Here a hermit crab has made good use of the shell of this characteristic snail. Note the extremely prominent ribs perpendicular to the whorls.

Distribution: Widespread around European coasts. From the Black Sea and the Meditterranean to southwest Norway.

Description: Characteristic elongated shape with conspicuous ribs. Shell usually greyish. Shell opening almost circular, closed with a dark, almost black operculum. Lacks umbilicus (central canal). Up to 40 mm in length. Note that there are four other species within this genus in northern European waters, but these are relatively rare.

Hermit crabs allow us to see shells that we otherwise probably would overlook.

Habitat: Muddy sand, most abundant between 5 - 20 m depth.

Biology: Predatory, feeding on sea anemones. The species is a consecutive hermaphrodite, changing sex each year (one year female, next year male…).

Euspira montagui (= *Polinices, Natica* and *Lunatia montagui*, amongst others)

This snail has an extremely large foot. The lack of brown stripes on the shell distinguishes it from a similar species, **Euspira pulchella.**

Distribution: Found from the Mediterranean north to northern Norway.
Description: Up to 8 x 8.5 mm, rounded shell, with 6 - 7 bulging whorls, of which the basal body whorl comprises most of the height shell of the shell. Shell apparently smooth, but with numerous fine growth lines. Large oblong umbilicus with hole at base). Pale brown in colour, lacking the brown band present on the basal whorl of the somewhat larger, but otherwise similar ***Euspira pulchella.*** Very large foot, parts of which cover both anterior and posterior parts of the shell in active animals.
Habitat: Sandy or gravely bottoms, usually from 30 - 200 m depth. *E. pulchella* is found slightly shallower in the same areas.
Biology: This snail belongs within the family Naticidae, which drill holes in burying bivalves using the radula. After the hole is drilled, a tentacle is pushed into it, and the bivalve flesh is sucked out. Bivalve shells found with one or more sharply defined round holes are usually the work of one of these snails. Characteristic, coiled egg masses mixed with sand and gravel grains are laid on the bottom.

BOTHERSOME SCIENTIFIC NAMES

The names of certain genera and species are frequently changed. This is due to unclear systematic relationships between the species. A prime example of this is the boring snail ***Euspira pulchella***. During the last decades both the generic and specific names have been changed several times. Amongst other names, it has been known as ***Natica alderi, Lunatia alderi, Natica nitida, Polinices polianus*** and ***Polinices pulchella*** - but currently (2003) it is referred to as ***Euspira pulchella*** (which incidentally is derived from the Latin word pulchellus, meaning beautiful).

Order Neogastropoda

The mantle of these snails forms a siphon (respiratory tube) which usually lies within a conspicuous siphon canal. The shell is often elongated, with a high spire and with a relatively long, oval shell opening. The shells usually are robust. All members feed on carrion and/or are predators. Separate sexes, with internal fertilisation. Males equipped with a penis. In most forms, females lay clusters of large egg capsules, which hatch into juvenile creeping snails.

Nucella lapillus - Dog whelk

Distribution: Widespread from Gibraltar to Greenland.

Description: Characteristic broad, conical, very thick shell. Up to 42 x 22 mm. Basal body whorl comprises more than 80% of the shell. Shell opening extended into a short siphon canal (respiratory canal). Outer lip evenly curved along its length towards the siphon canal. Shell often with both spiral striations and downward-pointing ribs, giving a chequered appearance. Shapes and colours vary; some pale, almost white, but yellow, red and brown variants occur. Colour influenced by prey organisms; individuals primarily feeding on blue mussels have a dark shell whereas those feeding on barnacles are pale. Brown or purple inner lip. Resembles ***Nassarius nitidus***, which also is common on the shore. However, the body whorl of the latter is not nearly as dominant and the spire comprises more than 1/3 of the shell's length.

Habitat: Very common on bare rock in the lower part of the tidal zone,

Dog whelks can occur in very dense aggregations.

Dog whelks occasionally have a number of smaller individuals perched on their shells.

FEEDS ON EGGS

In common with many other snails, the dog whelk has internal fertilisation. In early spring, it lays jug-shaped capsules, each containing 10 - 20 eggs. Only a few of these eggs hatch into 1 mm long juvenile snails. The rest of the eggs are eaten by the juveniles, until they grow large enough to start their predatory behaviour.

and most abundant in exposed areas.
Biology: The dog whelk is a ferocious predator which has barnacles and various other snails and bivalves on its menu, using its radula to drill a hole in the prey shell. This can take a few days. Individuals are sexually mature after 3 years, after which they cease growth. Can live up to 6 years. The flatworm ***Parorchis*** attacks dog whelks, destroying the reproductive organs, with the result that growth does not cease at 3 years of age. Dog whelks form dense assemblages in autumn, but the behaviour of infected individuals becomes disturbed such that they are late in joining these assemblages.

Redshank wading birds, which are the end-host for the parasite, mostly feed on dog whelks during their autumnal migrations. They never prey on dog whelks in assemblages, which means that solitary infected snails are at far greater risk of being eaten. This is one of many examples of a parasite causing behaviour modification in intermediate hosts (in this case the dog whelk) such that the chances of being transferred to an end-host are increased.

THE COLOUR PURPLE

In some languages, the dog whelk is called the purple snail because in previous times, purple dyes were made from snails in this family. The Romans were the first to exploit this.

Buccinum undatum - Common whelk

Distribution: In Europe from the Bay of Biscay to northern Norway. Common around most most of the British Isles.

Description: This large snail can become up to 110 mm high, and 68 mm wide. The basal body whorl comprises approx. 70% of the length of the shell. Shell surface rough with chequered appearance, due to prominent spiral striations, prominent downward-pointing ribs and crossing growth lines. Outer lip of shell opening evenly curved towards a short siphon canal. Inner and outer lips white; main shell body greyish-white, grey-green or grey-brown. In some areas, there are a range of species within the genus ***Buccinum***, which are difficult to distinguish from the common whelk.

Habitat: Occasionally as far up as the tidal zone, but usually from a few metres deep down to around 1200 m. Occurs both on hard and soft bottoms, but particularly common on soft bottom, in the vicinity of hard bottom. Tolerates brackish water down to a salinity of 14 ‰.

Biology: Spawns large aggregations of egg capsules, which are attached to hard substrates. From each capsule a number of juvenile snails hatch, which feed on carrion, various bivalves, polychaete worms and any other available organisms.

Commercial value: As for periwinkles, considered a delicacy.

Egg capsules of the common whelk, containing eggs, which have hatched into juvenile snails.

The common whelk is most commonly observed on soft sediments. Here showing its thick dark respiratory siphon.

Neptunea despecta

Distribution: Northern distribution. Large parts of the northern Atlantic. In the east from the Shetland Isles, Iceland, Arctic. Entire Norwegian coast.

Description: Our largest snail, together with the next species, ***Neptunea antiqua***. Characteristic, conspicuous, sometimes knobbled spiralling keel. Up to 200 mm in height, rough surface, often worn-looking. Colour dirty-white, yellow to pale brown. Eight whorls, of which the body whorl dominates, comprising approx. 75 % of the total length. Up to 120 mm wide. Shell opening a little less than half the total length of the shell. Wide siphon canal. Resembles, and by some considered to be a variation of *Neptunea antiqua*, which reaches a similar size, but is distinguished by the lack of a spiraling keel.

Neptunea despecta *is often seen carrying the sea anemone* **Hormathia digitata.**

Habitat: From approx. 10 - 1200 m depth around the coast or further off shore; prefers soft bottoms.

This photograph is taken at Saltstraumen in northern Norway at 12 m depth. The snail is in the process of laying egg capsules. Note the hermit crab **Pagurus bernhardus** *on the snail shell.*

Neptunea antiqua - Red whelk

Neptunea antiqua *is distinguished from* **N. despecta** *by the lack of spiralling keels.*

Distribution: Widespread from Gibraltar to the Arctic. Locally common around the British Isles, less in the south and off western Ireland.
Description: Large, up to 200 mm in height; resembles the previous ***Neptunea despecta***, but lacks the spiralling keels. Many, regular growth lines, but surface still even. Colour yellow, yellow-brown to grey. Thick outer lip, extending into a short, broad siphonal canal.
Habitat: Usually found off shore on soft bottoms, but also found in the kelp zone between 10 - 15 m. Down to approx. 1200 m.
Biology: Mostly feeds on carrion.

Colus gracilis

This snail resembles Neptunea antiqua *but is considerably smaller. Here inhabited by a hermit crab.*

Distribution: From the coast of Portugal and northwards; entire Norwegian coast. All British Isles, but less common in the south.
Description: Resembles the red whelk, ***Neptunea antiqua***, but does not reach nearly the same size; up to approx. 70 mm in height and 30 mm maximum diameter. Whorls almost straight (less outwardly bulging) than the latter. Siphonal canal relatively short and broad, markedly bent towards the left of the shell opening.
Habitat: Soft bottom. Usually occurs deeper than 30 metres, but shallower in the northernmost part of its distribution area.

Sub-class Heterobranchia

The Heterobranchia (Gr. hetero = varying + Gr. branchia = gills) include all the remaining snails that are not members of the sub-class Prosobranchia. The largest group is the supra-order ***Opistobranchia****; sea slugs (including nudibranchs and their relatives), within the infra-class* ***Euthyneura****. In addition, there is a group of snails belonging to the infra-class* ***Heterostropha****, which appear to resemble prosobranchs, but have the first body whorl in the opposite direction.*

The sea slug **Dendronotus frondosus** *prefers sheltered localities and is most often observed in wintertime. On the kelp blade (upper left) is an attached spiral egg mass.*

Supra-order Opisthobranchia

This kelp blade is covered by the bryozoan **Membranipora membranacea***. This is being grazed by members of the sea slugs* **Limacea clavigera** *(upper right; single individual) and* **Polycera quadrilineata***. To the left are several specimens of the snail* **Lacuna vincta***. Photograph from Arnøy, Troms in northern Norway.*

The name of this group (Gr. Ophisteu = behind, on the back + Gr. branchia = gills) reflects the posterior position of the gills, behind the heart. A few have a conspicuous exterior shell, others have a reduced shell enclosed in soft tissues, but the vast majority lack the shell altogether. Opisthobranchs include the beautiful ***nudibranchs***, ***sea hares*** *and* ***sea butterflies***. *Opistho-branch snails are all hermaphrodites, having both male and female sexual organs. When two such hermaphrodites mate, the individuals mutually exchange sperm.*

Eubranchus farrani *is a very characteristic opisthobranch. Most often seen in winter-time, grazing on the hydroid* **Obelia geniculata,** *or other members of that genus.*

Order Cephalaspida

Members of this group can possess or lack a shell. Where present, completely enclosed by soft tissues. Head flattened, often with extensions (parapodia), but lacking tentacles as such. Most with prominent extensions along the sides, folded round the shell. Most dig into soft sediments in search of prey.

Philine aperta

Distribution: Large parts of the northeast Atlantic. Mediterranean. Common around all British Isles.

Description: Internal, thin, brittle shell covered by soft tissues. Mantle folds divide the snail into four parts. Colour milky-white. Length up to 70 mm (shell approx. 30 mm), 35 mm wide.

Habitat: The latin name ***Philine*** is derived from ***Philaeni***, Carthaginian patriots who, out of passionate love for their homeland, allowed themselves to be buried alive. This species is found in soft bottoms, approx. 3 - 500 m depth.

Biology: Members of this genus lay spherical to oval egg masses. Several of the snails in this genus hunt for prey directly beneath the sediment surface, feeding on various polychaetes, bivalves and small snails. In common with various other apparently unprotected snails, it secretes a sulphuric acid solution as a weapon against predators.

Philine aperta *has an internal shell. These snails dig in the sediment in search of prey.*

This most likely is an egg mass of **Philine aperta**. *Photographed at Høgsfjorden at 15 m depth.*

Order Gymnosomata

Members of this group (Gr. gymnos = naked + Gr. somatos = body) are free-swimming and lack both shell and operculum (horny shield on the foot). Foot often with two prominent 'wings', which are used primarily for vertical movement. Two head tentacles.

Clione limacina - Naked sea butterfly

Distribution: Widespread throughout large parts of the North Atlantic.
Description: Cylindrical-shaped body, up to 40 mm in length. Near the head end are two relatively small 'wings', which are modified from the foot. Body almost transparent, but the red inner organs clearly visible. Head with sticky, conical tentacles that help to seize prey, together with hook-bearing sacs, which are everted from the proboscis.
Habitat: Lives in the plankton in shallow water. Follows the vertical migrations of its prey and rises to the surface at night, moving deeper during the day.
Biology: Mainly preys on the somewhat smaller, 'shell-bearing' winged snails (order Thecosomata) such as ***Limacina retroversa*** and ***Limacina helicina***.

One of the few molluscs that live as zooplankton.

These feed on single-celled algae and flagellates.
Commercial value: Important food source for plankton-eating fish.

Limacina helicina.

Order Aplysiomorpha

Most of these snails are herbivorous and can be relatively large. Some have a shell, either externally or internally, whereas it is lacking in others. This group includes sea hares.

BIG SNAILS

They vary in size, some no bigger than a grain of sand to the huge sea hares found in the tropical Pacific, which have a wet weight up to 2 kg!

*This sea hare (**Aplysia punctata**) is large, up to 20 cm in length. The species epithet **punctata** refers to the numerous pale spots that cover the outer surface of the reddish-brown body.*

Aplysia punctata

Photograph showing the mouth tentacles (lower part of picture; one damaged) and a pair of head tentacles (rhinophores; upper part).

Distribution: Widespread from the Canary Islands to Greenland. Norwegian coast. Mediterranean. Common around all British Isles.
Description: Characteristically appears to lack a shell, but in fact it is present, only surrounded by soft tissues. Can be very large, up to 300 mm long, but in northern waters seldom longer than 70 mm. Shell up to 40 mm in length. Colour usually reddish-brown with pale patches. Darker and paler variations can occur in purple, green and almost black colours. Colour influenced by diet, such that individuals primarily grazing on red algae become reddish, those eating green algae become greenish and those with a diet of brown algae are brownish in colour. Two anteriorly orientated, tube-shaped mouth tentacles located on the head, behind which are two head tentacles (rhinophores). Marked folding of the foot (parapodia). Two other species of ***Aplysia*** have been recorded from south west England.
Habitat: Occurs in shallow water amongst algae upon which it feeds. Often observed in early summer, when spawning occurs.

The sea hare **Aplysia punctata.**

Orden Sacoglossa

This group comprises snails with an internal or external shell, or no shell at all. The head lacks mouth tentacles, but usually has a pair of head tentacles (rhinophores). Most are very small in size and feed on algae in shallow water.

AMAZING!

A characteristic of this animal is that the chloroplasts (chlorophyll-containing cell organs) from the algae it eats are not broken down in its gut, but they continue photosynthesis in the outer cellular layer. The snail therefore receives carbohydrates from the photosynthetic activities of the chloroplasts!

Elysia viridis

Elysia viridis *can be difficult to detect due to its small size. Often occurs in large numbers amongst green algae.*

Distribution: Widespread across large areas of the northeastern Atlantic. Mediterranean. All around the British Isles.

Description: Small, but very attractive; up to 45 mm long, but usually smaller in northern waters. Shell absent. Foot modified into thin, broad folds (parapodia) extending almost the length of the animal, and which can be folded out to the sides or across the back. Colour varies according to the algae eaten; often bright green, but also brown, red and almost black variations are common. Usually with brilliant blue, green and red patches at the edges of the parapodia.

Habitat: Usually in shallow water, where it feeds on algae. Particularly common on the green algae ***Codium fragile***, and ***Cladophora rupestris***. Found in rock pools, but also down as deep as 80 metres.

Order Nudibranchia

Nudibranchs are often colourful and so characteristic that it is usually possible to identify them from a good quality photograph. Here **Coryphella pedata**, *photographed in Egersund harbour.*

Nudibranchs are perhaps the most widely photographed of all the snails. The reason is obvious. Most nudibranchs are beautifully coloured, which probably is a warning signal to their enemies -"eat me and you'll suffer".

This is the largest group within the opisthobranchs. In common with all opisthobranchs, nudibranchs arose from prosobranchs. However, they have lost both the shell and operculum (plate) during the adult stages (excepting the Pleurobranchomorpha). However, as larvae they posess a shell. All nudibranchs are **predatory**, feeding on various sessile animals, especially hydroids (**Hydroidea**), bryozoans (**Bryozoa**), soft leather corals (**Alcyoniidae**) and sponges (**Porifera**). Species within the same family will often eat the same type of organism. It has been shown that metamorphosis (larval development) with subsequent settling of pelagic larvae to the bottom is triggered by the presence of their prey. Those nudibranchs whose prey organisms have a short generation time, such as hydroids, often produce several generations within the same year. Those that prey on organisms that generally are multi-annual, such as sponges or dead man's finger (***Alcyonium digitatum***), also tend to live longer than a year.

Identification can be difficult within certain families, but many of the nudibranchs have such characteristic colourings, fringes, gills and the two usually upwardly orientated head tentacles (**rhinophores**), that they can be identified from an *in situ* photograph. In addition, the prey upon which the nudibranch is found also is a good indicator. To identify specimens without knowledge of their habitat, it is often necessary to examine the teeth of the **radula**.

DID YOU KNOW...

Mating between **herma-phroditic** nudibranchs occurs by the two individuals turning their right sides towards each other, such that sperm can be exchanged between the sexual organs located on the right-hand side. The sperm is kept in **sperm sacs** before the actual fertilisation takes place. Each of the mated individuals can lay eggs. The obvious advantage of such a system is that all sexually mature individuals of the same species can reproduce. Further, the snails can self-fertilise, but this is less common.

Sub-order Pleurobranchomorpha

Unlike most nudibranchs, most members of this group have an internal or external shell. The group only recently is recognised within the nudibranchs. The mantle is often hard due to the presence of calcareous spicules, and contains glands that secrete acid when attacked by a predator. Almost all are predators. Two species are presented here.

A representative of ***Cylichnidae*** *or* ***Philinidae****. These are small and often bury. The pictured individual is less than 10 mm long and difficult to see without magnification.*

Berthella sideralis

Distribution: Widespread in Europe from the Mediterranean to northern Norway.
Description: Up to 60 mm in length, oval shape with an up to 30 mm long internal shell. Colour lemon yellow to orange. A single,

SULPHURIC ACID

Berthella sideralis can secrete sulphuric acid if it is being attacked!

oblong gill situated posteriorly on the right-hand side under the mantle. Head with two tubular head tentacles, which can be withdrawn under the mantle. Foot extending posteriorly from the mantle.

Habitat: Feeds on various sea squirts from the lower part of the tidal zone down to a few metres depth.

Characteristic snail with internal shell.

Right: **Bertella plumula** *can reach up to 50 mm in length, and the internal shell is around half this length. It is found in shallow water down to around 10 m depth, usually amongst algae. Feeds on tunicates, often* **Botryllus**.

Below: **Bertella plumula**, *photographed at Egersund, southwest Norway.*

Sub-order Anthobranchia

*Most members of this group (Gr. anthos = flower + branchia = gills) have the gills positioned in a rosette around the anus on the posterior of the back. The mantle on the back is more or less covered in tubercles or spicules. These never contain nematocysts from prey or branches of the digestive canal, such as is often the case in the sub-order Cladobranchina. Most eat moss animals (**Bryozoa**), sea squirts (**Ascidiacea**) or sponges (**Porifera**). Exceptionally they eat other prey, such as barnacles (e.g. **Balanus**) but never cnidarians.*

Goniodoris nodosa

This sea slug is often observed in shallow water in the inner parts of fjords and inlets. To the right are two individuals preparing to mate.

Distribution: Distributed from Spain to northern Norway. All British coasts.
Description: Translucent white in colour, with a characteristically fringed mantle. Mantle along the back with small tubercles and patches of white and/or yellow pigment. Head tentacles (rhinophores) lamellate (shaped like stacks of discs), often with a faint yellow edge. Mouth tentacles somewhat compressed, visible from above. Prominent 'tail' protruding from behind the mantle. Mantle with faint yellow, almost transparent longitudinal low keel extending to the tip of the 'tail'. Up to 13 prominent gills, arranged in a bushy ring. Length up to 27 mm.
Habitat: Found shallower than 2 m, where the juveniles prey on bryozoans, particularly ***Alcyonidium diaphanum.*** Adult individuals prey on sea squirts such as ***Botryllus schlosseri, Dendrodoa grossularia*** and ***Diplosoma listerianum***. Recorded down to 120 m.

Adalaria proxima

Distribution: Northern distribution. From northern parts of the British Isles north to eastern Greenland and the White Sea. Norwegian coast.

Description: Up to 25 mm long, dorsally compressed (from the back). Colour usually yellow, but occasionally completely white. Dorsally (from the back) flattened. Gills and head tentacles (rhinophores) usually darker yellow than the rest of the animal. Up to 12 prominent gills around the anus on the dorsal (back) side. Entire back covered by round, relatively large, pointed tubercles, a feature distinguishing it from the very similar ***Onchidoris muricata***, which is discussed later.

Habitat: Usually found feeding on the bryozoan ***Electra pilosa,*** but also on other encrusting bryozoans such as ***Membranipora membranacea.***

This picture clearly shows the lamellate (like stacks of discs) head tentacles, or rhinophores. Members of this species are almost always found on kelp blades, feeding on bryozoans.

Onchidoris fusca

Distribution: From the Atlantic coast of France north to the Arctic circle.

Description: Small in size, usually white with a brown pattern on the back. Members of the genus are dorsally compressed. The back is densely covered with club-shaped, low tubercles. The two prominent head tentacles (rhinophores) are lamellate (like stacks of discs). Gills arranged in a horseshoe-shape posteriorly on the back (not easily visible).

Habitat: Found in very shallow water, preying on barnacles (Balanidae). Often found in aggregations.

These individuals are in the process of laying egg masses in characteristic ribbon-like coils.

The dispersed brown pigment is typical for this species.

Onchidoris inconspicua

DISCOVERY OF 'NEW' SPECIES

Leafing through this book, the immense diversity of animal life in the sea soon becomes obvious. However, shown here is only a tiny selection of our marine organisms. In the majority of species, the individuals are very small and need magnification to be visible to us. Further, many marine organisms live in deep water inaccessible to divers and less easy to sample. Because the seas are vast and support so many species, we do not know all of them. As a result, species new to science, or not previously recorded from our waters are constantly being discovered. This small nudibranch, recently found for the first time in Norwegian, waters is such an example. It may be a new arrival, or it may have been present in the area for a long time, just not detected until now.

Distribution: The distribution of this relatively rare species is poorly known. Recorded from parts of the British Isles and recently also from Normandy (France). Also recorded once in southwest Norway.

Description: Up to approx. 12 mm in length. Resembles the other, more common species within the genus ***Onchidoris***, but distinguished by being white of pale brown in colour, with a marked tint of violet. Dorsum (back) covered in very small brown patches. *Papillae on dorsum* very small, rounded and pale (not brown). Head tentacles bear approx. 14 lamellae (discs); up to 10 gills located posteriorly on the dorsum.

Habitat: Previously found on the bryozoan ***Cellepora pumicosa,*** but later also reported to feed on another bryozoan, ***Cellaria sinuosa***. Both species found on kelp stipes, various shells, stones or other hard substrates. Down to approx. 50 m.

Onchidoris muricata

O. muricata *feeds mainly on the bryozoan* **Membranipora membranacea.**

Recognisable by the flattened tubercles covering the back. In the lower left hand corner is an egg mass.

Distribution: From the west coast of France north to northern Norway. All British coasts, less common in the south.

Description: Up to 14 mm in length, usually white or yellow in colour. Occasionally with brown pattern. Back densely covered with large, round distally flattened tubercles. Probably often confused with the previously mentioned ***Adalaria proxima***, together with which it often occurs. The latter can be distinguished from *O. muricata* by its tubercles, which, although also round, are not distally flattened, but relatively pointed at their tips. Further, *O. muricata* is somewhat smaller. Because the radula (rasp-like tongue) is very different between the two species, these are placed in different genera.

Habitat: This species is found feeding on a variety of encrusting bryozoa. Most frequently on its menu is the common ***Membranipora membranacea***. This bryozoan is very common on seaweed and kelp along most of our coasts. Also commonly occurs on ***Securiflustra securifrons***, which grows somewhat deeper.

Biology: Eggs laid in ribbon-like formations.

Diaphorodoris luteocincta (= *Onchidoris luteocincta*)

This very small sea slug is easily recognised by its distinct colouring pattern.

Distribution: Reported from the Mediterranean to Norway.

Description: The colour of this nudibranch is so characteristic it cannot be confused with any other northern European species. Recorded up to 11 mm in length. Dorsal (back) side white, with a yellow band round the entire mantle. Reddish-brown pigmentation in the middle of the dorsum. Note that in the Mediterranean, a number of similar species occur.

Habitat: Often found on rock substrates covered with a fine layer of mud. Feeds on, amongst others, the bryozoans ***Smittoidea reticulata***, ***Crisia*** spp. and ***Cellepora pumicosa.*** Found down to 50 meters depth.

Diaphorodoris luteocincta *is usually found in fjords and inlets.*

Polycera quadrilineata

Note the four orange-coloured mouth tentacles.

Distribution: In Europe, reported from the Mediterranean to northern Norway. All British Isles.

Description: Body white, usually with yellow or orange patches, arranged to form five or more longitudinal dashed lines. Head with two lamellate head tentacles (rhinophores), and usually 4 (occasionally 6), anteriorly directed (forward pointing) 'mouth tentacles'. Two posteriorly directed (backwards pointing) papillae extend from beside the gills, and are somewhat longer than the gills. Papillae, gills and tentacles all with yellow or orange pigmentation at their tips. Length up to 45 mm. These features are so characteristic that the species cannot be confused with others in our waters. However, specimens from Norway tend to have more yellow pigmentation than individuals found around the British Isles; there are more than five lines of yellow patches. In some populations, considerable numbers of the larger individuals have small black pigment patches. A similar species, ***Polycera faeroensis***, is found from southern Spain north to British Isles, Sweden, mid Norway and the Faeroes. This lacks the yellow patches on the actual body, and it has 8 or more anteriorly directed mouth tentacles.

Habitat: Usually occurs on kelp laminae covered with the bryozoan ***Membranipora membranacea***, or on red algae covered with ***Electra pilosa***, both of which it feeds on. Large numbers of individuals often occur together.

Biology: Egg masses laid in half-moon shaped white masses, which are attached to algae.

Polycera quadrilineata *is a relatively common species, most often seen on kelp blades.*

Limacia clavigera

Distribution: From northern Africa to northern Norway. All British coasts, except the south-east.

Description: As in the previous species, white body with yellow pigmentation. Up to 20 mm in length. Distinguishable from ***Polycera quadrilineata*** by having a larger number of anteriorly directed tentacles, and these are rather 'bushy' at their tips. Further towards the rear of the animal are several branching growths, or protrusions. Lamellate head tentacles and the three small gills orange or yellow pigmented at the tips. On the back, or dorsal side, are a number of large conspicuous projections and smaller tubercles, also yellow-orange in colour.

Habitat: Occurs on its prey, which is various encrusting bryozoans, including ***Electra pilosa***, which grows on red or brown algae in shallow water.

Limacina clavigera, *like many sea slugs, feeds on bryozoans. Here attacking its favourite food,* **Membranipora membranacea.**

Almost always associated with kelp, on the holdfasts, stipes or blades.

Jorunna tomentosa

Distribution: Found from the Mediterranean to northern Norway. Common all around the British coasts.

Description: Unlike many of its relatives, this nudibranch is a single colour, light brown, often with darker brown patches along the mantle margin. The variant ***J. tomentosa lemechi*** lacks these patches. Back covered in small, soft, equal-sized, velvet-like tubercles. Rhinophores (head tentacles) lamellate (formed like stacks of discs). Up to 55 mm long.

Habitat: Down to approx. 400 m depth, exceptionally found in the lower part of the tidal zone. Occurs on sponges, upon which it feeds, such as ***Halichondria panicea***, but also ***Haliclona oculata*** and ***H. cinerea.***

Biology: Egg masses form a broad, tightly coiled band.

The body surface of this sea slug is smooth and velvet-like. It feeds on sponges.

Cadlina laevis

Distribution: Widely distributed. In Europe from Spain to the British Isles, Iceland, Norway and Greenland. All British Isles, but most common in the north.

Description: Flattened shape, translucent yellow-white, often with lemon yellow pigmented glands around the mantle edge. However, occasionally lacks this pigmentation. Dorsal (back) side covered in very small tubercles. Up to 32 mm in length.

This sea slug is very flat, almost leaf-like and usually has white or yellow pigment granules.

Habitat: Found on various sponges, including ***Halisarca dujardini.***
Biology: This is one of the few nudibranchs with direct development, and lacks the free-swimming larval stage. Small juveniles emerge directly from the egg capsules.

Cadlina glabra *is a small slender snail found on hard bottoms in shallow waters. Those individuals we have seen are around 3 cm long. This one from Egersund, south-west Norway, 8 metres depth.*

Archidoris pseudoargus - Sea lemon

The gills are located at the rear end of the animal, and are retracted if disturbed.

Distribution: Distributed from the Mediterranean to Iceland and northern Norway. All British Isles.

Egg masses are laid in spirals.

Description: One of our largest nudibranchs. Up to 120 mm in length. Colour usually mustard yellow, with patches in shades of brown, yellow green or violet. Colour varies with habitat, thought to be a method of camouflage. Body covered in tubercles of varying sizes. Body flattened, some individuals appear almost flush with the substrate, and blend into it colour-wise. Has 8 - 9 large bushy gills at the rear of the body, usually brown in

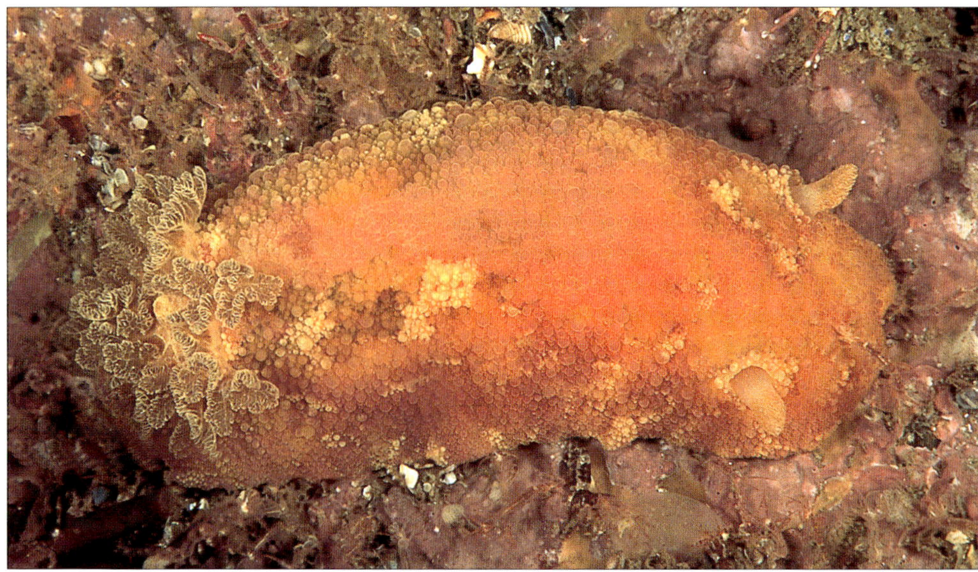

The most common colour is yellow but orange variants are also seen.

These two individuals have begun mating.

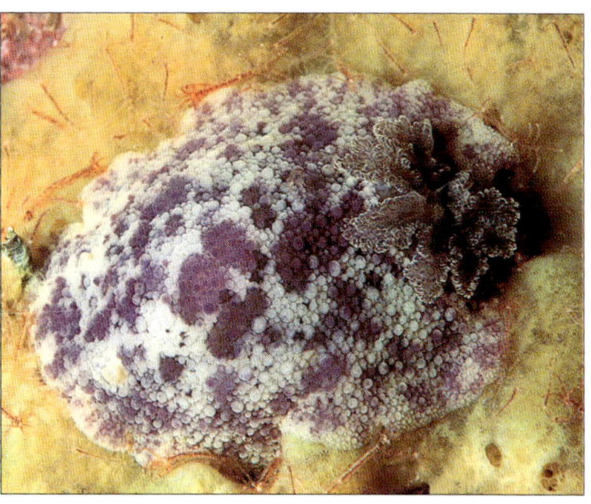

A violet coloured individual on the sponge **Halichondria panicea**. *Also visible are a number of skeleton shrimps, Caprellidae.*

colour. Can be confused with ***Discodoris planata***, which has star-shaped patches towards the rear of the animal, and dark brown patches on the underside of the mantle. This species is presented below.

Habitat: Very common on the sponge ***Halichondria panicea*** in shallow exposed areas. Also found on the sponges ***Halichondria bowerbanki*** and ***Suberites ficus***. Found down to approx. 300 m.
Commercial value: This nudibranch is used as bait by Shetland fishermen.

Discodoris planata

Note the scattered pale patches.

Distribution: Recorded from the Mediterranean north to the Norwegian coast. On the south and west coasts of the British Isles.
Description: Up to 65 mm in length; closely resembles the previous species ***Archidoris pseudoargus***. However, ***Discodoris planata*** can be distinguished by having dark brown patches on the underside of the mantle, and finger-like mouth tentacles. Further, it has up to 12 star-shaped patches radiating from a central papilla, which is paler in colour than any of the others. The papillae contain acid. The dorsal (back) side is further covered by small tubercles. Colour variable, and patchy as in ***A. pseudoargus***, often in tones of brown, orange and/or purple.
Habitat: Found in shallow water on sponges such as ***Hemimycale columella.***
Biology: Egg masses form a spiralling ribbon-like structure.

Sub-order Cladobranchina

Most members of this group (Gr. klados = branch, bud + branchia = gills) have either unbranched or branching fringes/outgrowths on the dorsal (back) side, which often contain poison extracted from its prey. The group includes three supra-families; ***Aeolidoidea, Dendronotoidea*** *and* ***Arminoidea****.*

Dendronotus frondosus

A beautiful, bushy sea slug that appears in a number of colours.

Even the head- and mouth tentacles are bushy.

Distribution: Northern distribution from the Bay of Biscay and northwards. All around the British Isles.

Description: Slender body form, up to 100 mm in length. Up to 9 pairs very bushy, large cerata on the dorsum (back side), which function as gills. Between these are small gills. Similar outgrowths present around the head end, in front of the bushy head tentacles (rhinophores). Adult individuals very characteristic, due to their bushy appearance, but the gills are less branched in juveniles. Colour variable, from lemon yellow to brown, red or even completely white. Even at the same site, individuals can occur in completely different colour variations. The marked variation in form, combined with the fact that some individuals spawn at a much smaller body size than others, suggests that these may represent more than one species.

Habitat: Very common amongst algae and in the kelp zone, usually in sheltered localities.

Biology: Small individuals mainly feed on hydroids within the genera ***Obelia***, ***Sertularia*** and ***Halecium***, whereas larger individuals appear to prefer ***Tubularia***.

Doto coronata

Distribution: Widely distributed along large stretches of the European Atlantic coast. From the Mediterranean north to Svalbard. All British Isles.

Description: This 'species' actually comprises a species complex within the genus *Doto*, characterised by a translucent white to yellow body, with reddish to violet patches at the tips of the club-shaped, usually 7 (5- 8) pairs of cerata ('gills'). Other recognised species within ***Doto*** have a dark pigmentation at the tips of the cerata, but only this species complex has pigmentation at the inner bases of the cerata. On the dorsal (back) side of the body itself are reddish to violet pigment patches. Length up to 12 mm. Two species, ***D. hydrallmaniae*** and ***D. sarsiae***, have recently been separated from this species complex by means of molecular analyses.

Habitat: Individuals within this species complex are found on a variety of hydroids, including ***Obelia geniculata.***

Biology: Eggs are laid in 'strings of pearls', wound around the hydroids upon which it lives.

The species complex **Doto coronata** *is characterised by violet or red spots all over the usually 7 pairs of spherical gills.*

This sea slug lays its eggs in 'strings of pearls'. Although a common species, it is often overlooked due to its small size.

Doto fragilis *feeds on various hydroids. Note the sheaths enclosing the head tentacles (rhinophores).*

Doto fragilis

Distribution: Widely distributed and common from the Mediterranean to northern Norway. All British Isles.

Description: This species is not unlike the previous species complex, but the body is pale brown in colour, often with white up to 10 pairs large, tube-shaped with large, up to 10 pairs, tube-like, dorsal (on the back) longitudinal cerata, without dark pigment patches. Considerably larger than the previous species complex; up to 30 mm in length.

Habitat: Occurs down to approx. 200 m depth, mainly on the hydroids ***Nemertesia antennina*** and ***N. ramosa.*** A smaller variant is found on the hydroid ***Halecium halecinum***, and a very large variant on ***H. muricatum.***

This most likely is **Doto millbayana** *creeping along a hydroid. There are 6 pairs of cerata, with dark pigment spots surrounded by smaller irregular red spots. The rhinophores have white spots. In front of the sea slug is an amphipod.*

Tritonia hombergi

This species has short bushy gills along the sides of the body.

Distribution: Recorded from the Mediterranean to Norway. All British Isles.
Description: Up to 200 mm, our largest nudibranch. Bushy gills all along the sides of the body Head tentacles bushy at their tips, encased in a sheath at their bases. Dorsal (back) side covered in round 'warts', which secrete an irritant compound which can cause fluid-filled blisters after contact with skin. Colour varies from completely white, pale orange to red and brown forms.
Habitat: Found from a few metres deept, usually on, or close to, the dead man's finger coral, ***Alcyonium digitatum***, upon which it feeds.

This individual is eating the dead man's finger coral **Alcyonium digitatum.**

Janolus cristatus

Distribution: Distributed from the Mediterranean to the Norwegian coast, also relatively common around the British Isles. **Description**: Relatively large, up to 75 mm in length, with numerous translucent cristata ('fringes'). Within each 'fringe' is a thin, dark brown digestive canal, terminating in a bright blue to white pigment patch at the tips. Similar pigment patches also occurring on the dorsum (back), and as a stripe posteriorly on the upper side of the foot. Head tentacles (rhinophores) lamellate (like stacks of discs), always with a pigment patch at their tips; usually white. Between the rhinophores is a raised comb-like sense organ, which is unique for this species. These latter two characteristics distinguish it from the aeolid nudibranch ***Eubranchus tricolor,*** with which it can be confused. Further, the rhinophores of ***E. tricolor*** are not lamellate and lack pigment, and it has two anteriorly directed mouth tentacles.

Long, partially transparent fringes (cerata).

This sea slug is almost transparent.

Janolus cristatus *is extremely characteristic and easy to recognise.*

Habitat: Usually found in shallow localities, especially sheltered rock substrates. Observed feeding on erect bryozoans around the British Isles, but we have seen it feeding on encrusting bryozoa in southwestern Norway.

Biology: Egg masses white, laid on the bottom in strongly waved coils resembling strings of pearls.

CNIDARIAN POISON AS DEFENCE

Many nudibranchs are elongate in shape, can move slowly and have numerous fringes (cerata) on their backs. These fringes contain sacs that store intact poison cells (nematocysts) from its cnidarian prey. In other words, the poison cells of the prey are passed through the gut of the nudibranch without being activated. The otherwise unprotected nudibranch can then use these as its defence mechanism. The tips of the cerata are often brightly coloured, to advertise the presence of poison to potential predators. Most fringed nudibranchs have hydroids on their menu, but some eat sea anemones. Shown here is ***Facelina bostoniensis*** (s. 345).

Coryphella verrucosa

Coryphella verrucosa *has relatively short fringes with thick orange-brown digestive canals. Upper right: egg masses are laid in spirals on hard substrata.*

Distribution: Northern distribution, from the northern parts of the British Isles to northern Norway.

Description: Closely resembles the more rare ***Coryphella browni***, with which it probably often is confused. However, *C. verrucosa* has thicker digestive canals and the white rings just below the tip of the fringes are narrower. Further, *C. verrucosa* has a central longitudinal white stripe along its back, but which often is hidden by the fringes. This stripe is absent in *C. browni*, but present in ***Coryphella lineata***, described below. However, the latter also has white stripes on the fringes and along the sides of the body. *Coryphella verrucosa* is usually between 15 - 25 mm in length but can reach a length up to 62 mm.

Habitat: Common, particularly in spring and early summer. Often found in exposed areas in shallow water, or in deeper water with tidal currents down to approx. 300 m. Usually feeds on ***Tubularia*** or other hydroids.

Biology: Eggs laid in a spiralling string.

Coryphella lineata

Distribution: Found from the Mediterranean to northern Norway. All British Isles.
Description: Closely resembles the previous species, but distinguished by the conspicuous white lines along the length of each of the fringes, tentacles and along the body. The translucent fringes have marked digestive canals, usually brown in colour, but can also be red or orange-yellow. Usual length between 20 - 30 mm, but can reach up to 40 mm.
Habitat: Found at the same kind of shallow, exposed localities as the previous species, or deeper water influenced by tidal currents, down to 400 m. This species also preferentially preys on the hydroid ***Tubularia indivisa*** and to some extent ***Corymorpha nutans***, but it also feeds on other hydroids within ***Coryne***, ***Hydrallmania*** and ***Sertularia***.

Characteristic longitudinal white stripes on the body and tentacles.

Biology: Eggs laid in waving spiralling ribbon-like masses.

Members of this species feed on hydroids. Here amongst the spines of a sea urchin.

Coryphella pellucida

Characteristic bright red fringes.

Distribution: Northern distribution, from northern parts of the British Isles north to mid-Norway.

Description: Body translucent white, with white pigment patches at the tips of the head tentacles, mouth tentacles and the tail. The approximately equally long fringes are filled with red digestive canals, and the tips are white. No pale rings just below the tips, as in most similar species. Usual length for fully-grown individuals is 30 mm.

Habitat: Typically found in sheltered areas exposed to tidal currents usually feeding on hydroids within the genus ***Eudendrium***. At some localities, large aggregations of individuals can occur.

Biology: Eggs laid in waved, spiralling threads either directly on the prey organism or on the bottom close beside it. Spawning observed around the British Isles in early summer, but in southwest Norway, we have observed spawning between October - November.

Coryphella gracilis *only reaches 15 mm in length and lives in current-rich areas. Can easily be confused with juveniles of* **C. pellucida**. *Both species feed on* **Eudendrium** *spp.*

Coryphella pedata

Distribution: From the Mediterranean to mid-Norway. All around the British Isles.

Description: Easily distinguished from other north European species by the characteristic violet-pink colour, both of the body and the fringes. Additional white pigments

on all outgrowths, including white rings at the tips of the fringes. Fringes arranged in bundles, fused together at their bases.
Habitat: Rather rare, occurs in ones and twos in partly exposed localities, feeding on hydroids within the genus ***Eudendrium***.
Biology: Eggs laid in thin threads, often wound around the prey hydroids.

In our waters, this species is unique in having violet fringes. Usually observed on kelp, feeding on hydroids.

Facelina bostoniensis

Robust horn-shaped mouth tentacles.

Here together with brittle stars.

Distribution: The species epithet refers to its north American distribution (Boston). In Europe widespread from the Mediterranean to northern Norway.
Description: Body translucent white, with red pigmentation on the head, particularly just behind the bases of the head tentacles. Mouth tentacles very long, with white pigmentation at their tips. Thin reddish to dark brown digestive canals visible through the fringes. Outer part of fringes with a broad white ring extending almost to their tips. Head tentacles bearing narrow rings. Up to 55 mm in length. A very similar species,

This species has characteristic red pigmentation between the head tentacles.

Facelina auriculata, is distinguished by having conspicuous blue-violet, not reddish, pigmentation behind the head tentacles.

Habitat: Lives and feeds on the hydroids ***Tubularia larynx*** and ***Clava multicornis.***

Cuthona caerulea

Distribution: Widely distributed around the coasts of Europe, from the Mediterranean to Scandinavia.
Description: Small body size, usually around 10 mm in length, but can reach 26 mm. Seen at close hand, this is a beautiful nudibranch. The body itself is more or less translucent, but the fringes have a dark green core, with blue pigmentation on the outer layer. The distal-most (outer) third of the fringes is orange. Occasionally also has a narrow orange ring directly below the blue pigmentation. Fringes club-shaped, arranged in approx. ten paired groups, on either side of the body.
Habitat: Shallow water, on various hydroids, often within the genus ***Sertularella***.
Biology: In common with its nearest relatives, it feeds on hydroids.

Cuthona caerulea *is a small sea slug around 1 cm in length. Here photographed in Egersund Harbour, southwest Norway.*

Tergipes tergipes

This small sea slug is rarely encountered. Being only 6 - 8 mm long, magnification or a strong macro-lens, is needed to see it properly.

Distribution: Common around the British Isles; also recorded along the Norwegian coast.

Description: Very small body size, only up to 6 - 8 mm in length; as a result often overlooked. Single row of swollen cerata (fringes) on each side, forming alternating pairs. Each of the cerata with large white cnidarian stinging cells at their tips. Body translucent whitish, with dark reddish spots on each side of the head, as well as at and immediately behind the bases of the relatively long rhinophores (head tentacles). Mouth tentacles short.

Habitat: As in other nudibranchs, strongly associated with its prey. Lives on hydroids within the genus ***Obelia***, particularly ***Obelia geniculata*** (p. 79), which in turn is common on seaweed and kelp.

Eubranchus farrani

Of the two colour variations shown here, this orange form is the most common.

Distribution: From the Mediterranean north to mid-Norway. All British coasts.
Description: Up to approx. 20 mm in length. The very characteristic swollen fringes can have different colours, and four different variants are recognised within the species. The most common of these is translucent white with dispersed orange or yellow patches on its back, on the tips of the fringes as well as the head and mouth tentacles. In another variant, the body has tones of dark violet, yellow-brown to orange, instead of white. A further variant common around the British Isles is completely white. Individuals seem only to mate with other individuals within the same colour variant, supporting the hypothesis that these represent separate species.
Habitat: Usually found in shallow water, feeding on hydroids within the genus ***Obelia***, including ***O. geniculata***.
Biology: Eggs laid in spiralling ribbon-like masses with two coils.

The characteristic swollen fringes leave little doubt as to the identity of this individual.

Eubranchus pallidus *can reach 23 mm in length. It is recognised by the yellow-white spots on the body and the prominent fringes. Adults eat the hydroid* **Tubularia**.

Cuthona viridis

This is one of the few sea slugs that has green digestive canals.

Distribution: From the northern Atlantic coast of France, the British Isles and north to Iceland and northern Norway.
Description: Recorded up to approx. 15 mm in length around the British Isles. Larger individuals reported from Iceland. Fringes filled with dark green digestive canals and with characteristic small green pigment patches. Pale yellow

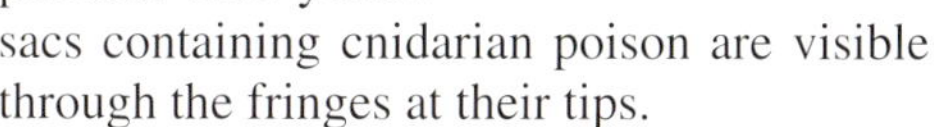

sacs containing cnidarian poison are visible through the fringes at their tips.
Habitat: Occurs and feeds on hydroids within the genus ***Sertularella***, particularly ***S. rugosa.*** Usually in exposed to partially exposed areas in shallow water.

Cuthona rubescens *is recorded as far north as the Orkney Islands and southwestern Norway (here photographed at Egersund). It is easily recognised by the characteristic band of reddish pigment half way up the rhinophores, which are white at the tips. Can reach 15 mm in length and feeds on the hydroid* **Halecium halecinum.**

Aeolidia papillosa

One of our largest nudibranchs, which has various sea anemones on its menu.

Distribution: Widespread in European waters, from Spain north to northern Norway. All British Isles.

Description: Up to 120 mm; as such the largest member of the group Aeolidiacea in our waters. Some variations in colour, but most commonly white with grey, brown and/or dark purple pigmentation. Occasionally with a white V-shaped pigment patch between the head tentacles, extending to the mouth tentacles. Both mouth and head tentacles wider at their bases, gradually becoming narrower towards the tips. Fringes somewhat flattened from the sides.

Egg masses are laid in long, folded 'strings of pearls'.

Habitat: Found in shallow water, both in exposed and shallow localities.

Biology: This is one of the few nudibranchs within this group that does not eat hydroids, but instead feeds on a wide range of sea anemones, including ***Actinia equina, Sagartia*** spp., ***Urticina*** spp. ***Metridium senile, Actinothoe sphyrodeta*** and ***Anemonia viridis***. Egg masses laid in characteristic thread-like, waved spirals.

Class Bivalvia

*Bivalves (Lat. bi- = double, two + Lat. valvae = valves) comprise molluscs with two more or less symmetrical, hinged shells. These are joined together with elastic bands (**ligaments**) and closed by means of an abductor muscle. The part of the paired shells that bears the hinge is called the **dorsal** side and the opposite end is the **ventral** side. In total, more than 8000 existing species are recorded, most of them marine.*

The shells are more or less curved, and the basic shape is oval or almost circular. However, some species have shells that are an exception to this basic form. Razor shells (**Solenidae**) are long and narrow, and in saddle shells (**Anomiidae**), one of the shell valves is much reduced. Ship-worms (**Teredinidae**) have a strongly derived, bullet-shaped shell that covers only the front part of the elongate animal. In bivalve molluscs, the head is reduced and lacks eyes and a radula. Eyes and balance organs (**statocysts**) can however be located throughout the body.

The vast majority of bivalves filter plankton and dead organic material across the gills. This is carried out in two ways. In mussels (**Mytilidae**), the water is transported into the shells through the mantle margins along most of the length of the shell opening, whereas digging bivalves, such as the **Myacidae**, have two separate or joined respiratory tubes called **siphons**, which extend above the sediment surface to varying degrees. In this case, water is pumped through the one siphon, and suspended food particles are captured. The water is then pumped out through the other siphon. Other digging bivalves have long siphons that extend relatively far above the surface, but can be bent over, allowing the animal to 'vacuum-clean' the sediment surface to obtain food particles. The 'primitive' **Nuculidae**, obtain food particles directly from the bottom by means of two 'lips' that protrude from between the shells. In the most derived, or 'advanced', forms, such as the shipworms, the transformed shell is used to bore through wood, which is digested by special enzymes that break down cellulose.

Most bivalves have separate sexes, i.e. individuals are either male or female. Reproduction occurs by eggs and sperm being released in 'clouds' directly into the water, where fertilisation occurs. The fertilised eggs develop into ciliated

The Icelandic scallop **Chlamys islandica** *may have a symbiotic relationship with a sponge. The sponge protects the scallop from predators, and in return it is lifted up from the substrate, where it can filter particles more effectively. Photograph from Tromsø, northern Norway.*

Bivalves have usually two shell halves that protect the soft tissues. Mytilus edulis.

trochophore larvae, which further develop into shell-bearing **veliger larvae**, which look like small bivalves. These live in the plankton for around 1 - 8 weeks, before settling to the bottom.

Relationships within the bivalves are not entirely clear, and there are various approaches to classification. Some classification systems are entirely based on the appearance of the shells, others focus on the form of the internal organs and still others use ecological characteristics, such as means of feeding or specialised structures selected for a particular habitat. For non-specialists, the shape, size and to some extent colour of the shells, together with the habitat are the most important features used in identification. The symmetry and lengths of the shells, together with their sculpturing, are also important for identification. Shell **length** is measured as the greatest distance between the outer points of the shell, in a parallel line to the dorsal side of the bivalve (the dorsal side is where the two valves are joined together). Shell **height** refers to the distance from the hinge/dorsal side to the outer edge of the valves (opposite or ventral side of the bivalve). The **anterior** (front end) of bivalves is where the foot sticks out; opposite to the **posterior** (rear end), from which the siphons emerge. The sculpture of the valves is often very characteristic. **Radial sculptures** radiate from the often strongly arched oldest part of the valves; the umbo, which often is bent forwards. **Concentric sculptures** are parallel to the edge of the valves. The terms left and right valves are used. When the dorsal side of the bivalve is turned upwards, and the anterior, or front end, of the bivalve turned away, the right-hand valve is to the right. The thickness of the sculpting varies from thin, barely visible **lines** to prominent **ridges** or **keels**. Further, the valves can be covered in sharp spines or rounded **knobbles**. In certain groups, more or less distinct growth lines can be seen, which can indicate the age of the animal. The shell grows more in summer, such that the summer growth lines will always be thicker and paler than the winter lines.

In some bivalves, the pallial line, which extends between the **abductor muscle** scars, is markedly invaginated to form the pallial sinus. When present, the mantle cavity is located in the rear part of the shells. The scars of the **abductor muscle** are usually largest in the posterior part. The shape of the abductor muscle scars, pallial line and the pallial sinus are important characters for identification, if the soft tissues are missing. The dorsally-located hinge consists of a specific number of teeth, ridges or plates that fit into each other and contribute to keeping the shells held together. There are two types of teeth: the **cardinal teeth** are located directly under the umbo and the **lateral teeth** are in front of or behind the umbo. The form of the hinge is characteristic and is also important in species determination. Using the characters described above, it should be possible to identify most of the bivalves commonly found along our shores.

Because this book presents animals photographed in their natural environments, generally only those bivalves which live as **epifauna** (on or above the bottom sediment) are included. However, many species live buried in the sediment, such that we usually only see them as dead shells washed up on the beach. Only a few of these are included here. There are 203 species of bivalves recorded from the Norwegian coast alone. The class Bivalvia is divided into the two sub-classes **Protobranchia**, with weakly developed gills, and the **Lamellibranchia**, which have well-developed and strongly folded thread- or disc-like gills, or where the gills take the form of a muscular, vertical perforated shield. Because of the discrepancies between different classification systems at the higher taxon level, we here present the species according to family membership.

In **Limaria hians**, *the soft tissues are larger than the shells can cover.*

Family Anomiidae - Saddle oysters

Saddle oysters are characterised by particularly one of the two shells being reduced to a thin, often transparent, very fragile shell. These bivalves attach themselves to stones, other bivalves, snails, crustaceans by means of calcareous byssus threads extending through the thin shell. The upper shell is more robust, often slightly curving, almost circular. The shape of the shell is formed to suit the substrate. Resembles small oysters.

Pododesmus patelliformis (= *Monia patelliformis*) - Ribbed saddle oyster

Distribution: Widespread along European coasts from the Mediterranean to northern Norway. Common at all British Isles.
Description: Up to 40 mm in diameter. Upper shell whiteish with reddish-brown to pale brown patterns or stripes. Radial ribs present, and concentric, blade-like sculpturing reminiscent of young oysters (***O. edulis***). Lower shell (only visible if the animal is prised off the substrate) very thin, almost transparent, circular or oval in shape, with a triangular hole through which the byssus threads pass. ***Pododesmus squama*** (= ***Monia squama***), is now considered to be synonymous (the same species) as ***P. patelliformis***. The other common saddle oyster, ***Heteranomia squamula***, is distinguished from ***Pododesmus patelliformis*** by its small size, only up to 10 mm in diameter. This little saddle oyster is often attached to the shells of crabs, lobsters and other shell-fish. Otherwise very similar to ***Pododesmus patelliformis***.
Habitat: Attaches to hard substrates, often stones or other shellfish.

Saddle oysters attach to hard substrates. On the shell are serpulid and spirorbid tubeworms.

Left: In **Pododesmus patelliformis**, *the hole from which the foot extends, is relatively large and triangular. Right: An individual covered with bryozoans.*

Family Mytilidae - Mussels

This family includes the blue mussel and the horse mussel. Mussels have oblong, symmetrical shells, where the umbo (the oldest part of the shell) is strongly displaced to one side. Mussels attach themselves to the substrate by means of byssus threads. The insides of the shells have a mother-of-pearl appearance.

Mytilus edulis - Blue mussel

Distribution: Widespread in Europe, from the Mediterranean to the Arctic. Common around all British Isles.
Description: Somewhat variable in form, but with a very characteristic deep blue colour, with purple tones, always with a brown mantle margin. The blue mussel usually reaches up to 100 mm in length, but can be up to 150 mm.
Habitat: Found attached to the substrate, often in large numbers, from the high tide mark down to approx. 10 m dept. Small individuals can move by means of the foot to deeper water as they grow. Tolerant of large variations in temperature and salinity, and occurs in brackish water in the inner parts of many inlets and fjords and all the way to Bottenviken in the Baltic Sea.

This blue mussel probably is very old.

Mussels attach to the substrate using byssus threads. Large individuals may be confused with horse mussels (next page).

Biology: During the first year, blue mussels generally grow to about 30 - 40 mm in length, but it takes approximately three years before they are of a commercially viable size. Blue mussels have separate sexes and spawning occurs in the spring. Individuals release eggs and sperm (milt) directly into the water, where fertilisation occurs. A single blue mussel can spawn 5 - 10 million eggs. The larvae will settle on almost any type of solid substrate in shallow water. By their filtering activities, a population of blue mussels can effectively clear the surrounding waters of nutrients and phytoplankton.
Commercial value: Because of their popularity as seafood, blue mussels are commercially cultured on a large scale in certain parts of Europe. Best eaten between September and April, when they are filled with roe and milt.

CAN BE POISONOUS...

Because blue mussels filter particles out of the water, they are sensitive to pollution, and heavy metals can accumulate in their tissues. For example, blue mussels can have a body load of lead that is 10,000 times that in the surrounding water. This means we should avoid eating mussels near industrial or other sites likely to release comtaminants. Another risk is toxic algae - blue mussels filter out all types of planktonic algae, including those toxic to humans. There are two main types of algal toxins. **Diaretic Shellfish Poison (DSP)** causes vomiting and diarrhoea within a short time after consumption of affected mussels. This poison is found in dinoflagellates within the genus ***Dinophysis***. These occur in blooms, usually in late summer and early autumn. **Paralytic Shellfish Poison (PSP)** is much more dangerous and is found in mussels that have taken up dinoflagellates within the genus ***Gonyaulax***. These algae usually occur in the springtime and early summer, especially May, but can occur throughout the year. PSP is a nerve poison that causes paralysis and which can be lethal in even relatively low concentrations. It is wise to wait at least 14 days after a toxic algal bloom before eating mussels. Health authorities carry out regular sampling in order to inform the public about the safety of mussels in the relevant areas. Such information often is available on the internet (see the web site of this book).

Modiolus modiolus - Horse mussel

Horse mussels resemble large blue mussels.

Distribution: Widely spread in northern Europe, from the Bay of Biscay to northern Norway. All around the British Isles.

Description: Adult horse mussels look like a larger version of the edible blue mussel (***Mytilus edulis***). Fully grown individuals are often 200 mm in length. The largest individuals are deep brown in colour, whereas smaller individuals are more blue. Older horse mussels have an eroded exterior, and generally look rather battered. Small individuals can be distinguished from members of ***Mytilus*** by the umbo, which in horse mussels is rounded and situated further back, and does not extend beyond the shell edge. However, small individuals can be confused with ***Modiolula phaseolina***, which only reaches up to 20 mm in

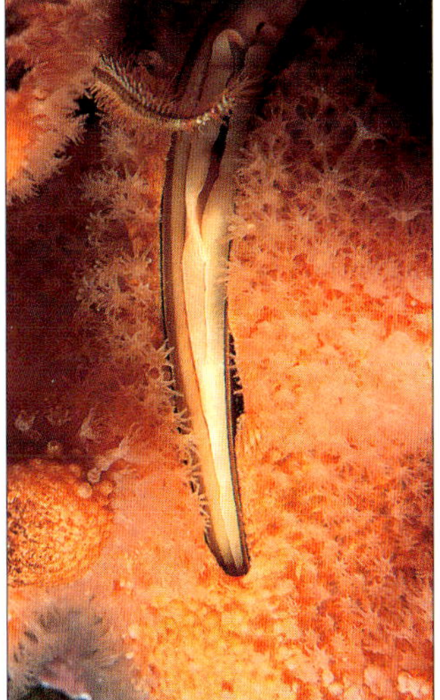

Covered by dead man's finger.

In mussels, it is important that the release of eggs and sperm is coordinated. During a dive in May, when these pictures were taken, all the horse mussles in the area spawned imultaneously. To the left are eggs being released; to the right, sperm. From Orkney Island near Hoy.

length. These have conspicuous hairs, especially on the posterior part of the shell. Small horse mussels also have such hairs, but not fully grown ones.

Habitat: Often found in large aggregations in hard bottom areas with some current, from a few metres depth, down to around 150 metres. Most abundant between 10 - 20 m. Often covered by other organisms, such as moss animals (Bryozoa), hydroids (Hydroida), barnacles (Balanidae), tube dwelling bristle-worms (Polychaeta) and saddle shells (Anomiidae).

Commercial value: Edible, commercially exploited in northern Europe. The mantle margin is removed before use, together with the gills, stomach and the byssus threads.

Modiolula phaseolina - Bean horse mussel

Distribution: In Europe distributed from the Mediterranean to northern Norway. All around the British Isles.
Description: The bean-like shape of this mussel has inspired its common name. Resembles the horse mussel (***Modiolus modiolus***), but only reaches around 30 mm in length. Bean horse mussels have characteristic hairs on the rear part of the shells. These also are found in young horse mussels, but not in such large quantities. These species are further distinguished by ***M. phaseolina*** which has a rough-textured, finely striped margin of the closing mechanism situated directly under the umbo (usually only seen under magnification). This margin is completely smooth in horse mussels (*Modiolus*).
Habitat: Usually occurs in deeper waters, from 50 m down to 1000 m. Found in shallow water around the British Isles, up to the lowest tidal level. Often occurs on coral reefs of ***Lophelia pertusa.***

The bean horse mussel has stout hairs along the outer margins. Note that small horse mussels also have hair, but all over the shell. Approx. 50 m depth.

Family Ostreidae - Oysters

In most northern European waters, this family is represented by a single species, the 'jewel' among molluscs; the native oyster, **Ostrea edulis**. *World-wide, the family comprises more than 100 species, mostly tropical forms. Oysters are characterised by their very robust, irregular and highly varied shell.*

Ostrea edulis - Native oyster

Distribution: In Europe, found from the Mediterranean to mid-Norway. Not particularly common. Fossil oyster shells are found as far north as Finnmark in northern Norway, where these oysters thrived during a time when the temperature was higher than today. All British coasts.
Description: Because of the very characteristic irregular and rough surface of the shell, fully grown oysters are not likely to be confused with other species. The lower, left hand shell is concave whereas the overlying right hand shell forms an almost flat lid. Greyish-white in colour, becoming darker with age. Can reach up to 150 mm in diameter, depending on the temperature.
Habitat: Usually found at only a few metres

Living oysters rarely occur naturally in our waters.

depth, usually partially partially attached to hard substrates. Due to the relative decline in water temperature, the distribution of oysters has decreased in northern Europe, making it now rather rare. However, it can be locally abundant in areas with high summer temperatures, such as shallow inlets.

Biology: Oysters usually are sexually mature after approximately 1 year, but this can vary with temperature. Can reach up to 60 mm in diameter after 2 years, but are usually not of an exploitable size until 3 - 4 years old. Can become up to 20 years old. Unlike most other bivalves, this species is a hermaphrodite and changes sex according to the water temperature. If the summer temperature is around 15 - 16 °C, it produces female gametes only once every four years, and remains male during the remaining times. In higher temperatures, the animals can become female once a year.

PARTIAL PARENTAL CARE

The around 1 million eggs spawned by the female are fertilised within the mantle cavity. Sperm cells from a nearby male swim into the female oyster. The hatched larvae live protected inside the mantle cavity of the female for an initial period. When they leave the mantle cavity of the mother individual, the larvae are pelagic (living freely within the water masses) for approximately 2 weeks, after which they settle on the bottom.

The upper shell is flat, whereas the lower shell is curved.

Commercial value: Oysters have been used as food for many centuries, and their shells have been found in garbage dumps from stone age times. Cultivation of oysters was carried out already in the middle ages and oyster farming is still carried out along many European coasts today. In northern areas, commercial farming is limited by the usually low winter temperatures, but some shallow, enclosed inlets are suitable. Oysters are highly prized by restaurants and connoisseurs, who usually prefer to serve them raw.

Family Pectinidae - Scallops

*This family includes a variety of species found in Europe. Scallops are characterised by their rounded, flat shells, usually with radial ribs or folds. The shell is extended into a pair of '**ears**' or '**wings**', one on each side of the umbo. The mantle margin is equipped with numerous sensory tentacles and eyes. Single large **abductor muscle**. The ligaments of the abductor muscle are clearly visible in the central part of the shell. Some species are hermaphrodites, whereas others have separate sexes. Scallops usually lie on the bottom, with the right hand shell facing downwards. In some species, the organisms are attached to the substrate, whereas others are unattached and can swim.*

Scallops have a row of rudimentary eyes along the mantle margins.

Pecten maximus - King scallop/ Great scallop

Distribution: Found from the Atlantic coast of Spain north to northern Norway. All around the British Isles.

Description: Up to 170 mm in diameter. The outline of the shell is almost circular. The right hand shell is strongly ridged, with 15 - 17 pronounced equidistant ribs. The left hand shell is almost completely flat. Reddish-brown to yellowish-brown in colour, often with irregular patches. Along the mantle margin are three rows of tentacles and 30 - 40 eyes. Unlikely to be confused with other species.

Habitat: Usually occurs on coarse sandy bottoms between 15 - 35 m depth, where it lies partially submerged in the sediment. Living scallops have been found as shallow as the tidal zone and down to around 150 m depth. The overlying shell is generally covered with a thin layer of sand, and lies flush with the bottom. The shells are orientated with the opening towards the main current direction, and the animals filter both living and dead organic material from the water. Scallops thrive especially well in areas of moderate current, particularly in relatively narrow sounds.

Shell opening with tentacles.

Biology: In common with other scallops, king scallops can swim. They have a rapid reaction to shadow, which is easily observed by passing a hand over a scallop that has its shell open. The numerous eyes

The king scallop is very characteristic, with a flat upper shell lying as a lid on the curved lower shell.

react to the shadow, and send impulses to the abductor muscle, which rapidly closes, propelling the scallop forward. This is obviously a defence against predators such as fish. Although spontaneous swimming is rarely observed, mass movements of scallops are known. Divers have reported areas that have been almost cleared of scallops to be repopulated by grown scallops within a short time. These adults must have moved themselves to the newly available area. King scallops are hermaphrodites and both the white milt and orange roe can be seen in mature individuals.

Commercial value: Scallops are considered a delicacy, and restaurants pay well. The adductor muscle is particularly tasty, and can be eaten raw, steamed or sautéd. The testes and roe are also edible. The mantle margin and the stomach are discarded. In certain parts of Europe, they are commercially exploited by divers or caught using special scallop dredges. There are great expectations for commercial culture of king scallops.

CAN SWIM

A reaction can be observed in scallops when the starfish come too close. The tentacles have organs that react to chemical stimuli from the starfish. When the starfish approaches, the scallop reacts by rapidly opening and closing its shell. Water is first pulled in then rapidly expelled, and the scallop 'swims' away from the starfish.

Aequipecten opercularis - Queen scallop

Both upper and lower shells are curved in this scallop. Note the strong ribs.

Distribution: In the northeastern Atlantic from the Canary Islands as far north as northern Norway. Mediterranean. Common around all British Isles.

Description: Lower (right-hand) shell valve distinctly flatter than the upper valve. Both valves with 18 - 22 radial ribs that are rough to the touch. The shell between the ribs is radially striped. Shell ears similar, although the ears of the right-hand shell overlap those on the left-hand shell. Colour usually reddish brown, yellow, violet or greyish, often with scattered patches of colour. Diameter up to 90 mm.

Habitat: From a few metres depth to approx. 200 m. Attaches to the substrate by means of byssus threads as a juvenile, but more motile and swims as an adult. Left shell lies uppermost. Prefers sandy or fine gravelly bottoms.

Pseudamussium perlutrae - Seven-rayed scallop

Characteristically with seven (5-9) radial ribs.

Distribution: Widespread from western Africa to northern Norway. West coast of Scotland, and coasts of Northumberland and Durham.

Description: Up to 55 mm diameter, recognisable by the 5 - 9 (usually 7) broad radial ribs. Between the ribs are indistinct concentric stripes, most prominent towards the edge of the shell. Shell colour differs between the valves; right-hand shell grey-white and left-hand shell usually reddish. Shell ears symmetrical. The few broad radial ribs makes it easy to distinguish this scallop from other northern European forms.

Habitat: Most common on sand and gravelly bottoms, down to 500 - 600 m.

Chlamys islandica - Icelandic scallop

The Icelandic scallop has fine radial ribs.

Distribution: As both its common and latin name suggests, this scallop has a northern distribution and is found all across the southern parts of the Arctic. In northern Europe from the Barents Sea and south to the coast of southern Norway.

Description: Up to 110 mm in diameter. Both shell valves similar; both concave, with numerous densely distributed radial ribs. Shells reddish or greyish, often differences in colour between the valves, with reddish left-hand shell and greyish right-hand shell. Shell ears markedly asymmetrical.

Habitat: In coastal areas usually found in current-rich areas from approx. 10 - 100 metres depth. In northern open waters, often found in high densities, in so-called "scallop banks".

Commercial value: Highly prized by restaurants and as gourmet food. Consequently, there has been considerable fishing activities for this scallop and dredgers have eradicated it from large areas of the sea floor. Due to the relatively slow growth rate of this valued scallop, some of the fishing grounds have been awarded protection (e.g. around Jan Mayen), whereas others are strictly regulated.

Chlamys varia - Variegated scallop

Chlamys varia *is recognised by the prominent spines on the outer parts of the shells.*

Distribution: Found from west-Africa north to northern Norway. Mediterranean. Off all British Isles.

Description: Up to 60 - 70 mm in diameter. Characteristic spines distributed especially along the outer parts of the 25 - 35 radial ribs. Shell ears asymmetrical; one pair markedly larger than the other but both with ribs. Colour yellow-white to orange, brown or violet, often with stripes or patches in darker colours. Distinguished from other scallops, such as the Icelandic scallop (***Chlamys islandica***) by the latter's lack of spines and denser distribution of radial ribs. Adult specimens unlikely to be confused with other species.

Habitat: Juveniles often attached to the substrate by means of byssus threads, but adults often unattached. Often found in association with kelp holdfasts. From the surface down to approx. 100 m depth.

Chlamys sulcata - Sulcate scallop

Chlamys sulcata *has many ribs, which vary in thickness, giving an irregular appearance.*

Distribution: Distributed across large areas of the northern Atlantic. Norwegian coast. Mediterranean. Off the coasts of Scotland.

Description: Small in size, up to approx. 25 mm in diameter. Shell height somewhat larger or equal to its length. Few but prominent radial ribs, 8 - 15 in number, between which are several less pronounced ribs. This gives the shell surface an uneven appearance. Ribs roughened by small scales, particularly towards the shell edges. Left-hand shell characteristically with flattened ribs, whereas those on the right-hand shell are high. Asymmetrical shell ears; anterior pair largest. Shell reddish in colour, with pale patches in between the ribs.

Habitat: Usually deep water, frequently occurring on coral reefs of ***Lophelia pertusa***. Found on *Lophelia* at 50 m in the Trondheimsfjord in Norway, but usually lives deeper than 150 m.

Palliolum striatum (= *Chlamys striata*)

Distribution: Widespread from the Mediterranean to northern Norway.

Description: One of several scallops not exceeding 20 mm in diameter. Fine radial ribs densely distributed, covered with fine conical projections or scales. One shell ear markedly larger than the other. These two characteristics distinguish it from two other species. ***Palliolum tigerinum*** (tiger scallop), lacks the conical projections, but has a red or yellowish colour, variously patterned. This species can reach up to 32 mm in diameter and is found in relatively shallow water down to several hundreds of metres depth. ***Delectopecten vitreus***, is distinguished by the shells being completely white and translucent. The latter scallop is found in relatively deep water, from approx. 50 to several hundred metres depth. Often found on reefs of the stone coral ***Lophelia pertusa***. These three scallops have very thin shells, but that of *P. tigerinum* is the most robust. Another species, ***Similipecten similis***, which only reaches up to 10 mm in length, is, in common with ***D. vitreus***, completely white or yellowish in colour, but totally lacks radial ribs. ***Similipecten similis*** is found in our waters, from approx. 10 - 1000 m dyp. ***Palliolum furtivum*** can reach up to 20 mm, and can be distinguished from the others under magnification; the apparently smooth shells have very indistinct ribs, which are granular on the left-hand shell. This species occurs most frequently between 40 - 60 metres depth, recorded from northern Norway, Swedish and Danish coasts. ***Chlamys sulcata*** (see previous page), resembles an up to 25 mm version of the Icelandic scallop, but is distinguished from juveniles of that species by its total lack of sculpture between the ribs. It is only found in deep water.

Habitat: *Palliolum striatum* is found from approx. 10 m down to several hundred metres depth, but most abundant between depths of 30 - 80 m.

Palliolum striatum *is often found under stones on hard substrates.*

Familie Limidae - File clams

These bivalves are closely related to the scallops. The left and right shells are thin and almost alike, with pronounced radial ribs and small shell ears. They are distinguished from the scallops by the more oval shape of the shells, and most are more or less white in colour. In common with the scallops, the mantle margin is provided with eyes and tentacles, which in file clams are very long. The shells are usually gaping, and the animal lies on the bottom, often in a 'nest' of byssus threads. File clams swim even better than scallops. Seven species within this family have been recorded in our waters.

Limaria hians - Gaping file clam

Limaria hians *has extremely long tentacles which are too large to be retracted into the shells. It therefore hides under stones or similar objects.*

Distribution: Northeastern Atlantic north to northern Norway. Mediterranean. Western and southern coasts of the British Isles.
Description: Rarely exceeds 25 mm in length. A very characteristic mussel, with both anterior and posterior parts of the shells widely gaping. Shells obliquely oval in circumference. Tentacles and mantle margins orange in colour, extending from the gaping shells, such that the shells cannot be closed. Distinguished from other small file clams by the non-translucent shell.
Habitat: Found from the lower shore down to approx. 700 m. Individuals often aggregated.

ACTIVE LITTLE CREATURE

In some languages, the gaping file clam is called the "nesting shell" because it often makes nests of small stones bound together with byssus threads. It has a large capacity for movement. When we lifted one of the stones, two shells shot out of the 'nest' and swam in rapid bursts across the bottom. This motility and its habit of building nests are rather uncharacteristic of bivalves, which we expect to be silent and unmoving. Norwegians have in fact coined the phrase "silent as an oyster".

Limea loscombi - Loscomb's file clam

Limea loscombi.

Distribution: Distributed from northern Norway to west-Africa. Mediterranean.

Description: Up to 20 mm in length. Shells white in colour, somewhat translucent, with very weakly defined ribs. Soft tissues orange, visible through the shell. Shells very similar to the gaping file clam, characterised by their asymmetrical, oblique oval shape. In life, the difference between the two is more obvious. In common with the gaping file clam, *Limea loscombi* (generic name changed several times, from the earlier ***Lima*** to ***Limaria*** and now to ***Limea***) has numerous strong long tentacles. However, when disturbed, *L. loscombi* can withdraw its tentacles into the shell and the shell valves can completely close. In the gaping file clam, the tentacles are always exposed through the gaping shell valves. Further, the shells of the gaping file clam are not translucent. The umbo of *L. loscombi* is large in comparison with two other similar, but even smaller species with more symmetrical shells; ***Limatula gwyni*** and ***Limatula subauriculata.***

Characteristic white, almost translucent shells.

Habitat: Found on soft bottoms, from mud to gravelly sand from 2 - 3000 metres depth.

Biology: Like the gaping file clam, this clam is a very 'active' bivalve. It can quickly move away from predators - or irritating photographers!

Acesta excavata - Giant file clam

Distribution: Northern distribution, from the coast of Sweden to the Arctic. Entire Norwegian coast.
Description: Up to 20 cm in length. As the name suggests, this clam is distinguished by its size. In common with other file clams, the shell height is greater than its length. Shells pale greyish-white. Tentacles of the mantle margins bright orange in colour and equipped with tentacles at their tips.
Habitat: Found on hard bottom substrates in relatively deep water, down to several hundred metres depth. Very commonly associated with the reef-forming deep water stone coral ***Lophelia***. The shallowest known occurrence of this clam is around the shallowest recorded *Lophelia* coral reef, at 39 m depth in the Trondheimsfjord in Norway.

The giant file shell can hardly be confused with other species. It can be up to 20 cm long.

*This bivalve (***Acesta excavata***) thrives together with* **Lophelia pertusa.** *From Trondheimsfjorden.*

Familie Arcticidae

This family comprises only one species in our waters, the Icelandic cyprine.

Arctica islandica - Icelandic cyprine

The Icelandic cyprine with the foot partially extended.

Distribution: In Europe from the Bay of Biscay to the southern Barents Sea. Entire Norwegian coast. All British Isles.

Description: Up to 120 mm in diameter. The extremely robust, wide, convex shells, together with its brown-black colour makes adults of this species easy to identify. Shells with very prominent umbo and numerous marked, fine concentric rings. Each shell with three stout main teeth. Two short respiratory siphons.

Habitat: Sandy or muddy bottoms with a high clay content; up to 100 per m^2. Common near the shore in a few metres depth, but individuals have been found as deep as around 500 m. Occasionally found between boulders on gravelly bottoms. This clam buries such that it barely projects from the sediment surface.

Biology: Adults have few enemies; the wolffish is probably the only fish able to crack the robust shells.

Commercial value: Harvested for food in USA (where it is known as the ocean quahog). In northern European waters and around Iceland, it is used as bait. In northern Norway, the shells are commonly hunted using underwater binoculars and a long-shafted 'spear'.

THE WORLD'S OLDEST?

Studies of the Icelandic cyprine have revealed that individuals often reach an extremely old age. The oldest recorded individuals are 220 years old or even older. In fact, living individuals are known to exist today that were hatched in the late 1700s.

The Icelandic cyprine usually buries such that only the shell opening projects from the sediment surface.

Familie Cardiidae - Cockles

Shells belonging to this family are very commonly seen washed up on the beach. When viewed 'side-on' (although actually the front/back), the shells have a heart-shaped outline. In addition, they are characterised by very pronounced radial ribs and a thick shell. Umbo prominent. Cockles live buried down to a few cm in sandy bottoms, often as shallow as the intertidal zone. Worldwide, the family comprises around 100 species and in Europe more than 20 species are represented. However, in northern European waters, two species are particularly common.

The common cockle has strongly curved shells, with relatively few, but prominent, spineless ribs.

ENORMOUS DENSITIES

In the Wadden Sea in the southern North Sea, it is not unusual to find around one thousand, approximately 10 mm large cockles per m^2. A redshank wading bird can consume 300 cockles a day. In this case, it would need more than three days to clear a square metre…

Cerastoderma edule - Common cockle

Distribution: Widespread from western-Africa to northern Norway. Common all around the British Isles.

Description: Up to 50 mm diameter. Shells white to brownish in colour with 22 - 27 prominent radial ribs without spikes, but often with weak perpendicular ridges. Ribs flattened, not pointed to form a keel, as in the very similar ***Cerastoderma glaucum***, which lives in brackish water (salinity 4 - 15‰).

Habitat: Common on European sandy beaches. Some beaches have such a high density of cockles, that the sand feels like it has been reinforced by cobbles. Found in the tidal zone down to a few metres depth. Tolerates salinities down to 10‰.

Biology: Buries to 1 - 2 cm below the sediment surface, with the two respiratory tubes barely projecting above the sand. The other end is anchored using the foot. If a cockle is dug up and placed on the sand, it will quickly bury itself again, using its foot to pull itself into the sediment.

Commercial value: This cockle is a popular food item in many European countries.

Acanthocardia echinata - Prickly cockle

The prickly cockle can be 70 mm in diameter. Here both the inhalent and exhalent siphons are visible.

Distribution: From the Canary Islands north to northern Norway. Mediterranean. All around the British Isles.

Description: Shells almost circular, up to 70 mm in diameter, usually brownish-white. Strongly resembles the common cockle, ***Cerastoderma edule***, but characterised by having prominent spines on its 19 - 21 radial ribs. This feature means it cannot be confused with other species in our waters.

Habitat: Unlike the common cockle, the prickly cockle lies only partially buried in sand or gravel. Found from approx. 2 metres depth down to around 150 metres. Usually deeper than 10 metres. Not found alive in the intertidal zone.

Biology: Observations of the prickly cockle in aquaria show that it can move relatively fast by using its strong, up to 100 mm long foot. If a starfish approaches, the cockles push the foot against the bottom and carry out a series of jumps.

The characteristic spines on the outer parts of the shells make this species easily recognisable.

Family Veneridae - Venus clams

This large family comprises 500 species world-wide. Dead shells within this family often are found washed up on the beach. The shells are solid, often with concentric sculpting. The two respiratory siphons are fused together and have a dense array of short tentacles at their openings. The foot is large and can be extended through a gap in the mantle margins. The shells are alike, but the umbo is displaced backwards. Each shell has three cardinal teeth, and occasionally also a lateral tooth. Only a small selection of the most common species is presented here.

Dosinia exoleta - Rayed artemis

Distribution: From western Africa to northern Norway. Mediterranean. Common around all the British Isles.

Description: Up to 60 mm in diameter. Shells almost circular, with fine but clearly-defined, blade-like concentric lines. On the insides of empty shells, it can be seen that the pallial line has a marked and pointed invagination. Shells white in colour, often with darker patches. Can be confused with the completely white ***Lucinoma borealis***, which reaches up to 40 mm in diameter. The latter has a relatively thin, oblong abductor muscle scar, and no invagination of the pallial line. *Dosinia exoleta* can also be confused with ***Dosinia lincta***, which has a similar distribution. The latter has a similar characteristic invagination of the pallial line, but is distinguished from *D. exoleta* by a more blunt posterior margin, and the even finer sculpturing of the concentric lines. In the latter, the lines become coarser towards the margins.

Dosinia exoleta *has a rounded form and is easily confused with* **D. lincta**. *Distinguishing between them requires examination of the interior of the shells.*

Habitat: Lives buried within the sediment from the lower part of the tidal zone down to approx. 100 metres.

Dosinia lincta

Distribution: As for the previous species, found from western Africa and the Mediterranean, north to northern Norway. Also Faeroe Islands and Iceland. All around the British Isles.

Description: Up to 35 mm in length. Resembles ***Dosinia exoleta***, but distinguished by the somewhat blunter and less round posterior margin. Shell edge in front of the anteriorly-orientated umbo is only slightly concave in *D. lincta*, whereas in *D. exoleta*, it is distinctly concave (curving inwards). Further, the sculpturing on the exterior of the somewhat thinner shells in *D. lincta* is even finer. Completely white, without any dark patches.

Habitat: Primarily clayey sediments, prefers finer-grained substrates than ***D. exoleta***. Usually also found at greater depths; most abundant between 20 - 50 metres.

Dosinia lincta *strongly resembles* **D. exoleta.**

Clausinella fasciata - Banded venus shell

Distribution: From the Canary islands to northern Norway. All around the British Isles. **Description**: Up to 25 mm in length. Common name inspired by the often very conspicuous dark radial stripes on the otherwise pale shell. Triangular outline with marked concentric sculpturing, as is typical for venus shells. *Clausinella fasciata* has approx. 15 broad, rounded concentric rings, or keels. Small U-shaped mantle cavity. Left-hand shell with a small lateral tooth situated behind the umbo. Other venus shells also exist in our waters, amongst these ***Chamelea striatula***, which resembles *C. fasciata* in having reddish-brown patches in three radial stripes. However, this clam has a greater number of concentric rings, which also are much finer, and reaches a larger size; up to 50 mm in length.
Habitat: Found on shell-sand and gravelly bottoms from around 5 metres depth to 100 metres.

This species is characterised by the prominent sculpturing and the faint bands of colour.

Venerupis pullastra

Venerupis pullastra *has fine concentric lines. Empty shells are often found on the beach.*

Distribution: From the Mediterranean to northern Norway. All around the British Isles.
Description: Length up to 70 mm. Clams within this genus are characterised by having oblong, almost rhomboid-shaped shells, with concentric and often also radial sculpting. ***Venerupis pullastra*** has fine concentric lines and has characteristic orange to dark brown zig-zag stripes, against a yellowish to brownish-white base colour. Umbo markedly displaced towards the front end; rear part of shell bluntly truncated. Another species that closely resembles *Venerupis pullastra* is ***Paphia rhomboides*** (= ***Venerupis rhomboides***). However, *P. rhomboides* is distinguishable because the rear part of the shell is not truncated.
Habitat: Buries into sand and living individuals are rarely seen by a diver. Usually shallower than 10 m, but occurs down to around 30 m. Empty shells commonly washed up on beaches.

Timoclea ovata - Oval venus

Timoclea ovata *resembles a small cockle when viewed from above. However, seen from the side, it is much flatter.*

Distribution: Common from western Africa to northern Norway. All around the British Isles.

Description: Relatively small; up to 20 mm in length. Shells triangular in shape. Sculpting in the form of 40 - 50 prominent radial ribs, crossed by 20 - 25 faint concentric lines. Resembles the common cockle ***Cerastoderma edule*** (p. 353), but distinguished by the lack of concentric lines in the latter. Further, *Timoclea ovata* has three cardinal teeth. Colour variable, most commonly brownish with irregular streaks and patches in brown, pink or violet. On the inside of the shells, the outer margins are finely granulated.

Habitat: Sandy and muddy bottoms from 2 - 200 metres depth.

Family Tellinidae

Clams in this family are characterised by their thin, flat, plate-like triangular or oval shells. The shell valves are similar, but the umbo is somewhat displaced towards the rear in most species. Each shell has two cardinal teeth and one lateral tooth on each side. Prominent mantle cavity. The animals dig into the upper layers of fine sand or mud.

Tellina pygmaea

The shells of this small bivalve often are found washed up on the beach. It is easily confused with other tellinids.

Distribution: Northeast Atlantic from western Africa to Troms in northern Norway.

Description: Up to 30 mm in length. Small, bean-shaped clam; up to only 10 mm in length. Distinguished from other members of the family by having a small lateral tooth at the anterior margin of the umbo on the left-hand shell; an essential character to separate it from the abundant ***Angulus tenuis***, which has the lateral tooth on the right-hand valve. The latter also has a weakly-defined keel. Umbo in *T. pygmae,* displaced further backwards than in most other tellinid shells. Can also be confused with ***Macoma balthica***, which is pink, yellow or white in colour. However, the umbo in *M. balthica* is located near the centre, the mantle cavity is very deep and the shell has weakly-defined concentric stripes. Further, *M. balthica* lacks lateral teeth.

Habitat: Shell-sand and mixed bottoms, deeper than 1 metre.

Family Psammobiidae - Sunset shells

This family is characterised by oval to rectangular, symmetrical shells, with the umbo located just in front of the center. Lateral teeth are absent, but each shell has one to three cardinal teeth. The shells are thin and flattened; in most species gaping at the posterior (rear) ends.

Gari fervensis

This species has a marked keel extending obliquely from the umbo to the outer shell margin.

Distribution: Widely distributed in the northeast Atlantic from the Canary Islands to northern Norway.

Description: Up to 50 mm in length. Bivalves within the genus ***Gari*** have very long, thin respiratory siphons; umbo located almost completely in the centre. Right-hand shell with 1 - 2 cardinal teeth; 2 - 3 in the left hand shell. Prominent concentric ribs, but most eyecatching is the marked keel extending from the umbo and diagonally down to the rear corner. The concentric ribs are crossed by radial ribs, which are particularly prominent in the rear parts of the shells and give a latticed appearance to the sculpturing. The front part of the shells are rounded, whereas the rear part is obliquely truncated. Shells gaping at the rear. Colour variable, often reddish, with darker concentric bands of colour and radial stripes. Shell insides pale pink to violet. With these characteristics, this species can hardly be confused with other bivalves. There are other species within this genus in northern European waters; ***G. depressa, G. costulata*** and ***G. tellinella***. These resemble *G. fervensis* but are distinguished by their lack of a clearly-defined keel, smoother shells and lack of lattice-like sculpturing in the rear parts. *G. depressa* can reach up to approx. 60 mm in length, whereas *G. tellinella* reaches up to 35 mm. The latter is more rounded at the rear and lacks a marked keel. The up to 25 mm long *G. costulata* has 12-20 strong, sharps ribs radiating from umbone to posterior margins of shell.

Habitat: Common on mixed substrates, often shell-sand. Occurs from the lower part of the tidal zone down to approximately 50 m.

Family Solenidae - Razor shells

Razor shells are easily recognisable by their very characteristic knife-shaped shells. The shells are open at both ends and left and right sides are similar. The umbo and ligament are markedly displaced towards the anterior end. These shells live in vertical burrows within sandy or gravelly sediments, from the lower shore and occasionally as deep as 100 m. They are anchored at the bottom by the foot which projects out of the shell, and at the top, the two short fused respiratory siphons barely project above the sediment. If disturbed, razor shells rapidly retreat down into their burrows by contracting the extended foot. The largest shells can retreat half a metre down into the sand. Razor shells are exploited for food and bait in certain parts of Europe, which has led to a marked decline in local populations.

Ensis arcuatus

Ensis arcuatus *has an almost straight dorsal side and a curved ventral side.*

Distribution: From Portugal to northern Norway. All around the British Isles.
Description: Up to 150 mm. Dorsal side almost straight, whereas the ventral side is evenly curving.
Habitat: Buried in sand or gravel, from the lower shore down to approx. 40 meters depth. ***Ensis ensis*** resembles *E. arcuatus*, but in *E. ensis*, both shell sides are curving. The even larger ***Ensis siliqua***, distributed from Norway to the Mediterranean and northwest Africa, is distinguished from both these by the dorsal and ventral margins being almost completely straight. Another razor shell in our waters is ***Solen marginatus,*** which resembles *E. arcuatus*, but is distinguished by a marked constriction or groove around the anterior part of both shells. The latter is recorded from southwestern Norway to western Africa. A common razor shell found between 15 - 100 metres depth is ***Phaxas pellucidus***, which reaches only up to 40 mm in length. In this, the ventral shell margin is curved and the dorsal margin is straight; notably more curving in the anterior region. The shells are greyish-white and partially translucent, and found at 15 - 100 metres depths. The American jack-knife clam, ***Ensis directus*** was introduced to European waters probably in 1978 through tanker ballast water. It is now distributed from Oslofjord in north to southeastern parts of England, and Belgium in the south, and is still spreading.

Ensis arcuatus. *Young individuals are much more colourful than older specimens.*

Family Myidae - Gapers/ soft-shell clams

Members of this family are characterised by their robust white shells, with an exterior yellowish-white to dark brown covering (periostracum). The respiratory siphons are large and fused, and in the largest individuals, cannot be withdrawn into the shells. They bury deep into the sand with the siphons extended up to twice the length of the shells, but barely projecting above the sediment surface. Shell valves similar, with the umbo somewhat displaced towards the rear. Posterior edges often widely gaping. The mainly internal ligament is supported by a chondrophore, usually prominent in the left valve. The palliate line is curved constructing a palliate sinus.

Mya arenaria - Sand gaper

The thick shells gape at one end.

Divers often see the respiratory siphons projecting from the sediment.

Distribution: From northern France to the Arctic. Circumboreal, all British Isles.

Description: Robust white shells; up to 150 mm in length. The left valve with a prominent projecting chondrophore. Resembles ***Mya truncata***, but both ends of the shells are rounded (*M. truncata* is truncated at one end). Can also remind of ***Lutraria lutraria***, but this lacks the characteristic chondrophore.

Divers can dig out **Mya arenaria** *by a vigorous waving of the hand.*

Habitat: Lives buried in sand, often muddy sand, from the lower part of the tidal zone and usually down to only a few metres depth. In the tidal zone, it buries down to approx. 30 cm into the sand, with only the siphons projecting from the sediment surface. Siphons surrounded by relatively long, light-sensitive tentacles. If disturbed, these are rapidly withdrawn into the sediment. The foot is small, and individuals do not move to any great extent.

Biology: Individuals can reach an age of 10 -12 years.

Commercial value: Commercially exploited, especially along the Atlantic coast of north America.

Mya truncata - Blunt gaper

Distribution: In common with ***Mya arenaria***, widely distributed in the northern part of the northern hemisphere. In Europe from the Bay of Biscay to the Arctic. All British Isles.

Description: Resembles ***Mya arenaria***, but distinguished by the truncated posterior, which is markedly gaping. Shells white in colour, with yellowish to brown external

'skin' (periostracum), which flakes off in dead shells.
Habitat: Similar habitat as ***M. arenaria***, but generally in deeper water, often between 10 - 30 metres.

The rear part of **Mya truncata** *is truncated, or 'cut off'. Here pictured with its long respiratory siphons.*

Family Hiatellidae

This group is characterised by very irregularly-shaped shells, often strongly gaping. In shallow waters, **Hiatella arctica** *is dominant, but there also are two other less common species.*

Hiatella arctica - Wrinkled rock borer

Distribution: In Europe from the Mediterranean to the Arctic. Common all around the British Isles.
Description: Shells irregularly rectangular in shape; rough surface, with prominent concentric ridges. Shells white in colour; gaping. Soft tissues larger than can be contained within the shells, such that some parts protrude. Umbo anteriorly displaced. Posterior, or rear part truncated; anterior part rounded towards the ventral side. Another species, ***Panomya arctica***, resembles *Hiatella arctica,* but is considerably larger, up to 80 mm in length. It can also be distinguished from *H. arctica* by the two abductor muscle scars being markedly different; the rearmost one is longer and thinner than the one in front of it. The cosmopolitan *Panomya arctica* lives in soft sediments down to several hundred metres water-depth. A third species, ***Saxicavella jeffreysi,*** only reaches up to 10 mm in length and can be distinguished by the dorsal margin being markedly oblique. The latter is found on sandy bottoms in most northern European waters, usually between 20 - 150 m depth.
Habitat: Hard substrates; often wedged into crevices in rocks. Very common in the spaces within the branched holdfasts of the cuvie kelp ***Laminaria hyperborea.***

Hiatella arctica *is often found wedged into rock crevices, between stones or among kelp holdfasts. The shells are irregular and basically white in colour.*

Class Cephalopoda - Squid, octopus and cuttlefish

We have all been impressed by tales and pictures of giant octopi (Gr. Kephale = head + Gr. podos = foot) up to 18 metres in total length. Specimens up to 10 metres long have been captured along the Norwegian coast. The cephalopods thus have the largest-bodied representatives among the invertebrates. Some giant octopi have been seen in battle with large toothed whales, particularly sperm whales, which are the only known enemies of adult giant octopi.

Sepietta oweniana. *We often see this squid in fjords during night dives.*

Cephalopods are highly-evolved molluscs, with a well-developed head with eyes, a radula and strong jaws. The body usually is long and more or less torpedo-shaped. The head, with the brain and mouth, is located on the ventral (front) side, and the dorsal side ends in a rounded posterior end. Most forms have 8 or 10 arms and there usually is a reduced shell inside the body. The body surface is formed by the mantle, which is folded on one side to form an internal mantle cavity. The mantle cavity has a funnel-like opening through which water can be ejected, thus propelling the animal through the water. Water enters the mantle cavity through gaps between the mantle margin and the head. A number of cephalopods also have fins, which are used for swimming backwards or forwards. Species which live in the water column have a more pronounced torpedo-shape than those with a more or less stationary existence on the bottom.

Cuttlefish have an inter-

Eggs from the squid **Todarodes sagittatus** *at around 60 m depth. Here attached to an anchor-rope.*

MATING LEADS TO AMPUTATION!

Cephalopods have separate sexes. In males, one or two of the arms, or tentacles, are transformed into a copulatory organ. Within this organ, sperm cells are stored in small vessels (spermatophores), which are placed inside the mantle cavity of the female. Once within the mantle cavity, the spermatophores release the sperm, and the female's eggs are fertilised. In some species, the outer part of the arm, containing the spermatophores, breaks off inside the female, leaving the male with a short stump. Previously these broken-off arm tips were thought to be distinct parasitic organisms within the female's mantle cavity, and were named ***Hectocotylys***. After the true connection was made, this name is now used to refer to the entire copulatory arms of the male. An arm modified in this way is said to be hectocotylised. Most cephalopods have a short pelagic larval stage, where the larvae look like miniature adults.

"INTELLIGENT"

The well developed brain of cephalopods is protected by a cartilaginous layer, forming a rudimentary cranium. The eyes are extremely well-developed and show many similarities with those of vertebrates. Studies in aquaria have shown the octopus to possess remarkable learning abilities. It is well established that octopi can select objects that give a reward in the form of food. Further, it has been shown that an octopus allowed to watch its peers conduct an exercise using coloured balls and rewards, will learn from the others experiences. If the observer octopus is given the same coloured balls, it will at once choose the one that contains the food reward. This shows that sight and colour discrimination are important in octopi. Picture: ***Eledone cirrhosa.***

nal shell, which is perforated by canals, which can be filled either with gas (N_2) or with water. Regulation of the relationship between gas and water allows the animal to adjust its swimming depth (as in the swim bladder of fish). The air-filled shells of dead/decomposed cuttlefish float to the surface and are often found washed up on beaches. Such shells usually belong to ***Sepia officinalis***, which has a rather southerly distribution, and is common around the British Isles, but rare in southern Scandinavia.

All cephalopods are predators. In some languages, cephalopods are known as "ink-squirts" because most forms have an ink gland that produces ink, which can be squirted out through the mantle cavity. This ink can confuse or blind its enemies. Not surprisingly, however, the ink sac is absent in deep-water cephalopods because colour tricks do not have effect in deep, dark waters. Another form of protection is the amazing ability for camouflage, by changing colour according to its background. Colour changes also occur in response to threatening or dangerous situations, and can take place in a matter of seconds.

The Cephalopoda comprises around 650 species, of which all are marine. Traditionally, the group is divided into two sub-classes and three orders. The first sub-class, Nautiloidea, comprises but one genus with living representatives, *Nautilus*, comprising five or six species from the Indian Sea and the Pacific. These have an external air-filled shell, several tens of tentacles and lack an ink sac. In members of the sub-class Coleoidea, the shell is absent, or reduced and enclosed by the mantle. This class comprises four orders, of which representatives from four are presented below.

Orden Sepiida

This is one of two orders of ten-armed cephalopods, whose members have a short body and fins along the sides. The shell is either absent or encased by the mantle. There are 8 short arms and two long tentacles than can be retracted. The two tentacles have suckers at their tips only, and lack hooks. The other order of ten-armed cephalopods, the order Teuthida, comprises forms with an elongated, torpedo-shaped body. Also these have lateral fins, eight arms and two long tentacles, but these are not retractable. The suckers are often equipped with hooks. Within this latter order is the squid, **Todarodes sagittatus**, *which migrates in large shoals towards the shores during autumn/winter in search of fish, such as herring. Squid is a popular European dish, but in Norway it is mostly used as bait, although the liver is salted and eaten.*

Sepietta oweniana

Sepietta oweniana *resembles* **Sepiola atlantica**, *but has a more elongate body. The outer parts of the fins in* Sepiola *are semicircular, not pointed as here.*

Distribution: Widely distributed in the northeast Atlantic from western Africa to northern Norway.
Description: Total length up to 80 mm. Body longer and more slender in appearance than the next treated species, ***Sepiola atlantica***. Lateral fins (along the sides) relatively short and usually rounded. Mantle margin fused with body wall for a short length between the eyes. The eight arms are of equal lengths; the two tentacles much longer. The two most dorsally-situated arms are partially fused, forming a membrane between them. The free ends of these two arms are therefore shorter than those of the other arms. Colour typically yellowish-brown with reddish-brown pigment patches (chromatophores).
Habitat: Primarily in deep water, usually between approx. 20 - 200 m. Often observed resting on soft bottoms, or swimming just above the sediment.

Sepiola atlantica - Little cuttlefish

Distribution: In the northeast Atlantic found from Morocco to Iceland and mid-Norway. All around the British Isles.
Description: Small in size; mantle up to 20 mm in length. Body rather squat, rounded, and with two prominent rounded, almost circular side fins. Fins thin and almost transparent, each approximately as wide as the body. Eight short arms and two long contractible tentacles. Arms with two to eight rows of suckers. More or less transparent body with red pigmentation. Ventral edge of mantle margin (opposite side to the eyes) with straight, thickened edge. Mantle margin fused with the head between the eyes. In males, one arm is modified into a copulatory organ. Light-sensitive organs present. Can be confused with the previous species, ***Sepietta oweniana,*** but this has a more elongate shape (see description above).
Habitat: Around the British Isles, individuals of this species have been found in very shallow water; even in tidal pools. Often caught in shrimp trawls together with ***Rossia macrosoma.*** Can partially bury into sandy sediments.

Biology: Mainly feeds on shrimps and other small crustaceans. Eggs large, with parchment-like outer layer, often laid on, or in, sponges or empty bivalve or gastropod shells.

Sepiola atlantica, *which appears to have fewer, and more dispersed chromatophores than* **Sepietta oweniana.** *Also, the body is shorter.*

Rossia glaucopis

Distribution: In our waters there are two species within this genus; ***Rossia macrosoma*** and ***R. glaucopis.*** *Rossia macrosoma* has a southerly distribution; from the Azores and Mediterranean north to mid-Norway and all British Isles, whereas *R. glaucopis* has a more northern distribution; along the Swedish western coast and the entire Norwegian coast.

Description: Both these species have a mantle up to 80 mm in length, which comprises around one third of the total length. Reddish-brown in colour. Mantle margin not fused to the body between the eyes. This distinguishes these two species from the other two, considerably smaller species ***Sepiola atlantica*** and ***Sepietta oweniana. R. macrosoma*** is distinguished from *R. glaucopis* by the latter having two rows of suckers along the inner side of the arms. *R. macrosoma* has four rows of suckers per arm. Further, the mantle margin of *R. macrosoma* is said to be folded to form a 'lip'; a feature absent in *R. glaucopis*. The latter has small triangular white papillae on the upper side of the head and mantle; most prominent between the eyes.

Habitat: Both species normally live in relatively deep water; *R. glaucopis* between approx. 50 - 600 metres; somewhat deeper than *R. macrosoma.* Usually occurring on soft bottom sediments, often appearing as by-catch in shrimp trawls.

Orden Octopodida - Octopus

Octopi lack an internal shell and the mantle also generally lacks fins. As the name suggests, there are 8 arms and no tentacles. Most seek refuge on the bottom and are poorer swimmers than their ten-armed relatives, the squids. They use the arms to creep along the substrate, but a number of octopi can move relatively quickly through the water by means of rhythmic contractions of the mantle cavity. Only one representative is found in our waters, which can be met occasionally in relatively shallow water.

Eledone cirrhosa - Curled octopus

Eledone cirrhosa *looks like the classic cartoon octopus. Usually reddish or yellowish brown.*

Distribution: Distributed throughout the northeastern Atlantic, lacking in the Mediterranean. All around the British Isles.

Description: Total length up to 500 mm, with an arm-span up to 700 mm. Arms long and thin, about twice as long as the body, usually inwardly curled while resting on the substrate. Each arm bears one row of suckers, whereas the common octopus, ***Octopus vulgaris*** bears two rows of suckers on each arm. The latter is distributed from the south coast of England and the southern North Sea coast to the Mediterranean and South Africa. Colour usually yellowish-brown, with reddish-brown patches. Possibility for large colour variations by means of the pigment cells (chromatophores) dispersed across the body, which can be contracted or expanded by muscular action.

Habitat: Usually lives on hard bottoms. Creeps along the substrate and swims only when necessary. Uses extremely effective camouflage, therefore probably often overlooked. Common.

Biology: Diving with this octopus is a remarkable experience. Its large eyes follow every movement and one really feels under scrutiny. It will react with a violent display of colour if divers come too close, and if it feels sufficiently threatened, it will squirt a good portion of poison at the offender.

Commercial value: Very tasty!

Class Scaphopoda - Tooth shells

The elongate form of tooth-shells comprises an incisor-like shell open at both ends. The shell usually is white and gently curving. The head and short cylindrical foot extends through the widest opening of the shell. The animal is orientated head-down in the sediment, with the narrowest opening barely projecting above the sand. Prey orgamisms are caught from around the buried head-end, using the tentacles around the head. Gas exchange is carried out without gills, and tooth shells also lack a partitioned heart. Sexes are separate, and there is a free-swimming larval stage lasting for around 5 days. Adults live buried in soft bottoms from around 5 - 4000 metres depth. Found exclusively in seawater, with a salinity greater than 30 ‰.

Antalis entalis

Tooth shells are difficult to identify without microscopy. Based on its habitat and the smooth surface, this probably is **Antalis entalis.**

Distribution: Widespread around large areas of the northeast Atlantic, from the Arctic to Portugal. Common.

Description: Up to 40 mm in length, with a maximum width of 5 mm at the head end. Shell smooth, white in colour and gently curving towards the narrowest end. This characteristic group of molluscs can hardly be confused with other groups, except for the calcarious tube-building serpulid fanworm ***Ditrupa arientina***. However, tooth shells can be distinguished from these by the wide opening at the head end of the shell, which is narrowed in the fanworm. Easily confused with the other 9 species found in our waters, but most of these live in deeper waters. Encrusting serpulid fanworms often settle on the shells. Empty shells are often green in colour and the echiurid ***Phascolion strombus*** often takes up residence inside them.

Habitat: Fine, often clayey sandy bottoms from 1 - 3200 metres depth; in our waters usually between 20 - 90 metres depth. Usually 2 - 4 individuals per m^2 (but up to 30 per m^2).

Order Phoronida

***Phoronids** (Gr. Mythical Phoronis: another name for Io, Inachos' daughter, who was turned into a white heifer by Hera) are very small, bilaterally symmetrical, worm-like, unsegmented and rather curious marine animals.*

The body itself consists of an elongated 'tube' ending in a mouth opening encircled by ciliated tentacles, which form a horseshoe-shaped lophophore (hence its common name). The gut forms a U-shaped passage within the body, with the mouth opening at the same end as the anus. The animals build tubes completely encasing the long body. The tubes are often buried within the sediment or the animal bores into chalk rocks (or other calcareous substrates such as mussel shells), such that only the anterior part of the animal projects above the substrate. Can be hermaphrodites or separate sexes. Reproduction both sexually, via planktotrophic, so-called **actinotrocha** larvae, or also asexually, usually by budding. Phoronids are filter-feeders. The phylum comprises but a single family; the **Phoronoiidae**, which contains two genera and around 20 species world-wide. In our waters, there are at least 4 species, but their distribution is poorly known. Due to their boring activities, phoronids play an important role in breaking apart chalk and calcareous structures, including dead shells, thereby facilitating the further breaking-down processes by other organisms.

Phoronis hippocrepia

These small creatures are found in very shallow water. Here photographed in Jøsenfjorden.

Distribution: Widely distributed in the northeastern Atlantic.

Description: Tubes up to 10 cm in length and around 1 mm in diameter. The tip of the tube, together with the horseshoe-shaped lophophore projects above the surface. Lophophores with between 50 - 150 ciliated tentacles, each up to 3 mm in length. Body tube yellowish-white, with the dark gut clearly showing through it. Lophophore white. A similar species, ***P. ovalis,*** is less common, but registered sporadically in northern European waters. This also bores into calcareous structures, but is distinguished from ***P. hippocrepia*** by having an oval-shaped lophophore and only up to 20 tentacles. The latter favours deeper water, between 20 - 50 metres.

Habitat: Bores into calcareous material, from the tidal zone down to approx. 50 metres (most common shallower than 10 metres).

Brachiopods had their peak in diversity around 300 million years ago.

Phylum Brachiopoda

*Brachiopods resemble bivalves and, in common with the moss animals (phylum **Bryozoa**), they can be regarded as 'living fossils'.*

WRONGLY NAMED...

The name Brachiopoda (Gr. brachion = upper arm, Gr. podos = feet), is in fact the result of a misunderstanding. In 1805, the french paleontologist Cuvier thought that the arms (lophophores) could be extended from the shell opening and used for locomotion.

OLD GROUP

Brachiopods were in existence already 600 million years ago, and reached their peak around 300 million years ago. At that time, no less than 30 000 species existed. Fossil shells of some of these have been found to be up to 40 cm long. Now they have been outcompeted by bivalves, and are much less abundant or diverse (around 300 living species), most of these live at intermediate depths on continental shelves.

Brachiopods are most commonly seen below 30 m depth. Here **Terebratulina retusa**.

Externally, brachiopods show a high resemblance to bivalve molluscs, but in terms of evolutionary relationships, they are not close at all. Like bivalves, they have two shell valves, and some attach themselves to the substrate by a fleshy **stalk**. Others directly attach to the substrate while still others bury. The shell halves are called **ventral** and **dorsal** shells (compare with left and right shells in bivalves). The shells are always dissimilar; the ventral shell is larger than the dorsal one. In most species, both shells are convex, and many have an 'underbite'–the lower shell can protrude beyond the upper shell, forming a beak. In stalked species, the dorsal shell often faces the substrate, whereas those directly attached have their ventral shells attached to the substrate. The main part of the shells is filled by two spiralling crowns of 'arms' (**lophophores = tentacular** crowns), equipped with rows of cilia. The cilia create water currents bringing food particles towards the shell openings. The tentacles extend beyond the shell opening and food particles are caught by their mucus, and passed unto the mouth by means of a special furrow. The mouth ends in a '**gullet**' which extends to form a **stomach**. A rudimentary heart lies above the stomach and there is an open bloodstream.

They usually attach to steep rocky faces, either via a fleshy stalk, or by cementing the lower shell to the substrate.

Most brachiopods have **separate sexes**, use **external fertilisation** and produce ciliated **planktonic larvae**. In our waters, the planktonic phase lasts from a few hours to several days. The life expectancy of adults is from around 20 months to 10 years. There are more then 10 species recorded in northern European waters within two classes. Most of these are relatively small; only a few millimetres in length.

Class Inarticulata

Members of this group (Gr. In = not + Gr. articulatus = jointed) lack a hinge- the shell halves are held together purely by muscles. A calcareous skeleton (lophophore skeleton) is absent. Only a single species occurs in our waters.

Crania anomala

A common species, usually below 30 m depth. The vestigial ventral shell is attached directly to the rock surface.

Distribution: In Europe from the Canary Islands to the Arctic.

Description: Up to 15 mm in diameter. Vestigial ventral shell thin, stalk absent. Dorsal shell almost square in outline and domed. Colour brown, often with a darker spiralling stripe from the centre. When feeding, stiff bristles can be seen emerging from the shell margins. Can be confused with saddle-oysters (bivalve family **Anomiidae**), which also attach to the substrate. However, these are almost circular, not squareish, as here

Habitat: Attached by its ventral shell to rocks, stones or other shelled animals, usually between 15 - 80 metres depth. Very common, but often overlooked due to its habitat and small size.

Class Articulata

Members of this group (Gr. articulatus = jointed) have hinge teeth on the ventral shell, and most have a calcareous skeleton. All occur as epifauna (living on the substrate, not buried).

Terebratulina retusa

The shells are often covered by a spongy layer. Also visible are three arms of the brittle star **Ophiothrix fragilis.**

Distribution: In Europe from the Mediterranean north to northern Norway.
Description: Up to 33 mm in length. 'Classic', almost pear-shaped appearance, with two white convex shells. Larger ventral shell with a small hole (**umbilicus**) near the hinge, from which a small fleshy stalk extends and attaches the animal to the substrate. From the umbilicus and outwards are 11 - 14 longitudinal stripes. The deep-living unstalked brachiopod, ***Terebratulina septentrionalis,*** is similar, but more densely striped (16 - 21 stripes). When alive, the shells are covered in a spongy, proteinaceous layer, giving it a rough appearance. Can also be confused with another species common along our coasts, but this has completely smooth shells and is yellow/ yellowish-brown in colour.
Habitat: Usually on relatively steep rock faces below 30 metres depth. Recorded depth distribution 15 - 1500 metres.

Phylum Bryozoa - Moss animals

Bryozoans (Gr. bryon = lichen, moss, seaweed + Gr. zoon = dyr) are evolutionarily very old marine animals, which usually form colonies consisting of small zooids, each individual usually less than 0.5 mm in length. Almost everyone who has been on the shore has seen these animals, but perhaps without being aware of what they are. Bryozoans are almost always found on the blades, fronds or stipes of kelp and seaweeds, especially older plants. They are visible as encrusting patches with a fine mesh-like structure, often white in colour.

Colonies are founded by a single individual (**ancestrula**), which was formed by sexual reproduction. Colonies grow from this individual by asexual budding. Therefore, all individuals within the same colony have an identical genetic composition, and often are identical in appearance. Each zooid secretes a small surrounding 'box' (**zooecium**), which usually has a lid (**epistome**). This 'box' is made of horny or calcareous material and protects the soft tissues inside. Bryozoan individuals have a ciliated crown of tentacles (**lophophore**), consisting of a ring of feeding tentacles surrounding the mouth with which they catch food particles. The tentacles are arranged such as to form a funnel down to the mouth. Rhythmic movements of the cilia cause plankton and particles to be passed through the lophophore. The mouth is connected

The shape of bryozoan colonies varies greatly. This type of bryozoan forms hard, coral-like colonies. Pictured is **Porella compressa.**

SPECIALISED INDIVIDUALS

In many species, certain zooids have specialised functions within the colony. Some are specialised for defence (**avicularia**). Avicularia lack a feeding lophophore, but the zooecium and its 'lid' form a beak-like structure. In such colonies, food particles are transported between individuals, so that also those without lophophores can feed.

LIVING FOSSILS...

These little animals already existed 600 million years ago, and have not changed notably during the last 300 million years.

Omalosecosa ramulosa *is common on rocky substrates. Here the feeding tentacles (lophophores) are clearly visible as 'down' on the colony surface.*

to the stomach and a gut that ends in an anal opening at the basis of the lophophore. Organs for excretion, circulation and respiration are absent (with some exceptions); the small size of the zooids makes **diffusion** sufficient for transporting substances.

Most bryozoans are **hermaphrodites**, producing both eggs and sperm. In many species, pelagic larvae are released, which feed on phytoplankton, until they **metamorphose** and found new colonies.

Colonies often consist of several thousands of individuals. Some form a membrane-like crust on suitable substrates, whereas others form erect, bush-like colonies. Some colonies are hard, almost coral-like, whereas others are soft. In a few species, the colonies have such characteristic forms that they can easily be recognised ***in situ*** or from photographs. However, in most cases, a dissecting microscope is needed to examine the individual zooids. For this reason, many of the photographed bryozoans presented here are not identified to species.

Also red algae eventually can become completely covered by bryozoan colonies. Here **Electra pilosa.**

Bryozoans are permanently attached to the substrate during their entire adult life, although some encrusting colonies actually can 'glide' across the substrate. Colonies are found on a wide range of substrates, from pier pilings, kelp and other algae, crustaceans and snail shells to stones and rock surfaces. In hard-bottom areas where there is insufficient light for algae, the sea floor can be covered by bryozoans.

It is a diverse group, with more than 4000 recorded species within approx. 1200 genera, divided into three classes, of which one (**Phylactolaemata**) comprises fresh-water species. The two other classes comprise around several hundred species recorded in northern European waters. The most common forms encountered in shallow waters are presented here.

Class Stenolaemata

This group (Gr. stenos = tight, squeezed + Gr. laimos = throat + Lat. -ata = equipped) comprises exclusively marine bryozoans, with cylindrical zooids and circular tentacular crowns. Each zooid is enclosed in a ***tube-shaped, calcareous*** *'box' (****zooecium****), usually covered by a thin membrane with a central opening, the orifice. The calcareous 'boxes' make the colony hard. Some colonies are bush-like, whereas others form partially erect, plate-shaped colonies. Some of the 'boxes' are specialised brooding chambers (****gonozooids****), and these are enlarged, often pear-shaped. In our waters, most members of this class belong to the families* ***Crisiidae*** *and* ***Tubuliporidae.***

Crisia eburnea

Distribution: Widely distributed in the north-east Atlantic. Mediterranean, western Baltic, Iceland, Arctic, all Norwegian and British coasts.

Description: Bryozoans within the family Crisiidae, of which this species is a member, form white, erect colonies, comprised of thin branches. Colonies of the pictured species up to 20 mm high. Branches formed of tube-like structures arranged in alternating pairs, and characteristically curved inwards. Directly beyond the point of branching are often enlarged brood chambers (gonozooids). Branches with gonozooids usually carry between 8 - 12 zooid pairs, whereas in branches without gonozooids this number varies; usually between 5 - 7.

Habitat: Found on almost all types of hard or firm substrates, including plants and animals, cliffs, stones and pier pilings. Very tolerant; found from 0 - 300 m depth, from hypersaline to brackish waters.

The latin species epithet eburneus is derived from the word for ivory, referring to the whitish colour of this bryozoan. The best macroscopic character for identification is the characteristic inward-bending of the branches.

Complicated species identification

Species determination of many bryozoans usually requires study under a dissecting microscope, although very high quality photographs can suffice in some cases. For the colony pictured here, we are not completely sure of its species identity, because we lack details of the structure and positioning of the individual zooids. However, it is almost certain that it belongs to the class ***Stenolaemata****. Further, we believe it to be a member of the family* ***Horneridae****, possibly* **Hornera lichenoides**. *This species is normally found deeper than 50 metres, although it has been recorded as shallow as 3 metres. The colonies are white, and reach up to 65 mm in height. It is distributed from the Bay of Biscay to the Arctic. Here from Egersund, southwest Norway, at 40 metres depth. Within the branches we can see the arms of the brittle star* **Ophiopholis aculeata**.

FURTHER READING...

Further information on most Scandinavian bryozoans can be found in "Danmark's Fauna" (1940), volume 47. This volume is devoted to the Bryozoa, with all of 401 pages on the subject.

These individuals within the class Stenolaemata, in common with the species on the previous page, belong to the suborder ***Tubuliporina****. This bryozoan occurs frequently in shallow waters at exposed localities and often is attached to the kelp cuvie,* **Laminaria hyperborea**.

Class Gymnolaemata

This group(Gr. gymnos = naked + Gr. laimos = throat + -ata = equipped) is the largest of the two classes of marine bryozoans, and includes most of the species in our waters. The zooids can be box-shaped, sack-like or cylindrical, arranged with their longitudinal axis more or less parallel to the direction of the colony's growth. The exterior 'boxes' are membranous, gelatinous or partially calcified, but never tube-shaped as in the class ***Stenolaemata****.*

This unidentified species forms a characteristic, often highly colourful crust on kelp holdfasts and various hard substrates. Colonies are often violet, white and/or orange. The zooecia ('boxes') have oval openings considerably smaller than the total surface of the 'box'.

Membranipora membranacea - Sea-mat

Many bryozoans typically form characteristic white, mesh-like mats on substrates including algae. Here the commonest of the encrusting forms, **Membranipora membranacea.**

This species is particularly common on kelp blades, but also on other algae. The zooid 'boxes' are rectangular.

Distribution: Widely distributed in the north-east Atlantic, including all around the British Isles and the Norwegian coast.

Description: Colonies of this species form white, lattice-like crusts on brown algae, primarily kelp (***Laminaria spp***.). The individual 'boxes' are rectangular, arranged side-by-side in long rows, and with a regular displacement of the rows (like bricks in a wall). Some specialised zooids, the avicularia, have beak-like outgrowths. Colonies can reach up to 1.5 m in length and 20 cm in width. It usually is possible to identify using the above characteristics.

Habitat: Grows mainly on the kelps cuvie and tangle, ***Laminaria hyperborea*** and ***L. digitata***, respectively, but can occur on other brown algae. Occurs both on the blades and stipes. Very common; found down to 500 metres depth. Tolerant of polluted water.

Biology: Colonies have a fast growth rate of several mm per day, in response to grazing by various gastropod snails and particularly sea-slugs.

Colonies can cover entire kelp blades.

Electra pilosa - Hairy sea-mat

Distribution: Widely distributed from the Arctic to south Africa. Entire Norwegian coast and British Isles.

Description: Also this species forms crusts on algae, however mostly on other species than ***M. membranacea***, but also on hydroids, other bryozoans, polychaete worms and various shellfish. Zooids rounded, squarish or pear-shaped. Very wide opening for the tentacular crown (lophophore), extending over ¾ of the length of the 'box'. Further, there are relatively large pores in the calcareous part in front of the opening. In some colonies, the 0.5 mm long zooids have two relatively thick spines at one end and a prominent, often long spine at the other, as well as two small spines around the edge of the opening. Can be confused with other species. Colour silver-grey. Colonies with very characteristic star-shape when growing on algae.

Habitat: Forms encrusting colonies on epiphytous red algae (red algae attached to the stipes of brown algae, especially the cuvie, ***Laminaria hyperborea***). Can eventually outgrow the epiphytes to form fringe-like colonies.

Electra pilosa *is especially common in exposed areas, where it covers epiphytic red algae growing on cuvie stipes.*

Above: *Lophophores extending from the 'boxes' (see lower part of picture).*
Right: *The species sometimes grows on brown algae, where it forms characteristic star-shaped colonies.*

Flustra foliacea - Hornwrack

Distribution: From the Bay of Biscay north to the Arctic. Kola coast, White Sea. All Norwegian coast and British Isles.

Description: Forms up to 200 mm high, branching, leaf-like colonies. Branches with a layer of zooids on each side, and broader towards the outer parts. This distinguishes it from a similar species, ***Securiflustra securifrons***, whose branches have a constant width towards the tips. Colonies grey-white to brownish-grey. Each zooid with four or five short spines at the edge of one side of the opening. 'Boxes' pear-shaped, not angular. In between the many identical zooids, are a few transformed zooids, avicularia, which have a beak-shaped lid and serve to protect the colony. Colonies have a characteristic smell; some claim a rosemary scent whereas others detect a lemon smell.

Characteristic bush-like colonies with very broad, leaf-like tips.

Detail of the zooids.

Habitat: Often forms continuous carpets on cliffs and overhangs; also found on gravel. Found from 2 - 220 metres depth. In shallow water, as for some sponges and other bryozoans, most often found in shaded places.

Biology: Colonies can reach at least 12 years of age.

Securiflustra securifrons

Colonies are strongly branched and, unlike the preceding species, has only a single layer of zooids. Branches of similar width towards the tips.

Distribution: Widely distributed in the northern part of the Atlantic. Mediterranean, Arctic. Norwegian coast and British Isles.

Description: Resembles the previous species, ***Flustra foliacea***, but distinguished by the zooids not being pear-shaped, but rectangular. Each 'box' approx. four times as long as

wide, and arranged in a regular pattern. Colony branches of even width towards the tips. Colonies slender; up to 150 mm in height.
Habitat: Attached to hard substrates, from the tidal zone down to around 500 m.

The individual 'boxes' surrounding each of the zooids are rectangular. On this colony is a juvenile common starfish **Asterias rubens.**

Caberea ellisii

Distribution: Northern distribution; from the Arctic south to France. Southwest to northern Norway. In British waters there are records from Northern Ireland, west coast of Scotland and south coasts of England.
Description: Forms characteristic, rust-brown or greyish brown, fan-shaped colonies. Branches relatively slender, rounded and often prominently forked at the tips. 'Boxes' almost rectangular, with a large elliptical opening; arranged in 2 - 4 rows of alternating orientation. Zooids form a single layer, with all the openings on the same side, such that the branches have a 'right' and 'wrong' side. Colonies reaching up to 40 mm in height. Very characteristic habitus (appearance), such that photographs are sufficient for species identification.
Habitat: Depth distribution 10 - 300 m. Usually deeper than 30 m, found attached to rock and stones.

This species forms very characteristic rust-brown, fan-shaped colonies and is unlikely to be confused with other bryozoans.

Porella compressa

Distribution: Northern distribution, in northern Europe from the Bay of Biscay north to the Arctic. Swedish west coast. Probably also entire Norwegian coast. Distributed along the west coasts of Ireland and Scotland.

Description: Colonies hard, bushy in appearance, with somewhat flattened branches arising from a short, cylindrical central base. Bushes mainly arranged in a single plane, such that the colonies form fan-like structures. Up to 80 mm in height, approximately similar width. Base diameter approx. 7 - 9 mm; branches approx. 2 - 4 mm in width, blunt at their tips and comprised of many irregular rows of zooids. Colour variable, often yellow or orange; can also be lightly flesh-coloured. When brought to the surface, it quickly loses its colour, becoming flesh-coloured. A number of other species form similar colonies, such as ***Tessarodoma boreale***, but this is almost white, the zooids on each branch have only four, but regular longitudinal rows (seen under magnification) and the branch tips are narrowed. The outer, new-formed parts of the branches are smooth, whereas older parts are rougher, due to the transformed, enlarged zooids. A close relative, ***Porella laevis***, is distinguished from this by, amongst others, having cylindrical, not flattened branches.

Habitat: Found at 7 - 323 metres depth. Often attached to rocky walls, with the stem at right angles to the substratum.

This bryozoan is reminiscent of a small coral, due to its shape and very hard branches.

Reteporella beaniana (= Sertella beaniana)

Distribution: Widely distributed in the northeast Atlantic north to the Arctic. Entire Norwegian coast and British Isles.

Description: Very characteristic colonies; hard, coral-like, primarily white, leaf-like with perforations. Some attractively violet coloured. Colonies up to 30 mm high; and 50 mm broad; often shaped like a funnel or beaker, with a more or less waved edge. Colonies extremely brittle and supported by a short stalk. There are at least three species in our waters, and certain identification requires detailed examination of the zooids.

Habitat: Often found in exposed, current-rich hard-bottom localities deeper than 20 metres, but sometimes shallower. Recorded down to 1000 m depth.

This bryozoan forms very characteristic, leaf-like, perforated colonies (see also photograph at page 392. To the left of the centre, a juvenile of the same species has attached itself.

Omalosecosa ramulosa (= Cellepora ramulosa)

The colonies are small. This one is attached to the tube of a polychaete worm.

Distribution: Southern distribution, recorded from the northern African coast, Madeira, north to northern Norway and British Isles.

Description: Colour pale yellowish-white; hard. Colonies rising from an attached crust, often with numerous, cylindrical, irregular branches. Branches becoming narrower toward the tips. In the older, inner parts of the colony, the barrel-shaped zooids are protruding and irregularly arranged. In the younger, outer parts, the zooids are more regularly arranged, and less prominent. The largest recorded colony is 76 mm in height.

Habitat: Found on hydroids, algae, stones and rock faces. Below 14 metres down to 400 metres.

ECHINODERMS

Phylum Echinodermata - Echinoderms

*Echinoderms (Gr. ekhinos = hedgehog + Gr. dermatos = skin) are exclusively marine animals with pentamerous radial symmetry and a skeleton of calcareous plates. They lack segmentation and they have no head. The morphology is complex, with a 'true' body cavity (**coelom**), and a body composed of three basic layers. Virtually all members are benthic in their adult stages, either, like feather stars (class **Crinoidea**), living attached to surfaces, or, like starfishes (class **Asteroidea**), brittlestars (class **Ophiuroidea**), sea urchins (class **Echinoidea**) and sea cucumbers (class **Holothuroidea**), crawling on hard bottoms or on and in soft bottoms.*

Echinoderms have a unique hydraulic vascular system (**the ambulacral system**) that regulates the tube-feet (**ambulacral feet**). The ambulacral liquid consists mostly of seawater. Apart from regulating movement, it transports nutrition and excretory products, oxygen and carbon dioxide. It shows many similarities to our own blood system, and contains amoeboid cells similar to our leucocytes. These **coelomocytes** are used both for excretion, wound healing and various reparations, including regeneration of body parts. The echinoderm skeleton consists of plates of calcium carbonate, providing a tough and hard body surface in most members. It is hollow and appears as a meshwork, thereby decreasing the weight without loss of strength. Echinoderms have a special nervous system, and lack a differentiated head and brain. Nerves extend along the arms, near the ambulacral canals, and join together in a nerve ring that surrounds the mouth. The different parts of the animals are coordinated from this ring. Sense organs are generally not highly developed. Apart from balance organs (**statocysts**) in some sea cucumbers, simple eyespots at the tip of the arms in some starfishes, and smell/taste organs on some tube-feet in sea urchins, they seem to lack specialised sense organs. However, they do have dispersed sense organs on the body wall, which react to touch and smell.

The common starfish, **Asterias rubens**, *is probably our most widely known starfish. Here feeding on blue mussels.*

Most echinderms have separate sexes and external fertilisation. Eggs and sperm are simply discharged in the water, events that are regulated by water temperatures and chemical signals between the animals. In most echinoderms the fertilised eggs will develop into pelagic larvae, and go through a series of larval planktonic stages before settling and metamorphosing into small adults. Brooding occurs in some cold-water species. Echinoderm larvae are highly characteristic, bilaterally symmetrical, and often provided with long spines. The phylum is subdivided into three subphyla and five classes. The sea feathers belong to **Crinozoa**, starfishes and brittlestars to **Asterozoa**, and sea urchins and sea cucumbers to **Echinozoa**.

Class Crinoidea - Feather stars and sea lilies

I northern European waters there are less than 10 different species, mostly feather stars (fam. ***Antedonidae****). Feather stars are strikingly beautiful echinoderms with five principal arms, each of which branches into two near the basis. The arms are thin, rather long and feather-like, and carry a large number of short appendages,* ***pinnulae****. In some taxa the arms do not subdivide, in others there are a larger number of arms. Crinoids differ from other echinoderms in having both mouth and anus situated on the upper side of the* ***central plate****. From the underside of this plate there is a stalk, which distally is provided with 'feet' (****cirri****). These are simple, unbranched extensions that are used for attachment. Crinoid tube-feet serve both for gas exchange and food collecting. They are provided with glands that produce mucus and serve to trap drifting food particles. These are then transported in longitudinal furrows along the arms, situated in between the tube-feet rows. Crinoid tube-feet, in contrast to other echinoderms, lack distal suckers and are not used for locomotion.*

The larvae develop from attached eggs, or from eggs that are released into the water, and obtain their nutrition from the yolk. After a few days in the plankton, the larvae settle on the bottom. Initially, they will be completely sessile, whereas adults are partly sessile and attached with the cirri, but can also let go and swim with long, elegant arm strokes.

Antedon petasus

Distribution: West and northern coasts of the British Isles, up to Iceland, the Faeroes, and northern Norway.
Description: Ten arms, evenly distributed in a ring around the central plate, often with relatively large red and white spots (picture p. 406), or one-coloured brown, rose, orange, yellow, deep purple or dark pinkish-red. Number, form and length of cirri are useful features for identification. ***Antedon petasus*** has 50-100 articulated, long (usually 70-100 mm) cirri. The similar-looking ***A. bifidus,*** instead has 25

Above and below: **Antedon petasus** *is a common crinoid, or feather star.*
Right: *The cirri ('feet') are used in identification.*

***Above:* Antedon bifida.**
Inset: *In feather stars, the mouth faces upwards.*

shorter (usually 50 mm) cirri, is generally less neat in appearance, and has arms of unequal length. Both species have cirri with 15 segments.

Habitat: In sheltered localities from approx. 20 metres and down to several hundred metres depth. ***Antedon bifida*** tends to occur in more shallow water.

Class Asteroidea - Starfish

Starfish, or sea stars, (Gr. aster = star + Gr. -oideos = kind/sort), with their characteristic shape and wide distribution, are among the best known animal groups in the sea. There are more than sixty species recorded from the area, including a number of deep-water and Arctic ones. Most are flattened, star-shaped echinoderms with five arms. They can be separated from brittle stars in that the arms are much broader at their point of insertion, and in lacking a distinct delineation between arms and central disc.

The tube feet are directed downwards. They are impressingly powerful and are, among other things, used to open mussels. **Marthasterias glacialis.**

The mouth, as well as the **tube-feet**, are situated on the underside of the central plate. The tube-feet have suckers, and sit in rows of **two** or **fours** in a furrow along the underside of the arms. A **filtering plate** (madrepore plate - p. 423 is situated on the dorsal side of the central disc, displaced to one side. It communicates with the **ambulacral system**, and it is also likely that it helps the animals in their orientation. The arms contain **gonads** and **blind sacs** of the gut. The anus, when present, is situated dorsally on the central plate. Some species (such as ***Luidia*** spp.) lack both blind sacs and anus, such that mouth and gut represent the complete digestive system. The tip of each arm is provided with a nerve, ending in a red/violet spot that constitutes a simple eye with a lens. The dorsal side of starfish is covered by a skeleton of **calcareous plates**. These usually have **spines**, and are sometimes arranged in rings, then referred to as **paxillae**. Between the spines there are fine **extensions**, **papulae**, serving as **gills**. In some taxa, such as ***Asterias***, the papulae are very dense, giving a furry appearance to the skin. Different starfish are separated by the shape of their arms and the presence of various kinds of calcareous plates and spines. Some of our starfishes, such as ***Leptasterias***, ***Solaster*** and ***Henricia***, brood their young and have direct development; other taxa have pelagic larvae that go through a series of different planktonic stages.

The common starfish, **Asterias rubens**, *has a varied diet.*

The majority are predators, and several feed on other echinoderms, including starfishes. Many prey on bivalves, although snails and polychaetes are also common food items. The prey can even include animals that are larger than themselves. Starfishes have excellent regeneration capacities, and can let go of arms if damaged or under threatening conditions. Arms will regenerate as long as the central plate remains, and some tropical species can even regenerate from single arm pieces. In several taxa, especially ***Solaster*** and ***Crossaster***, the central plate can subdivide into two, and thereafter regenerate both the missing side and arms.

Astropecten irregularis

Astropecten irregularis *usually is found on sandy bottoms, but occasionally is seen on coarse shell sand or gravel.*

Distribution: From Morocco and the Mediterranean to northern Norway. Common.

Description: A stiff, flattened and highly characteristic starfish. The five arms are short and provided with two series of large lateral plates, of which one can be seen from the dorsal side. Each plate on the underside of this upper row is provided with a short spine, whereas in the lower row there are 4-5 longer spines per plate. The animals are pale yellowish, often with violet arm tips and a violet spot in the middle of the central plate. The upper side is near smooth and leather-like. The diameter is up to 200 mm. It is unlikely to be confused with other species.

Habitat: Often partly burrowed in fine sand or muddy sand, and emerging during the night. Usually occurring at depths of 10-30 metres, although there are records down to 1000 metres.

Biology: The prey consists of various benthic animals, with a preference for bivalves, other starfishes, and brittlestars, all of which are consumed in a single piece.

The individual is in the process of eating a small feather star, **Virgularia mirabilis**. *From Sognefjorden, south-western Norway, 20 m depth.*

Luidia sarsi

Luidia sarsi. *This starfish tends to break apart if handled.*

Distribution: From Cap Verde and the Mediterranean to northern Norway.
Description: Five evenly pointed arms with smooth velvet-like upper sides, central plate small and flattened. Arms laterally with spines, sitting in vertical rows of three (rarely four). Upper side yellowish. Diameter usually not over 200 mm, although there are records up to 350 mm. Another species in the genus, ***Luidia ciliaris*** (picture below), is similar but has seven arms. Both species are very fragile and are rarely collected intact by dredging or trawling.
Habitat: Usually below 20 metres depth down to the continental slope. Less common.
Biology: Often burrowed in mud bottoms during the day, emerging during night in search for polychaetes, bivalves, small crustaceans, and other echinoderms. The favourite food consists of brittlestars. Starfishes of this genus produce an unusually large number of eggs, and *L. ciliaris* can generate over 200 million eggs per individual.

Luidia ciliaris *has 7 arms. Picture from 25 meters deap in the Hardangerfjord. It can grow to more than 60 cm in diameter, and can move very fast.*

GASTRIC ACID

When a starfish (***Asterias***) attacks a mussel, it uses the feet suckers, or tube-feet, to break it open. The mouth is everted in between the shells, secretes gastric juice (hydrochloric acid and enzymes), and will then proceed to take up the decomposed nutrients. Some species (***Henricia*** spp.) are filter feeders and take up microorganisms or detritus from the bottom or suspended in the water.

Ceramaster granularis

Ceramaster granularis *has a highly characteristic appearance.*

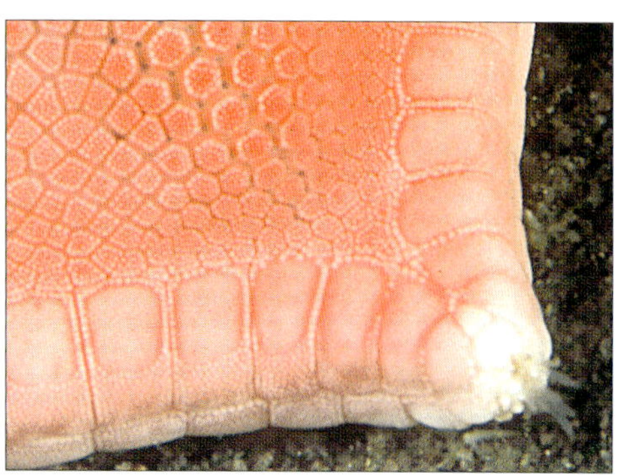

Detail of one of the arm corners.

Distribution: From Morocco to northern Norway and Greenland.
Description: Highly characteristic, with very short arms that appear only as corners of the central plate. Distinct and partly naked marginal plates. Upper side red, underside white. Up to 80 mm in diameter.
Habitat: On various kinds of bottoms at 20-1400 metres depth.

Pseudarchaster parelii

Distribution: Northern distribution, from Ireland to Murmansk and south Greenland. Less common.
Description: Arms about three times the length of the central plate. About 100 mm in diameter. Distinct marginal plates, covered by small granules. The lower row of marginal plates (not visible dorsel view) has small spines. Colour reddish brown or blood red.
Habitat: Occurs at 15-2 500 metres depth, usually below 75 metres.

This starfish also has marginal plates. Occurs in relatively deep water, usually below 75 m.

Hippasteria phrygiana

Distribution: Northern distribution, from Ireland to northern Norway and Greenland. Relatively common.
Description: Large central disc with short arms. Stiff. Upper side light red, densely covered with short white spines. Green spots are sometimes present dorsally, due to the presence of the symbiotic green algae, ***Coccomyxa***. Each marginal plate is surrounded by a ring of granules, and has a few, medium thick, bottle-shaped spines. Up to 200 mm in diameter. Unlikely to be confused with other species.
Habitat: Usually on hard bottoms, often on vertical walls, but also on various bottom types, at about 10-100 metres depth.
Biology: Feeds on other echinoderms, bivalves and polychaetes.

This starfish is easily recognised by its spines and short arms.

Pteraster militaris

Its uneven surface gives this starfish a rather 'untidy' appearance.

Right: **Pteraster obscurus**

Distribution: Circumpolar distribution, in northern Europe south to Skagerrak.

Description: Species within this genus are characterised by having the upper side covered with paxillae (rings of small spines) that are united by a skin, forming a roof-like covering over the rings. This 'roof' contains a series of pores that can be opened and closed. Medially on the dorsal side there is a larger pore that is surrounded by stouter spines, and the margin of the central plate has a fin-like edge composed of longer spines, also united by thin skin. The species ***P. pulvillus*** has paxillae with 8-15 spines on the upper side, whereas ***P. militaris*** has paxillae with 2-4 spines only. The specimen depicted furthest down seems to have about four spines and would then belong to *P. militaris*; the upper picture is not detailed enough for species identification. A third member of the family, ***Diplopteraster multipes***, is characterised by having tube-feet in four rows, in contrast to ***Pteraster***, which has two rows. *D. multipes* is supposed to have musculature forming a net on the upper side, although this is not visible on the illustrated specimens. It is also previously known from larger depths in the North Atlantic (>500 metres). *P. pulvillus*, according to the literature, is pale yellow with a diameter up to about 40 mm, whereas *P. militaris* **is** yellow to yellowish red, with a diameter up to 85 mm.

Habitat: There are three species in the genus Pteraster distributed in the northernmost part of Europe. ***Pteraster obscurus*** has 6 - 8 arms and an Arctic distribution, whereas *P. pulvillus* and *P. militaris* both have circumpolar distributions, in northern Europe south to Skagerrak. *P. militaris* occurs at 10 - 1100 metres, and *P. pulvillus* at 15 - 400 metres depth.

Biology: The skin pores in the family Pterasteridae serve for respiration, with water entering the small pores, and exiting through the large central one. Unusual for starfish, where the larvae mostly have pelagic development, members of this group have brooding chambers under the skin where the larvae develop, and the pores also serve for ventilating these.

Porania pulvillus

The surface of **Porania pulvillus** *is densely covered in papulae (gills).*

Distribution: In Europe from Bay of Biscay to northern Norway.
Description: ***Porania pulvillus*** is sometimes called the red cushion star, and both this and its Latin name (Lat. pulvillus = pillow) refers to the shape and consistency of the animal. It is perhaps our most beautiful starfish. In contrast with most of our other starfishes, the dorsal surface is smooth. It is deep red on the upper side with white, almost transparent retractable papulae (gills), and yellowish white on the underside. About 120 mm in diameter. Unlikely to be mistaken for other species.
Habitat: Often on semi-exposed or more sheltered bottoms, at about 10-250 metres depth.
Biology: Feeds on the leather coral dead man's finger, but also on detritus. Spawning takes place in late winter.

Distinctive red velvety surface, with numerous papulae (gills).

Solaster endeca - Yellow Sun star

Solaster endeca *can be yellow, orange or even a deep violet colour.*

A well-fed example of the violet variety.

Distribution: From the British Isles north to Spitsbergen and Greenland.

Description: Highly characteristic, usually with 7-13 (9-10) arms. In our waters only ***Crossaster papposus***, easily distinguished by its red colour and shorter arms, have that many arms. Apart from the number of arms, *S. endeca* reminds of the common starfish, ***Asterias rubens.*** The colour of *S. endeca* is yellow to deep violet. The surface is granulated with small spines. The arm tips are usually lighter in colour, often curving upwards. A large starfish, up to 350 mm in diameter. Another, arctic species, ***S. glacialis***, can be distinguished by the dense paxillae (rings of spines) with truncate spines. The paxillae in *S. endeca* are less densely distributed, and lack the truncate tips.

Habitat: Usually deeper than 30 metres, down to about 450 metres, usually in reasonably sheltered localities.

Biology: The development is direct, and the larvae lack pelagic stages. Breeding takes place in the spring.

Crossaster papposus - Red Sunstar

Crossaster papposus *can be up to 34 cm in diameter.*

Juvenile (young) individual.

Mouth region, with part of the stomach everted.

Distribution: Widely distributed in the northern parts of the North Atlantic, from the British Isles to the Arctic.

Description: Sometimes referred to as the rose sea star or the sun star. Beautiful, red, with 11-13 (8-14) arms. Dorsal side covered by rather long, slender spines. Colour usually red, with yellowish red areas in concentric bands. The only similar species is ***C. squamatus*** with a more northern distribution, and which has papulae (gill papillae) dispersed singly, whereas they are arranged in groups in ***C. papposus.***

Habitat: Common on partly sheltered to exposed localities, usually at 10-40 metres depth. Mainly on hard bottoms, often on rock faces.

Biology: Predator, mainly on other echinoderms, such as the common starfish (***Asterias rubens***) and sea cucumbers of the genus ***Psolus***, but also on bivalves, sea anemones, etc. It may attack starfishes of its own size. The larvae have a short pelagic stage, during which they do not feed.

VORACIOUS PREDATOR

Crossaster papposus is a greedy predator that often attacks and consumes other starfishes of its own size. Here it is feeding on the red cushion star (***Porania pulvillus***).

Henricia spp.

Distribution: There are several species in this genus. Their distribution is unclear due to difficulties in species determination. The most common species in the genus with a northern distribution are ***H. perforata, H. pertusa*** and ***H. sanguinolenta. Henricia oculata*** have a southern distribution.

Description: All ***Henricia*** species have five arms. The dorsal spines are very small and arranged in small groups, and the surface reminds of sandpaper. They are uniformly coloured, often violet, but can also be yellow, orange or red. Yellow specimens seem to be more common in deeper water. Species identification often requires microscopy. Rather small, most having a diameter less than 120 mm.

Habitat: Very common, often among algae in shallow water, from the intertidal to deep water. Yellow specimens are commonly associated with the coral ***Lophelia pertusa***.

Biology: Species of ***Henricia*** brood their larvae.

This is most likely **Henricia sanguinolenta.**

The upwardly curled arm tips are typical for **Henricia**.

Stichastrella rosea

Distribution: In Europe from the Bay of Biscay to northern Norway.
Description: An orange starfish with five arms. Central plate rather small, arms long, stiff and cylindrical. Dorsal side without protruding spines, but the skeleton forms a mesh. Habitus is characteristic, and it is unlikely to be confused with other species. Diameter about 150 mm.
Habitat: Both on hard and sandy bottoms, from a few down to 350 metres depth. Rather common on exposed localities, especially among kelp.
Biology: Reproduce in August-September. The larvae have pelagic development.

This starfish is stiff and hard to the touch.

An individual pictured together with **Luidia sarsi** *(left).*

Asterias rubens - Common starfish

Distribution: In the eastern parts of the North Atlantic from Senegal to the White Sea.

Description: Medium-sized central plate, arms rather stout, evenly tapering. Dorsal side with weakly developed skeleton, appearing soft in adults (small specimens are harder). Each arm provided with a central row of white spines. Other spines more irregular in distribution and number, occasionally completely lacking. Usually orange to yellowish brown, sometimes with additional violet colouration, and dark green, almost black, specimens also occur. Maximal size depending of food availability, approx. 300 mm in diameter. May be confused with small specimens of ***Marthasterias glacialis***, and the much smaller ***Leptasterias muelleri***. The former has more conspicuous spines, which are arranged in three rows, and the latter has thinner and less tapering arms, covered with spines, and with white tips.

Malformed individuals are occasionally observed - an error during regeneration or a mutant?

The pelagic larvae hatch in the summer and are seen all through autumn until they settle. Larvae swim with worm-like movements. Note the small starfish.

A juvenile, less than 1 cm in diameter.

Habitat: Widely distributed on various kinds of bottoms, from the intertidal down to about 200 metres depth.

Biology: Produces large numbers of pelagic larvae that occur in the water in the summer. During breeding time, the larvae of ***A. rubens*** may completely dominate the zooplankton. Under optimal food conditions they may become reproductive already after one year.

If the central disc remains intact, even a one-armed survivor of a brutal attack can survive, provided it can obtain food with its one arm.

Spawning

With a piece of good luck (combined with regular diving and keen eyes) one may come across events like this one: spawning ***Asterias rubens***.

The openings of the gonads are situated dorsally at the basis of the arms, and are the place where eggs and sperm are emitted. Most echinoderms have separate sexes and external fertilisation, where eggs and sperms are released into the surrounding water. The timing is regulated by temperature and chemical signals between the animals. The fertilised eggs develop into pelagic larvae, and these go through a number of different stages before they settle and take up the adult life on the bottom.

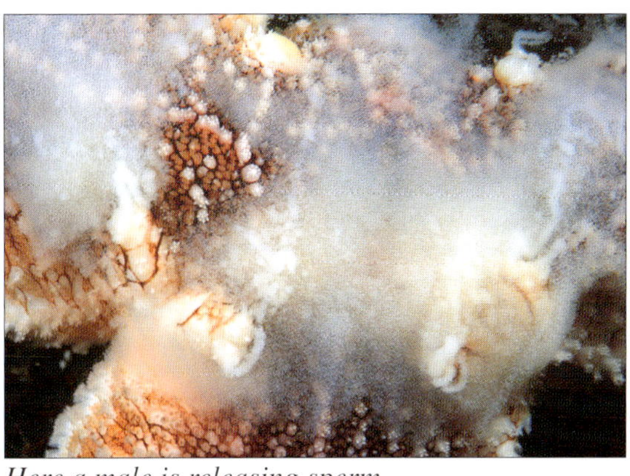

Here a male is releasing sperm.

A female spawning eggs.

FEEDING ON MUSSELS

The most frequent prey of the common starfish is the mussel ***Mytilus edulis***. They open the mussel by means of the tube-feet suckers. As the mussel gets tired and opens the shells, the starfish will evert its stomach between the shells and secrete gastric juices, thereby commencing the digestion of the prey. It also preys on barnacles, polychaetes and other echinoderms (including its own species in times of food shortage), and may feed on dead animals.

*This individual is attacking a razor shell (**Ensis** sp.).*

Leptasterias muelleri

A particularly attractive starfish, found from shallow to deep waters.

Distribution: In the North Atlantic from the British Isles to Spitsbergen.
Description: This small starfish resembles small specimens of the common starfish, ***Asterias rubens***, and ***Marthasterias glacialis***. The spines on the dorsal side appear in star-shaped groups that are arranged in more or less distinct longitudinal rows. The dorsal skeleton is well developed. A single papula (gill papilla) is present between each group of spine. Colour variable, including green, pink or violet, with light arm tips. Usually up to 60 mm in diameter, although it can be considerably larger.
Habitat: From the intertidal and usually in shallow water, although there are records down to 800 metres depth. Prefers moderately to strongly exposed rocky localities.
Biology: ***Leptasteris muelleri*** is a brooder, and carries the developing larvae around the mouth, at the basis of the arms. The larvae will not leave the mother until they have developed the first three pairs of tube-feet, and there is no planktonic stage at all. Reproduction takes place during the winter, and the mother will eat nothing as long as she carries the young.

SYMBIOTIC WITH GREEN ALGAE

The colour of intertidal specimens is often green, and comes from unicellular green algae, which live as symbionts in the skin.

Several colour variations exist.

Marthasterias glacialis

Because of its hard 'skin' and long spines, this species has few enemies. Easily recognised as adults.

Distribution: Widely distributed, from the Azores and the Mediterranean to Iceland and northern Norway.
Description: May in extreme cases reach a diameter of 80 cm, and is our largest starfish. Highly characteristic, with a small central disc and long arms provided with stout spines arranged in three main rows–it is sometimes referred to as the spiny starfish. Spines usually white with violet tips, colour otherwise brownish to greenish grey.
Habitat: Common both on hard bottom among algae and on various kinds of soft bottoms. The largest specimens live on soft bottoms where bivalves are abundant. From the intertidal down to a few hundred metres depth.
Biology: Like many other starfishes, this species is a voracious predator, and eats most of what comes in its way of live or dead animals, although mainly snails and bivalves. It has a good regeneration capability, and specimens with regenerating arms are commonly encountered.

The madreporeplate is connected to the water canal system.

Ophiothrix fragilis *is one of the most common brittle stars seen by divers.*

Class Ophiuroidea - Brittle stars

Brittle stars (Gr. ophis = snake, + Gr. oura = tail) are characterised by long, thin arms, which are distinctly delineated from the plate-like ***central disc****. Most have five arms, branching in some taxa. Each arm segment is provided with a pair of* ***tube-feet****, emerging through pores on the ventral side. These pores are usually surrounded by small scales, often used in species identifications. The tube-feet lack suckers, but are important for respiration, feeding and as sensory organs. In contrast to starfish, the arms do not contain internal organs.*

In brittle stars, the central disc is distinct from the arms. Here **Ophiopholis aculeata.**

The surface of the central disc is commonly covered by plates and scales. On the central disc, at the basis of each arm, there is always a pair of large plates called radial shields. They are usually visible, but can in some groups be covered by skin. Also the arms are covered by plates, the shapes of which are helpful in identifications. Between each arm segment there are a number of spines, more or less well developed, directed towards the tips of the arms, and the number and shape of these also provides useful information.

Brittle stars commonly appear in aggregations. Here **Ophiothrix fragilis** *and* **Ophiopholis aculeata**.

A number of brittle stars are predators, and feed on, for example, polychaetes and small molluscs. Others are filter-feeders, and live on detritus and plankton. Places exposed to strong currents can harbour enormous densities, up to 10 000 specimens per m^2, catching plankton with their arms raised up into the water. Some live partly buried in sediment. Several species are used by biologists to characterise different kinds of habitats. A few are luminescent.

Growth studies of brittle stars (***Ophiura ophiura***) indicate that they reach maturity at the age of two years, whereas it takes three or more before they have reached full size. They have well-developed regeneration capabilities, and even the central disc, containing the gut and the gonads, can be replaced. Some taxa have asexual reproduction, where the central disc divides into two, and the corresponding parts are regenerated. Several species are **hermaphrodites** and **viviparous**. The eggs hatch and develop within the parent, emerging as small adults. However, most species have pelagic development.

LARGE GROUP

Ophiuroidea is the largest class of echinoderms, with more than 1800 species worldwide. They are found from the lower shore to the deepest oceans.

NINE LIVES...

Commenting on a brittle star, Th. Mortensen (1924: Danmarks Fauna vol. 27) noted: *"The author has observed that an* ***Amphiura chiajei*** *, which was taken out of the gut of* ***Astropecten irregularis*** *18 hours after it had been consumed, was still alive, even though the skin and the tube-feet were partly digested. A few days later it was still alive and showed signs of regenerating the tube-feet, where after the experiment had to be aborted due to travels."*

Arm tip of brittle star. Many species use the arms to filter particles from the water.

Ophiothrix fragilis

Ophiothrix fragilis *is recognised by its marked colouration and long arm spines.*

Distribution: Widely distributed in the northeast Atlantic, from western Africa and the Mediterranean to northern Norway. Very common.

Description: A characteristic and colourful brittle star with long, numerous spines. Each arm segment is provided with seven, long spines each side, and also the central disc has with spines. The colour is variable, but it usually has transverse red or orange stripes on the arms. Rather large, with the diameter of the central disc of about 20 mm, and the arms approx. five times longer.

Habitat: From the low intertidal down into deep water. Usually occurring singly among other animals on mixed bottoms.

Biology: The sexes are separate and the larvae are planktonic. Reproduction takes place in the summer. They catch food with the arms extended into the water, and collect plankton, especially copepods, with the long spines and the tube-feet.

ENORMOUS DENSITIES

In deeper water this species can reach densities of 10 000 specimens per m^2.

Ophiocomina nigra

This brittle star often is almost completely black.

Distribution: From west Africa and the Mediterranean to mid-Norway.
Description: A relatively large brittle star, uniformly coloured in dark brown to almost black. Less pigmented, almost white specimens, sometimes occur. Large central disc, up to 25 mm in diameter, smooth. Arm length about five times the diameter of the central disc. Arm segments with 5-7 rather long spines each side, neatly arranged like the teeth of a comb.
Habitat: Usually appearing in rather sheltered localities, with some current. It is common to find many together, with up to 100 specimens per m^2. Occurs also in brackish water. Found on mixed bottoms, often openly displayed on the bottom, from the lower intertidal down to at least 400 metres.
Biology: Larvae with pelagic development. Food highly variable, including various algae, plankton, detritus, carrion, etc.

Ophiopholis aculeata

Ophiopholis aculeata *is often highly colourful, with reddish tones.*

Distribution: Widely distributed in the northeast Atlantic from Madeira to northern Norway, more common north of the English Channel.
Description: Diameter of central disc up to 20 mm, arms four times as long. Arms often curled, banded in red or brown. A characteristic feature is that each dorsal arm shield is completely surrounded by a single row of small scales. The arm segments have 4-5 robust, bottle-shaped spines each side. Central disc with distinct incisions at the insertion of the arms, dorsally covered with shields.
Habitat: Often in crevices, or between stones, with only the arms emerging. Sometimes on soft bottoms, but then always attached to solid objects, such as empty bivalve shells. Most common at 10-60 metres depth.
Biology: Food dominated by detritus. The most common predator is cod.

Ophiura albida

Distribution: From the Azores and the Mediterranean to northern Norway.
Description: Dorsal side usually reddish brown with white spots (therefore the name Lat. albidus = white). Diameter of central disc up to 15 mm, arm length to 60 mm. Disc more or less pentagonal.
Habitat: Occurs on various kinds of bottoms, but most common on muddy sand. From a few metres down to about 500 metres depth.
Biology: Larvae present in August-November. The adults feed on detritus.

This species has characteristic white patches on the disc at the arm bases. Here on a dead cod.

Ophiura ophiura

Distribution: In Europe from Morocco and the Mediterranean to northern Norway.
Description: Characterised by a large central disc, up to 35 mm in diameter, arms 3.5 times that length. Arms evenly tapered, straight and stiff. Arm segments with three short spines. The colour is reddish brown or brown, often with some lighter spots. Can be confused with other species within the genus. The preceding species, ***O. albida***, has a pair of white spots dorsally at the insertion of the arms.
Habitat: On soft bottoms, usually mud or

Unlike many of its relatives, **Ophiura ophiura** *lacks colour patterns.*

muddy sand, but also on shell sand. Generally below 10 metres, and down to at least 200 metres depth.

Biology: Pelagic larvae, which are common in the water throughout the summer. Feeding varies and it can be a predator, filter-feed, or feed from carrion.

The arm plates have a typical shape.

MATURITY

Ophiura ophiura reaches sexual maturity at about three years, and fully grown specimens are likely about 5-6 years.

Gorgonocephalus caputmedusae - Basket star

Distribution: Northern distribution, from the west coast of Ireland to middle Norway.

Description: Highly characteristic brittle star. The five arms are repeatedly branching, and a fully grown specimens may have 5000 terminal arm tips. This group of brittle stars are often referred to as basket stars. The central disc is large and densely covered with short, stout bottle-shaped spines. Arms densely covered with small granules. Colour usually pale yellow or pale orange. The diameter of the central disc is up to 90 mm. A similar species, ***Gorgonocephalus lamarcki***, occurs from the west coast of Norway and further north and can be distinguished by the central disc that has very fine and densely distributed spines, appearing almost smooth. It occurs at about 75 - 800 metres depth. Two other species also

Juvenile (young) individual.

The five arms are highly branched. The largest individuals have up to 5000 arm tips. Individuals are often found attached to soft corals in current-rich areas.

THE SPECIES NAME

Caputmedusa refers to Lat. caput = head, and Lat. from Greek mythology medusa: a beautiful gorgonian that seduced Neptune with her golden hair.

appear in more northerly waters, ***G. arcticus*** and ***G. eucnemis***. Both lack spines on the central disc, with the exception of the radial shields (at the insertion of the arms), which in the former has large elongated spines, and in the latter rounded spines.

Habitat: *Gorgonocephalus caputmedusae* is mainly found attached to deep-water corals, including ***Paragorgia arborea*** and ***Paramuricea placomus,*** although specimens on branching sponges and on rock surfaces also have been observed. Usually at 50-1200 metres depth. In the Trondheimsfjord they appear more shallow, and we have observed specimens on rock surfaces at 20 metres depth. On *Paragorgia arborea* we have recorded it at 30 metres, and on one occasion we encountered one of these gorgonians with over 15 basket stars. A common feature for all our observations is that the localities are exposed to strong currents. The specimens usually sit distally on the corals, with the arms directed against the current.

This species is seldom found shallower than 50 m depth, but we have seen it at 20 m.

Biology: The branching structure of the arms of this brittle star, as well as the preference for localities with strong currents, clearly designate it is a filter feeder. The development is unknown, but the large number of small eggs indicate that the larvae have a pelagic development.

Class Echinoidea - Sea urchins

Sea urchins (Gr. ekhinos = hedge hog) are more or less spherical echinoderms without arms, usually with a hard, inflexible skeleton consisting of calcareous plates. They are covered by long, moveable spines, which are expanded basally at their point of attachment. The skeleton is perforated by pores through which the tube-feet emerge. All have the mouth on the underside, and all have an anus.

The group is subdivided into two subclasses, **Cidaroidea**, mainly with fossil members and a single, recent species in our waters, ***Cidaris cidaris,*** and **Euechinoidea**. Previous, but now obsolete, classifications instead divided sea urchins into Regularia, with spherical body, and Irregularia, with flattened body. Those previously referred to Regularia now are referred to the order **Echinoida**.

Among the irregular sea urchins, our smallest member, ***Echinocyamus pusillus***, belongs to the order **Laganoida**; the remain-

The edible sea urchin, **Echinus esculentus** *(red; left in picture), together with* **Strongylocentrotus droebachiensis**, *both common in our waters.*

The Edible sea urchin, **Echinus esculentus.**

ing ones to the order **Spatangoida**. ***Echinocyamus pusillus*** is distinguished by having a maximum length of 15 mm only, and in that the anus is on the underside, just behind the centrally positioned mouth. Members of these two orders have a flattened shape. The Spatangoida lack a well developed jaw apparatus, and are adapted to a life within sediment. The empty skeletons of these animals often remain on the bottoms or are washed ashore on beaches.

Sea urchins move slowly, mostly by using the long spines. The regular urchins use the spines almost like stilts, and the tube-feet suckers are used when moving across vertical surfaces. Algae are important as food, but they may also feed on animals. The irregular sea urchins that lack the stout jaws will dig into the sediment down to about 20 cm, and feed on small bivalves, snails and foraminiferans. Identification of sea urchins is not always easy. For some taxa, one has to scrape away tube-feet and spines, and chemically remove the skin with chlorine solution, in order to examine the pores and the plates of the skeleton.

Order Echinoida

Regular sea urchins. Spherical shape, with the skeleton perforated with five double-rows of pores. Both mouth and anus are situated centrally, the former on the underside and the latter on the top. Most feed on algae and sessile animals, including barnacles and bryozoans.

STOUT JAW APPARATUS

The mouth is called **Aristotle's lantern**, and consists of five stout, calcareous jaws, which are centrally situated on the underside.

Psammechinus miliaris

Distribution: In the northeast Atlantic from Morocco to northern Norway.
Description: This small, slightly flattened sea urchin, sometimes referred to as the green sea urchin (although this name is also used for ***Strongylocentrotus droebachiensis***; see p. 437), attains a diameter of about 50 mm. They are green with violet spine tips. There are two kinds of spines of different length, the smallest ones very short. Spines rather stout, very sharp, up to about 15 mm long. Tube-feet arranged in three rows, visible as three pores on each plate of the naked skeleton. This species may be confused with small specimens of ***Echinus***, but is distinguished in having thick and tightly arranged plates on the membrane surrounding the mouth. It may also be confused with *S. droebachiensis*, which, however, has green spines with white tips, and tube-feet arranged in 5-6 rows.

Typical violet spine tips.

Habitat: Varying bottom types, with a preference for rocky, exposed localities, often among calcareous algae and other organisms covering the rocks. May bore into softer kinds of rock. From the intertidal down to about 100 metres depth.
Biology: Reproduction takes place in early summer. They are omnivorous, and the food will depend on the bottom type, mostly consisting of various kinds of benthic algae and animals.

Echinus esculentus - Edible sea urchin

The Edible sea urchin, **Echinus esculentus** *is common in our waters.*

Distribution: In the northeastern Atlantic from Portugal to northern Norway and Iceland.

Description: This, together with ***E. acutus***, is our largest sea urchin, with a diameter up to 160 mm, although rarely exceeding 100 mm. Shell spherical, red to rose, densely covered with rather short, equally sized spines. The spines have violet tips and white insertion points on the plates. ***Echinus esculentus*** hybridizes with *E. acutus.* It is very similar to ***Echinus elegans***, but the latter occurs in deeper water (50-2000 metres), it has spines of different lengths and that are arranged in more distinct rows, and it has a more flattened shape. *E. elegans* is also smaller, rarely attaining a diameter exceeding 50 mm. Another, rare species in the area is ***E. tenuispinus***, recorded from western Norway and the outer west Shetland shelf. It is virtually spherical and have a near-white colour.

Habitat: Among algae on rocky bottoms, from the intertidal down to about 40 metres depth, although there are records down to 1200 metres depth. Very common.

Biology: Mainly grazes on larger algae, especially on those covered by bryozoans, and with

FEEDING ON FAECES

The polychaete ***Flabelligera affinis*** and the isopod ***Astacilla intermedia*** sometimes occur among the spines of this sea urchin. The polychaete feeds on the faeces of the sea urchin.

a preference for ***Laminaria***. May also feed on a variety of sessile animals. Reproduction takes place in the early summer. Growth patterns have been recorded from Scottish waters, and the first year they are up to 40 mm in diameter, second year 40-70 mm, third year 70-90, and fourth year up to 110 mm. The largest specimens are likely to be 7-8 years old.
Utility: The gonads are eaten in several countries, including UK and Portugal, and are best before the reproduction in the early summer.

In fjords, the Edible sea urchin often has brighter colours than in other localities.

Echinus acutus

Distribution: From West Africa and the Mediterranean to Barents Sea.
Description: Characterised by sparsely distributed spines of unequal length, including very long ones. Shape not spherical as in the previous species, but with a more pointed top. Diameter up to 160 mm. The colour is reddish, often with white bands, and the spines are red with white tips. There are several varieties, differing in shape and spine length. Hybridises with ***E. esculentus.***

A juvenile individual from Lysefjorden, southwest Norway.

Echinus acutus *can appear in a wide range of colour variations.*

HYBRIDS

Hybridisation means reproduction between two different species. The progeny are called hybrids.

Habitat: Relatively deep water. It has previously been considered as rather uncommon, because it is not easily caught by dredging. However, since diving became common, we know that it can be abundant, especially on rock faces below 20 metres depth. Depth distribution 15-1000 metres. It often comes up into more shallow waters in winter and early spring.

Echinoderms also feed on dead animals, such as this fish.

Biology: Food consists of various benthic animals, but also sediment occurs in the gut.

This species has extremely long spines. It usually occurs on steep rocky walls below 20 m depth, but can also be seen grazing on algae, as here.

Strongylocentrotus droebachiensis

This sea urchin can occur in large aggregations, particularly in northern waters.

DELICACY

In later years there has been extensive fishing for ***S. droebachiensis***, especially in Norway. They are collected by divers and exported to Japan, where the gonads are considered a delicacy.

Distribution: Widely distributed across the Northern Hemisphere (circumpolar). In the northeast Atlantic it occurs from the English Channel to the Arctic.

Description: A very common sea urchin that attains a diameter of 80 mm. Sometimes referred to as the green sea urchin (although this name is also used for ***Psammechinus miliaris***; see p. 433) or the northern sea urchin. The colour is green, with greenish, reddish or violet spines with white tips. Shape usually somewhat flattened. The skeleton is covered with densely distributed tubercles, of which the largest, primary ones, form distinct rows. Also the smaller, secondary ones may form rows in larger specimens. Each plate in the skeleton has 5-6 pores. A congeneric species, ***S. pallidus***, occurs from the west coast of Norway and further north. It has slightly shorter spines, and reproduces later in the year, June to July.

Habitat: A typical shallow water species, often on rather sheltered localities. However, there are records down to 1200 metres depth.

Biology: Feeds on algae, especially ***Laminaria***, but also on various sessile animals. Now and then it can appear in large densities, and may cause great damage to kelp forests, especially on the cuvie, ***Laminaria hyperborea.*** Reproduction takes place in late winter to early spring.

Order Spatangoida - Irregular sea urchins

More or less flattened and elliptical, approaching heartshaped. All species have mouth and anus situated ventrally on opposite sides. The well-developed jaw apparatus-Aristotle's lantern-that occurs in other sea urchins, is lacking. About 5 - 7 species are common in our waters. All our species live buried in sand, mud or gravel. Apart from the illustrated ones, there is also **Brissopsis lyrifera**, *provided with a dorsal, highly characteristic lyre-shaped dark furrow, together with a smaller circular one under the anus, and a deep anterior incision. Rather common below 50 metres depth.*

Spatangus purpureus - Purple heart urchin

Distribution: In the northeast Atlantic from the Azores and the Mediterranean to northern Norway.

Description: An irregular sea urchin with a highly characteristic purple colour. As other burrowing forms it is flattened and elliptical. Diameter up to 120 mm. It has five rows of tube-feet, of which the anterior most ones are longest and situated in a shallow furrow leading to the mouth. Spines rather short and fine, giving a furry appearance to the animal. Deep anterior incision. Due to its colour it is unlikely to be confused with other species in the area. A congeneric species, ***S. raschi***, occurs below 150 metres depth.

The characteristic purple colour distinguishes this from other irregular sea urchins.

Habitat: Digs in shell sand and coarse sand, at 5-900 metres depth. Usually completely buried in the sediment, although the dorsal spines may sometimes be visible. Can be traced when diving by the elongated depressions it leaves behind in the sand.

Biology: Transports food particles from the sediment into the mouth by means of the brush-like tube-feet surrounding the mouth. It feeds on dead or live organic matter in the sediment. The special larvae of this species have an extremely prolonged tail-spine, and occur in the plankton in late summer and autumn. The adults often have a small bivalve attached to them, ***Montacuta substriata***.

Echinocardium cordatum - Sea potato

Distribution: A southern species that occurs in large parts of the Atlantic and the Mediterranean, north to northern Norway.

Description: This is one of three species in the genus that are present in our waters. Common for them is the lack of the anterior furrow that is present in all other members of Spatangoida, and also the absence of the lyre-shaped dorsal furrow that characterise ***Brissopsis*** and ***Brisaster***. Colour light, pale yellowish brown. The species ***Echinocardium cordatum*** is distinguished from the two other by the distinct anterior incision. It reaches a length of 90 mm. Both

Echinocardium flavescens *has grey-greenish spines of varying lengths.*

Echinocardium cordatum.

Echinocardium flavescens (illustrated above) and ***Echinocardium pennatifidum*** (illustrated on the page 440) lack the anterior incision. *E. flavescens*, also called "yellow sea potato", have a few dorsal spines that are longer than in *E. pennatifidum.* The latter has stouter and more evenly distributed spines than in the two

This most likely is **Echinocardium pennatifidum**, *but identifications from photographs alone are uncertain.*

other species. *E. pennatifidum* reaches a length of 70 mm, *E. flavescens* 50 mm.

Habitat: Echinocardium cordatum occurs from the intertidal down to 250 metres depth. It is very common, buried rather deep (10-15 cm) into clean sand, with a channel leading to the surface. The other two species prefer coarser sand and shell sand, *E, flavescens* mainly from 20 metres depth, and *E. pennatifidum* at 5-150 metres.

Biology: Food is captured by very long tube-feet that reach the sediment surface; they collect particles, and transport them down to the mouth. Accordingly, they do not simply ingest sediment as they dig, but actively select food items.

Dead sea potatoes easily lose their spines and the shells are brittle. They are often washed up on beaches.

"LOVELY LOVENIA"

The heart urchin ***Lovenia cordiformis***–found around the Channel Islands, only partially buries itself, leaving its long, backward pointing spines largely exposed.

Class Holothurioidea - Sea cucumbers

Sea cucumbers are often found in fjords. Most prefer current-rich areas but avoid wave-exposed sites.

*Sea cucumbers, or holothurians (Gr. holothourion: kind of sea cucumber), are distinguished by their lack of arms, and, for the most part, by having a sausage-shaped body. The calcareous skeleton is usually reduced to microscopic plates dispersed in the skin, and they tend to be soft bodied and lack a stiff surface (***Psolus*** being one exception).*

The mouth is situated at one end, surrounded by branching and sticky **tentacles**. They have a transient gut and an anus situated at the opposite end. As in other echinoderms, they have a pentamerous symmetry with five rows of **tube-feet**, although these may be lacking or dispersed in some taxa. Sea cumbers have '**water-lungs**', which are branching organs inserted on the distal most part of the gut, and take up water through the anus. The active pumping in and out of water serves for an efficient supply of oxygen. Most species live more or less buried in the sediment, from which they also feed. The organic content of the sediment is digested, whereas most part of it is just transported through the gut. Other species use the branched tentacles to filter plankton and dead organic matter from the water. One and one the arms are bent into the mouth and licked off. Some filter feeders live buried in sediment with only the tentacular crown emerging, whereas others are fully exposed. A common feature for all these is the preference for localities exposed to currents.

A majority of holothurians have separate sexes. Some brood their young, and in some the emerging larvae resemble the adults. Many have pelagic larvae (**auricularia**). The regeneration ability, as in most other echinoderms, is well developed. Under threatening conditions (as when caught in a trawl or dredge) they will contract and evert the whole gut and other internal organs through the anus. These will then regenerate within a few weeks. In our waters there are more than 30 species. The two mentioned subclasses below, **Dendrochirotacea** and **Aspidochirotacea**, have waterlungs, whereas these are lacking in **Apodacea**, a subclass of mainly deep-living, burying forms. Most of our species are small forms that live in the sediment in deeper water, and identifications of these usually involve microscopic examinations of the calcareous spicules in the body wall. The most common one on shallow hard bottoms from Scotland and further north is ***Cucumaria frondosa***; on soft bottoms ***Stichopus tremulus*** which can also occur in shallow waters. Members of ***Psolus*** have the underside formed as a sucker and sit attached to stones and rocks, often on rock faces, and mainly below 30 metres depth.

Subclass Dendrochirotacea

Species in this group use their tentacles to collect food particles. The arms are rhythmically moved into the mouth and licked off, one by one. They can also be completely retracted into the mouth.

Some species sit more or less attached to surfaces by means of their tube-feet, others are partly buried in sand, gravel or among stones, with the tentacles emerging into the water.

GUT AMPUTATION...

Carl Dons, a Norwegian marine biologist, wrote in "Sjøen" (1927): "*...some of these cucumbers clearly demonstrate their unhappiness-it sort of flows out of them as they evert their guts in protest, a habit they have, among other reasons, in order to get rid of a parasitic snail that lives in their guts, a long vermiform animal (***Enteroxenos oestergreni***) that looks more like a worm than a snail. Nevertheless, such gut amputation-autotomy as it is called-represent a highly incredible event. The Japanese attempt to imitate it in a ceremony called hari-kiri, but generally do not survive the exercise, at least not when properly performed. As stated by an old Roman, "Quod licet Jovi non licet bovi", meaning: what is fitting for a sea cucumber, may not be fitting for a Japanese."*

Cucumaria frondosa

The tentacles are darker than the body itself.

Distribution: Northern distribution. Occasionally from the Portuguese coast to the northern parts of the British Isles, common on the Faeroes, Iceland and Norway.
Description: Highly characteristic large, almost black holothurian, with leather-like skin and long, branching, bush-like tentacles. In extended condition it may reach a length of 50 cm. The tube-feet are arranged in distinct rows, running from the gut to the anus. When contracted and with the tentacles withdrawn it looks very much like an American football. Unlikely to be confused with other species in the area.
Habitat: May occur in thousands in localities with strong currents, sitting attached to the

surface with the tentacles extending into the water. From a few down to about 200 metres depth.

Biology: The red, barrel-shaped larvae are released in late winter and spring, and may occur in such abundances that the water turns red. Feeding in adults is accomplished with the long, branching tentacles.

White colour variants also occur.

FREELOADERS

A number of organisms live as parasites in the gut of holothurians. Strongly modified snails without shells are the most common ones, together with various unicellular animals (protozoans), but also flatworms, nematodes, copepods, bivalves, and even crabs live there, mainly in tropical regions. Even small fish of the genus *Fierasfer* live in the gut and the water-lungs, swimming in and out through the anus.

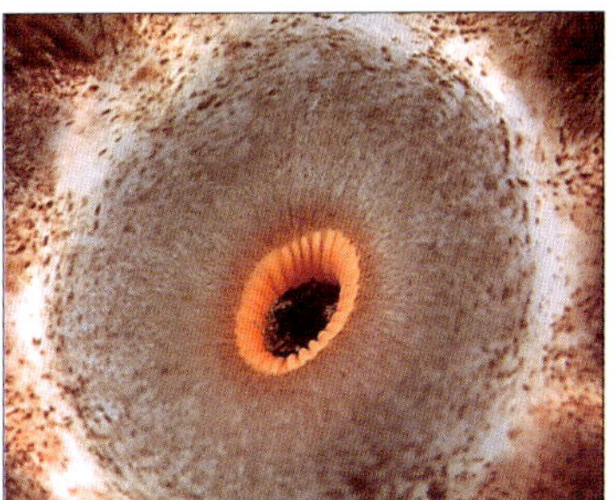

The mouth has an attractive orange ring. The tentacles bearing food particles are inserted into the mouth and licked clean.

LICKING THE FOOD

It is worthwhile when diving to observe the slow, rhythmical movements of holothurian tentacles. One by one the arms with the captured particles are bent into the mouth and licked clean. The next arm will begin its movement already before the preceding one is fully extended.

Thyone fusus

Distribution: From Madeira and the Mediterranean to northern Norway.
Description: Barrel-shaped with pointed ends, colour white or light brown to pink. Body surface delicate and easily damaged. Often covered with shell fragments. Tube-feet diffusely distributed.
Habitat: Bottoms with shells, mainly at 10-150 metres depth.

This sea cucumber was dug up to be photographed.

Thyone fusus *lives buried within the sediment.*

Psolus squamatus

Distribution: Of the two species of the genus that are present in the area, ***Psolus phantapus*** has a northerly distribution and occurs in the northern parts of the British Isles to Spitsbergen and the White Sea. The other, ***Psolus squamatus***, also has a northerly distribution in Europe, but has also been recorded from diverse places, including California and Japan.

Description: Both these ***Psolus*** species have the ventral surface modified into a sucker used to attach to the surface. The upper side is covered by thick, imbricated plates. Mouth and anus are situated dorsally on each side. All ***Psolus*** have ten tentacles. On the ventral side, the middle row of tube-feet in *P. phantapus* extends all across the sucker. The sucker is more or less rectangular and smaller than the circumference of the body. Length up to 20 cm. Colour variable, often with orange tentacles and pink spots on the body, but can also be golden brown, or sometimes almost black. ***Psolus squamatus*** is more flattened, without a marked 'tail'. The median tube-feet in the sucker are apparent only anteriorly. The plates are large, smooth, and only with small, dispersed granules. Length up to 9 cm. Body yellowish, often with reddish tentacles.

Habitat: Mainly in rather deep water, rarely above 20 metres, down to about 500 metres. Adult *P. phantapus* bury partly in sand and gravel. *P. squamatus*, and young *P. phantapus*, attach to hard surfaces by means of the sucker.

These sea cucumbers are attractive animals that occur in relatively deep water, seldom above 20 m. Here **Psolus squamatus**, *which often attaches to steep rocky slopes.*

Psolus phantapus

Psolus phantapus *usually lived buried within the sediment. Here photographed at 20 m depth, but it usually occurs deeper. Note the beautiful pigmentation (see facing page).*

Distribution: From the British Isles and north, probably circumpolar.
Description: Up to 20 cm long. Can be separated from ***Psolus squamatus*** by the following features: body with a tail-like, upward pointing extension, more flattened in ***Psolus squamatus***. The dorsal plates in *P. phantapus* are covered by the skin, and only visible anteriorly and posteriorly. The sucker is rectangular and comparatively small, whereas in *P. squamatus* it is larger and elliptical, almost circular. All ten tentacles in *P. phantapus* are of the same length. The colour is age-dependent, although most specimens tend to have red or red-spotted tentacles. Younger specimens are light red, older ones yellowish to brown, almost black for the largest ones.
Habitat: Small specimens live as epifauna (attach to solid surfaces), whereas larger ones bury in sediments with the tentacles emerging. It may be found as shallow as at one metres depth, although usually below 20 metres and down to about 400 metres.
Biology: Rather large (0.6 mm in diameter), pelagic eggs. The larvae do not feed before settling. Initially following the settling they develop five tentacles (at this stage labelled pentactula), which are used to crawl to a suitable point of attachment, where they will develop till adults.

DID YOU KNOW...

In ***Psolus phantapus***, the posterior end forms a tail-like extension, directed upwards, which explains the scientific generic name; Gr. psolos = Lat. psoleos = circumcised penis.

Thyonidium drummondi

This sea cucumber often occurs in fjords, often from 15 m depth and below.

Distribution: Limited to the north-east Atlantic, from the British Isles to the Arctic.

Description: Species of this genus are distinguished by having 20 tentacles sitting in two rings. The outer ring has five pairs of large tentacles, whereas the inner one has five pairs of small ones. In this species the elongated, rather stout animal is up to 25 cm long. The body is yellow to light orange, with red/orange tentacles. One variety, previous known as ***Thyonidium hyalinum***, is virtually identical. As indicated by the scientific species name, it is paler and more transparent (hyaline). It is currently uncertain whether or not they are conspecific.

Habitat: On various kinds of soft bottoms. Body buried with the characteristic tentacles emerging. From 5 to about 1000 metres depth.

The arms typically are arranged in pairs; with 5 pairs of large tentacles in the outer ring and 5 pairs of small ones in the inner ring.

Subclass Aspidochirotacea

Species in this group move along the bottom, and have short tentacles that they use to 'lick' sediment.

Stichopus tremulus

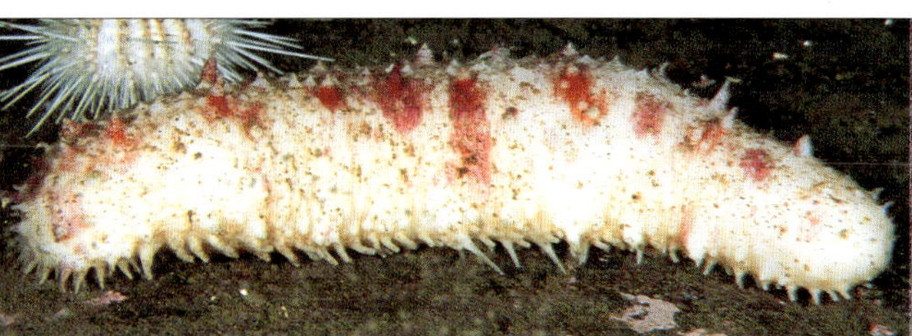

Some individuals are mostly white.

Distribution: From Bay of Biscay to northern Norway.
Description: Large holothurian, up to 50 cm long. Dorsal side coloured red with tiny black spots; ventral side white. Some specimens are near white or greyish. Body slightly tapered towards the ends. Dorsal side covered by rather long papillae. Tentacles short. Adult specimens are unlikely to be confused with any other holothurian.
Habitat: Originally described by the Norwegian bishop Gunnerus from the Trondheimsfjord. It is common in fjords, from 20 to about 1200 metres depth. Occasional specimens may appear at only a few metres depth, usually in combination with upwelling of cold water. Commonly caught in shrimp-trawls.
Biology: This species has a number of internal parasites. The highly modified and worm-like snail, ***Enteroxenos oestergreni***, is the most common, but also five different flatworms. In the southerly parts of its distribution it is also inhabited by fishes of the genus ***Fierasfer***. The holothurian feeds on deposited organic matter (detritus), and the larvae have pelagic development.

This species has such a characteristic shape that it cannot be confused with any other.

Mesothuria intestinalis

Mesothuria intestinalis, *photographed at 50 m depth.*

Distribution: Widely distributed along European costs, from the Azores and the Mediterranean to northern Norway.

Description: Highly characteristic, with shell fragments and other particles attached to the surface. Near cylindrical, underside slightly flattened. Rather large, up to 30 cm long and 6 cm wide. The mouth is ventrally directed, but the anus is terminal. Adults have 20 fully retractable tentacles. The tube-feet are small, diffusely distributed along the body, and provided with a small plate that is used to attach foreign particles. Body wall very thin and soft, and the animals are difficult to conserve. Colour pale greyish or brownish, often with tints of red, violet or blue.

Habitat: ***Mesothuria intestinalis*** lives on various bottom types, from sand and gravel to mud. Depth distribution 20-1500 metres.

Biology: Hermaphroditic. The eggs are large, over 0.6 mm in diameter.

Here on a soft substrate at 35 m depth.

CHORDATES

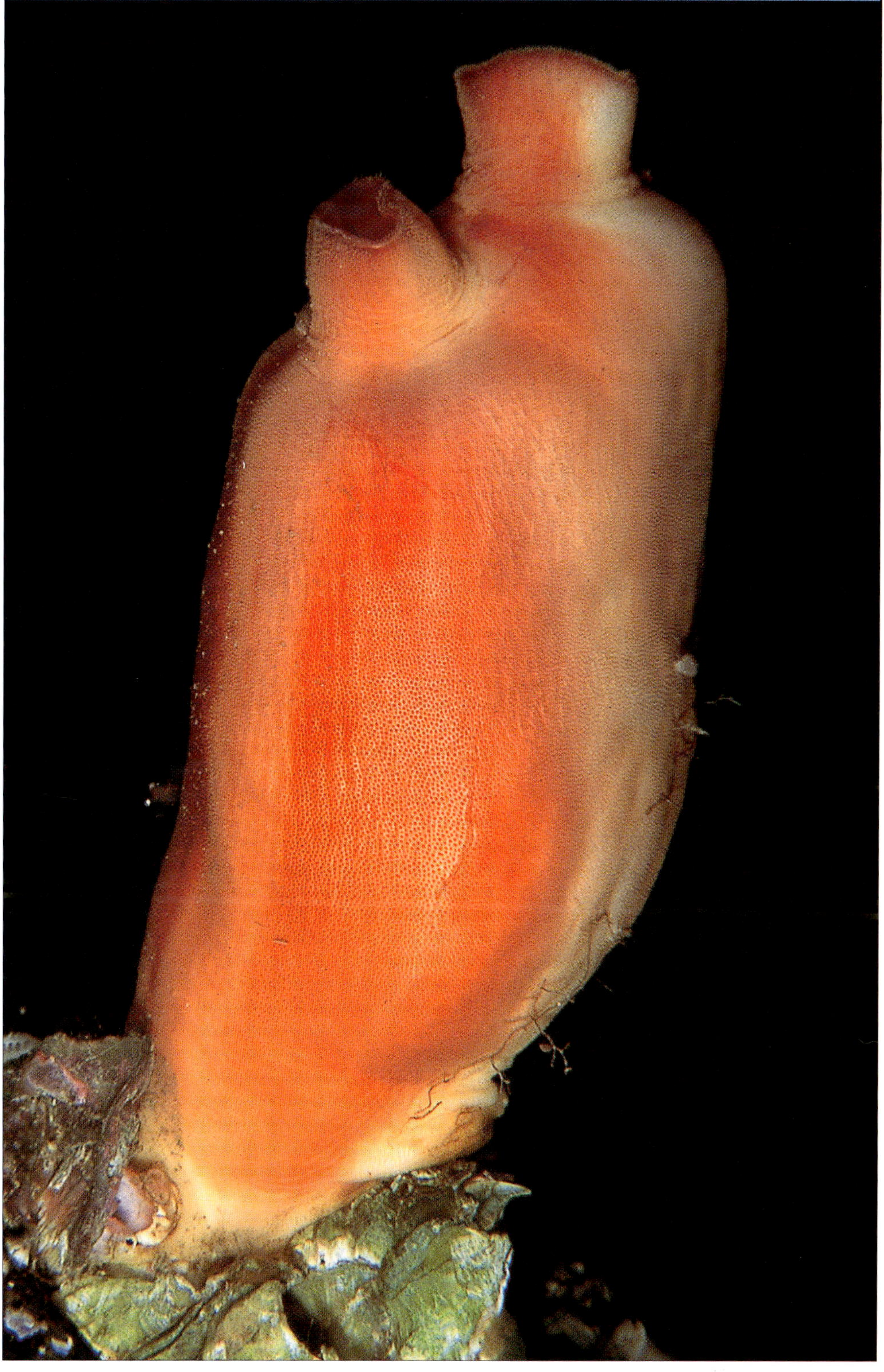

Phylum Chordata

This phylum (Lat. chorda = chorde + Lat. -ata = with-) includes tunicates, lancelets, and vertebrates. Chordates are bilaterally symmetrical, with an internal, longitudinal rod, the notochord, on the dorsal side (absent in adult tunicates). Fishes, which no longer are treated as a single taxonomic entity, belong to the chordates, and are presented in a chapter of their own.

HUMANS - *HOMO SAPIENS SAPIENS*

Man, as well sea squirts, belongs to the phylum Chordata. Our own species is labelled with the generic name *Homo* and the specific name *sapiens*. Often the subspecific name *sapiens* is added in order to separate us from the Neanderthals, *Homo sapiens neanderthaliensis*, that became extinct 35 000 years ago. The label subspecies indicates that *H. s. sapiens* and *H. s. neanderthaliensis* likely could reproduce with each other and obtain fertile offspring.

Subphylum Tunicata

Tunicates are unsegmented and exclusively marine chordates. The ***tunic****, or test, consists of a cellulose-like substance and covers all internal organs. The* ***chorda****, except in appendicularians, is present only in the larvae. It is reduced when the pelagic larvae metamorphose into the adult stage, which may be sessile or also pelagic, and solitary or colonial. The* ***branchial sac****, usually used for filtration, dominates the inside. The test has two openings, siphons, and the water enters through one and exits through the other. Some colonial tunicates have a common siphon for exiting water.*

Tunicata includes three classes. **Thaliacea**, where the **salps** (order **Salpida**) belong, are pelagic and have the siphons at opposite ends of the body. Some are solitary at some stage in their life cycle, but most are entirely colonial. The colonies are transparent, up to many metres in length, and occur near the surface. Most live in warm areas and are rare in our waters, although they may appear in connection with intrusions of more southerly water masses. **Appendicularia** also are pelagic, but solitary. They are 2 - 10 mm long, display many similarities with the larvae of other tunicates (they are sometimes referred to as Larvacea), and retain the tail with the chorda also as adults. Many produce a gelatinous house, which serves as a filtering device for plankton feeding. The house is shed after a few hours and replaced with a new one. Only a few appendicularians are common in the area. The third and largest group, **Ascidiacea**, is treated below.

Solitary sea squirts usually have a classic beaker-shape, with two siphons.

Class Ascidiacea - Sea squirts

Sea squirts (Gr. Ascidion = earthenware, - bag, - bladder) are sessile, benthic animals with the two siphons situated close together. As indicated by their Latin name, a thick, gelatinous or leather-like ***tunic*** *surrounds them. The* ***oral siphon*** *(for ingoing water) is situated on top in most species, and the* ***atrial siphon*** *(for outgoing water) on the side. Water, together with detritus and plankton, is inhaled into the* ***branchial sac*** *and trapped and concentrated by* ***cilia****. The food particles are subsequently transported to the gut through* ***ciliated furrows****. Some sea squirts catch larger prey, including amphipods and larger copepods. The water transport then continues out through the* ***atrial siphon****. Waste is discharged from the anus and follows the outgoing water. Some colonial taxa (such as* **Botryllus** *spp.) have a common opening for outgoing water, and each individual specimen in such colonies is referred to as a* ***zooid****. The branchial sac also serves for uptake of oxygen. A* ***heart*** *is positioned near the gut loop, connected to a simple* ***vascular system****.*

The vast majority of ascidians are **hermaphrodites**. Eggs and sperm are discharged near the opening of the atrial siphon. In order to avoid self-fertilisation, the male and female

Colonial ascidian - **Botryllus schlosseri**

The tunic often is transparent. Here the gut is visible, and the individual is defecating. At the base are the branchial sac and a small heart.

gonads mature at different times. The sperm is discharged into the water, and will in some species enter other animals and fertilise the eggs internally. The larvae may then be protected through the early stages of their life. In other species, both eggs and sperm are discharged. In addition, many colonial ascidians also have asexual reproduction.

The larvae are planktonic. There is a well-developed tail and chorda, which is resorbed and reduced when they settle and become sessile. Reproduction usually takes place in summer, and in most species the adults die soon afterwards. Some northern species, though, can attain an age of 2 - 3 years.

Many ascidians have a characteristic appearance (**habitus**), making identifications easy. Some genera, however, include a series of closely related species that can only be separated by closer examination.

PARASITES AND COMMENSALS

Here we can observe a specimen of **Ascidia virginea**, *attacked by two specimens of the snail* **Velutina velutina**. *These have deposited a number of eggs (white blotches) on the tunic of the sea squirt. The eggs soon become overgrown by the tunic, but with a canal through which they will escape in due time.*

Sea squirts have few predators, as the cellulose in the tunic is tough to digest. A few fishes feed on some species, and snails of the genus ***Velutina*** live by sucking body fluids from ascidians. As if this was not enough, they also deposit their eggs in the tunic of ***Ascidia virginea*** and ***Styela coriacea.*** The tunic grows to encapsulate the eggs, leaving only a small channel open where the larvae can crawl out. Many ascidians have parasites or commensals, mainly a number of copepods and amphipods, but also the crab ***Pinnotheres pinnotheres***, and nemerteans and shrimps may be present in the branchial sac. The mussel ***Musculus marmoratus*** often is attached to the outside of the tunic on species of ***Ascidia, Ascidiella*** and ***Denrodoa***, and after some time becomes completely surrounded by the tunic, with only a small opening for water exchange.

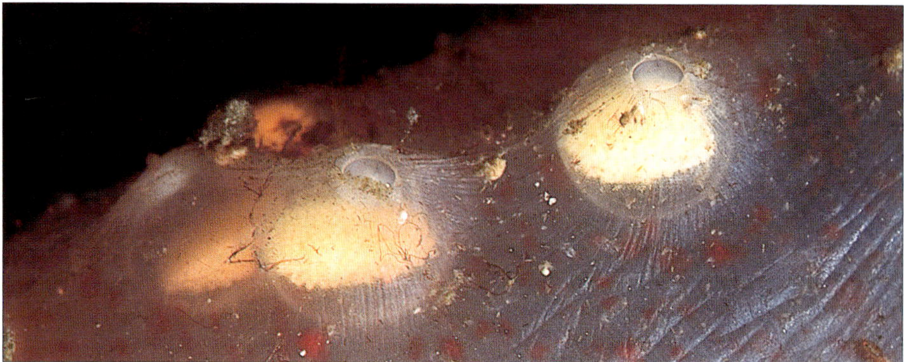

Clavelina lepadiformis

The small coiled individuals form characteristic colonies.

Distribution: Mediterranean, western Europe, north to mid-Norway.
Description: Characteristic colonies, specimens with long thin abdomen and wider thorax. Usually standing in small bouquets loosely connected to each other by a basal common cuticle. Solitary specimens also exist, then often larger, up to 40 mm. Tunic fully transparent, with white, light red or lemon yellow stripes. Oral and atrial openings circular.
Habitat: Colonies on various types of hard surfaces, often on rocks or large stones, but also on kelp, wrack and mussel shells. Common in sheltered places where summer temperatures are higher. Rarely below 20 metres.
Biology: Sexual reproduction takes place in July-August. The larvae swim for a few hours before settling and founding a new colony. In both new and older colonies, all zooids will be dispelled, and only tiny 'winter buds'–little larger than pin's heads but filled with nutrition–will survive the winter. These often occur 2 - 5 together in small groups. New development of zooids does not start until May, when water temperatures have risen. From this it takes about two months before they are reproductive.

Two individuals, with developing eggs. These will be fertilised by sperm from another individual. Self-fertilisation is avoided by the eggs and sperm maturing at different times within the same individual.

Sidnyum turbinatum

Colonies are formed of circular clusters of usually 6 - 12 elongate individuals.

Distribution: From western Mediterranean to northern Norway.
Description: Colonies with characteristic, elongated zooids, usually 6 - 12 together. Transparent, orange or light red, with white or yellowish white longitudinal stripes. Zooids 1 - 2 cm long. The eight lappets on the oral siphon are characteristic.
Habitat: From the intertidal and (rarely) down to 200 metres. Attached to various substrata, often rock, stones, mussel shells and different algae.
Biology: Reproduction takes place in summer. The parent colony dies when the larvae have left it. The larvae will usually settle and found new colonies in the nearby vicinity.

Some colonies are transparent and white.

This most likely is **Aplidium punctum.**

Diplosoma listerianum

Colonies form layers reminiscent of sponges.

Distribution: Mediterranean, northwestern Europe, north to northern Norway.
Description: Colonies with 1-2 mm long zooids, forming a thick, gelatinous and smooth cover on hard surfaces. Often on kelp stalks, or on flat surfaces where it will spread in all directions in an irregular fashion. More or less transparent, with green, brown or greyish yellow spots, zooids white internally.
Habitat: Covering algae, hydroid stalks, eelgrass or other surfaces.

Didemnum albidum

These colonies probably are **Didemnum albidum.** *Certain identification is difficult, and requires detailed examination.*

Distribution: Northerly distribution, from the Faeroes and western Norway to Spitsbergen and Greenland.

Description: Colonial, forming white crusts with characteristic, star-shaped oral siphons. May also be pink or yellow. Colonies grow in an irregular fashion, and can have a diameter of 70 - 80 mm, thickness rarely over 2 - 3 mm. The colony is hard and brittle, due to the presence of dense calcareous spicules.
Habitat: From the intertidal down to 1500 metres. Usually on stones or rock, but also on other organisms, including mussel shells, barnacles and other ascidians.

Ciona intestinalis

The siphonal margins are characteristically lemon-yellow.

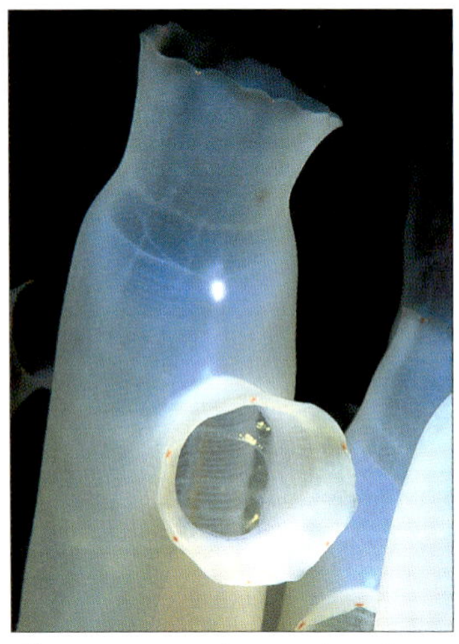

The orange patches on the edges of the siphons are sensory organs called ocelli.

In some localities, large aggregations of individuals may form.

DID YOU KNOW...

This species is one of the most studied sea squirts. Specimens have been kept for up to four years in aquaria, but will rarely become older than one year in nature. They increase rapidly in size, and go from larvae to a length of 40 mm in two summer months. For food they filter small plankton and detritus, and a medium-sized *Ciona* can filter 2 - 3 litres of water per hour. Observations show that it actually can 'creep' across surfaces, although very slowly and limited to very short distances.

Ciona intestinalis *usually has an obvious green tint. This individual clearly has grown over several seasons. Small sea anemones, probably* **Protanthea simplex**, *are growing on its surface. Here at 12 m depth in Hafrsfjord, Norway.*

Distribution: Very common in large parts of the northeast Atlantic, from the Black Sea to Spitsbergen.
Description: Solitary, even though many specimens often occur together. Tunic long, tubular, soft. Large, up to 150 mm long, rarely to 270 mm. The siphons are large and situated terminally, close together and with the atrial siphon slightly below the oral one, usually at an angle. Tunic rather transparent and internal organs visible. Colour bottle green, yellowish green or yellowish, occasionally bluish, rarely uncoloured. Siphons usually yellow. The oral siphon has eight, and the atrial siphon six incisions, each of which has a yellow or orange sense organ (ocellus) of uncertain significance. Body contractile, and will withdraw if touched.
Habitat: Often in large aggregations on hard bottoms, including vertical rock surfaces, stones, jetties, large kelp, etc. Can live in brackish waters down to salinities of 11‰, and is found from the intertidal down to over 1000 metres depth.

The species thrives on vertical cliff faces, often together with the fanworm **Sabella pavonina.** *Here together with* **Protanthea simplex**, *at 50 m depth in Høgsfjorden, southwestern Norway.*

Corella parallelogramma

Distribution: Northern Europe, from the English Channel to northern Norway. There are also some Mediterranean records.

Description: The tunic is completely transparent, and the internal organs, including the white branchial sac, are fully visible. Up to 50 mm high and 30 mm wide. Oral and atrial siphons inserted distally, rather close to each other. There are eight red pigment spots surrounding the oral siphon, and six around the atrial one. Usually standing up from surfaces. The gut is situated on the same side as the atrial siphon. The characteristically pigmented branchial sac makes this an easily identified species, and confusions are unlikely.

The branchial sac has a very distinct lattice-like appearance.

Habitat: Very common on hard bottoms from a few down to a few hundred metres depth. Often in large numbers on walls protected from light. Common on large brown algae.

Biology: Mature in spring, 9 - 12 months after hatching. Reproduction takes place during summer. It is likely that none of the animals survive reproduction, such that the life span does not exceed 15 months.

As in other species, the branchial sac is used both for oxygen uptake and filtering of food particles.

Ascidia mentula

Distribution: Mediterranean and northern Europe, from the English Channel to northern Norway.

Description: Relatively large species, usually lying on its 'side' with the atrial siphon pointing upwards. The two siphons are far from each other, with a distance exceeding half the length of the animal. Body shape elongated, rounded in transsection, slightly pointed towards the oral siphon, and largest slightly below the middle. Colour usually red, and the tunic is semitransparent. Other colourations occur, including green, olive, light brown and, rarely, fully transparent. Recorded length to 300 mm, although usually up to 180 mm.

This species usually lies on the bottom, with the atrial and oral siphons widely spaced.

Oral siphons with characteristic incisions.

Habitat: Attached with the lower part of one side, from the intertidal to 150 metres depth. Occurs also with shell gravel, given the presence of any larger shells for attachment.

Biology: Reproduction in northern waters takes place in summer. In the Mediterranean and around the British Isles the branchial sac of this sea squirt is commonly inhabited by the small crab ***Pinnotheres pinnotheres***.

This sea squirt is rectangular in shape, and usually has dispersed reddish-pink patches on the tunic.

Ascidia virginea

Distribution: Mediterranean and northern Europe, from the English Channel to northern Norway.

Description: A beautiful species, easily recognised by its even rectangular shape and the pale pink tunic, often with sharp rose-red spots. Shape elongated, more than twice as long as wide, slightly tapering towards the siphons. Oral and atrial siphons not prominent, appearing mostly as pores in the tunic, situated at a distance from each other representing 1/3 to 1/4 of the total length of the animal. Length up to 95 mm, although rarely exceeding 60 mm.

Habitat: Usually attached basally and standing erect on rocky bottoms. Mainly below 10

These individuals are almost completely covered with reddish-pink pigment spots.

metres depth, down to 3000 metres. Also on mixed bottoms with shell gravel and mud.

Biology: Often attacked by the snail ***Velutina velutina***, which is one of the few animals that feeds on ascidians. Not only does *Velutina* suck the body fluids of the ascidian, but it also deposits its eggs on the tunic. The tunic will soon overgrow the eggs, although, by an unknown mechanism, a small channel to the outside will be kept open. In due time the eggs will hatch, and the snail larvae emerge.

Ascidiella aspersa

Distribution: Mediterranean and northern Europe, from the English Channel to northern Norway.
Description: Large, up to 13 cm long, with a coarse, greyish and finely knobbed surface. The oral siphon is inserted terminally, whereas the atrial siphon is inserted at a distance corresponding to 1/3 of the total body length. Inside the oral siphon there is a row of 30 - 40 tentacles. Similar to the co-occurring ***A. scabra***, which does exceed 5 cm in length. The latter also is more transparent, has up to 100 oral tentacles, and, in contrast to *A. aspersa*, has internal red pigmentation.
Habitat: Attached basally, or with 1/3 of the side, to hard surfaces, from the intertidal down to about 60 metres depth.

Ascidiella *scabra* *often has a reddish pigmentation on the inside. Here at 10 m depth.*

This sea squirt has a characteristic knobbed surface.

WATER PISTOLS

The species epithet *aspersa* is derived from the latin asperio, meaning "squirting". This phenomenon is familiar to anyone who has picked up this creature from the shore and squeezed it - it squirts like a water pistol.

Dendrodoa grossularia

Distribution: Widely distributed in the northern parts of the North Atlantic, from the English Channel to Spitsbergen and Greenland.

Description: Exists in two forms, one solitary and one forming dense aggregations. In northern areas the solitary form dominates. It is inflated, spherical, with two short siphons. The aggregating form is dominating around the British Isles. Diameter of solitary specimens up to 11-12 mm. Colour light red, brick red, or rusty red, deeper specimens yellowish brown.

Habitat: On all kind of surfaces, especially stones, empty mussel and various other shells, wrack (***Fucus*** spp.), and on kelp holdfasts. From the intertidal down to 600 metres depth.

Biology: Viviparous, with large red eggs inside the tunic, where the larval development takes place during about two weeks (temperature dependent). Hatching and development can only proceed at temperatures of 8-15°C. A mature specimen (about 5-6 mm wide) can have 20 - 100 eggs, rarely up to 200. The eggs are fertilised in intervals, and the same specimens will therefore contain eggs at different developmental stages. The released larvae are 2.4 mm long. Metamorphosis and attachment to a surface will take place only a few hours after release, where after the larvae starts its sessile life. Specimens that settle in spring can be mature already in the autumn. An average lifespan is about 18 - 24 months.

This colonial ascidian forms attractive layers, usually on kelp stipes or blades.

DID YOU KNOW...

Dendroa grossularia around the British Isles occurs in high densities, with up to 50 - 60 000 specimens/m^2. It is a tough species that can stand several hours of desiccation, and salinities down to 13‰.

Botryllus schlosseri

This colonial ascidian forms attractive layers, usually on kelp stipes or blades.

Distribution: Widely distributed in the northern parts of the North Atlantic, from western France to Spitsbergen. Also the Black Sea and the Mediterranean.

Description: A beautiful, colonial species, forming 2 - 4 mm thick covers on kelp, wracks or other surfaces. The individual zooids are arranged in star-shaped clusters, and have a common atrial siphon. The colonies tend to be thicker when they grow on thinner algae or hydroid stalks. There are many colour morphs, but are often visible as white or yellow stars against the darker background. Blue, green, violet, brown or red morphs also occur, although all zooids in the same colony will have the same colours. Each star-like colony consists of 4 - 16 zooids, each of which is 3 - 4 mm long. The species may be confounded with the quite similar ***Botrylloides leachii***, which also is colonial. In the latter, however, the colonies are not typically star-shaped, but instead arranged in elongated double rows, interrupted by branching atrial siphons.

Habitat: Grows on hard surfaces, such as kelp stalks, algae or stones, sometimes also on other ascidians. From the intertidal down to 20 - 30 metres depth.

The communal atrial siphon is in the middle, surrounded by a ring of oral siphons.

Biology: Reproduction, at least in northern areas, takes place in May-August. The larvae will settle and metamorphose within a few hours after leaving the parent, and found a new colony by asexual reproduction. A colony that is founded during summer will in the autumn have attained a size of 8 - 20 mm, and include up to 12 zooids. Subsequently, growth decreases and stops due to low temperatures and food availability. It is resumed in April, and they will reach their maximum size in the following autumn. After that they die, at an age of about 18 months.

Botryllryllus leachii

This species occurs as colonies, usually seen on kelp blades in sheltered localities.

Distribution: Widely distributed in the northern parts of the North Atlantic, from western France to northern Norway. Also the Black Sea and the Mediterranean.

Description: Similar to the preceding species, ***Botryllus schlosseri***, but the colonies are not star-shaped (with the exception of young ones). The zooids are instead arranged pair wise in elongated, sometimes meandering, clusters now and then with common, branched atrial siphons. Colour highly variable, dominated by brown, yellow, violet or orange.

Many colours occur.

Habitat: As with the previous species, *B. leachii* grows on kelp, wracks and other surfaces. Large number of colonies may occur together on old kelp in more protected areas. From the intertidal down to 20-30 metres depth.

Biology: Each colony stems from a single larvae, which after metamorphosis subdivides asexually, thus founding a colony of genetically identical individuals.

Halocynthia pyriformis

Distribution: Mainly Arctic, from western Norway and northwards.
Description: A beautiful, large, barrel-shaped, solitary ascidian, up to 100 mm high. The tunic is leather-like, with minute spines sitting singly or in small groups. Both siphons similar, oral siphon near terminal, atrial siphon situated slightly further down.
Habitat: Occurs in shallow water. Some specimens may be completely white, although they are usually reddish brown with yellowish stripes. Lives both in areas with strong currents and in more sheltered places.

A magnificent sea squirt, which has a northern distribution.

Large aggregations are occasionally seen. Here at 20 m depth in Narvik harbour, northern Norway.

The colour also can be white.

Right**: these individuals belong within the genus* **Molgula, which includes about 10 species in our waters. They are more or less rounded in shape, with a thin, often transparent tunic which also can be covered by sand or mud. Usually attached to kelp stipes; here on red algae.*

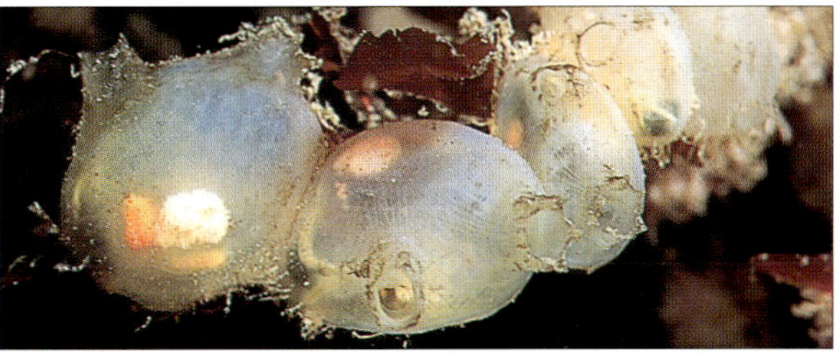

A few unidentified sea squirts from western Norway

Boltenia echinata *is common along the entire Norwegian coast.*

Subphylum Cephalochordata - Lancelets

Lancelets (Gr. Kephale head + Lat. chordata = chorde, synonym ***Acrania****, Gr. a- = not, and Gr. crania = skull) are small, free-living and benthic chordates with a fish-like body that tapers evenly towards both ends. The chorda runs through the whole body. A* ***nerve cord*** *runs dorsally to the chorda, and is anteriorly slightly widened, but shows little specialisation. They have a* ***closed vascular system*** *but lack heart. However, some gastric vessels are contractile and pumps blood through the vessels. They lack eyes and jaws. An* ***unpaired fin*** *extends along venter, dorsum and tail. The mouth opening is located ventrally on the 'head', leading into a large branchial sac or pharynx, provided with a large number of lateral* ***gill bars****. Lancelets have* ***separate sexes*** *and* ***external fertilisation****. The simplicity of the lancelet makes it well suited as a model organism for studies of vertebrate morphology. Although it does not actually belong to the vertebrates, it is considered as their closest relative. Adult lancelets burrow in coarse sand and shell sand, while the larvae are planktonic. They occur in coastal areas in tropical and temperate regions. The group only includes two genera, with less than 20 species in all. A single species occurs in the area.*

Branchiostoma lanceolatum - Lancelet

In evolutionary terms, the lancelet may be thought of as the precursor to the invertebrates.

Distribution: In the Mediterranean and Black Sea, and from Portugal to northern Norway.

Description: Bilaterally compressed, fish-like chordate. Chorda extending through whole animal. Eyes lacking. White to greyish white, semitransparent. Cannot be confused with other species. About 40 - 50 mm long.

Habitat: Adults live buried in medium fine sand or shell sand, with only the mouth in contact with the surface. Occurs from a few metres depth, in shallow water. Careful digging into bottoms as described above will frighten the lancelets to swim up and then quickly dig into the sand again. After a few times they will be exhausted, allowing them to be examined and photographed.

Biology: Filters small organisms from the water. They become mature only after three years, and may live about eight years. The larvae look like small adults, and can be found in the plankton in July-October. For a few months the larvae lead a double-life, spending the days on the bottom, but the nights as plankton.

FISHES

Fishes

Few, if any, animal groups can boast such a wide variation in form, colour, size and distribution amongst its members as is found in the fish. The colour display of fish around a tropical coral reef can be overwhelming. In this book, we show that also the fishes in our waters have a fascinating range of shapes and colours.

Fish are vertebrates (sub-phylum Vertebrata), but there is no single systematic taxon that includes only fish. The super-class **Agnatha**, the jawless fish, includes lampreys and hagfish. The super-class **Gnathostoma**, the jawed fish, comprises the **Chondrichthyes**, the cartilaginous fish, such as sharks and skates and the **Osteichthyes**, the bony fish, which includes our most familiar fish. Most fishes have a scaly, torpedo-shaped body with fins, a small two-chambered heart and take up oxygen through their gills.

MANY SIZES

The world's largest fish is the whale shark, from tropical waters. The largest recorded individual was 13.7 m long and weighed 15 tonnes. Whale sharks produce very large eggs, 30 x 14 x 9 cm; the largest of all currently existing organisms. Even though it looks dangerous, it is in fact harmless to humans, because it feeds solely on plankton. For comparison, the worlds smallest fish is the 10 mm long goby, ***Pandaka pygmaea***.

Haddock - *Melanogrammus aeglefinus*

MANY VARIETIES

There are currently 28 400 recorded fish species, and more are described each year. There are more than 300 living marine fish species recorded from Northern European waters.

This book presents 105 marine fish species, which can be encountered in relatively shallow coastal waters in various parts of northern Europe. There are a large number of deep-water, or rare fish that are not dealt with here.

Class Agnatha - Jawless fishes

This is a so-called primitive, or early evolved group of fish, including two orders; ***Petromyzontiformes****, the* ***lampreys*** *and* ***Myxiniformes****, the* ***hagfishes****. In our waters, there are four species of lamprey, and two of hagfish. Lampreys reproduce in freshwater, using external fertilisation, but three of the species spend a lengthy period of their lives in the sea. Jawless fish have typically slender, eel-like bodies lacking paired limbs and scales. Further, their digestive system lacks a stomach, but they have taste, smell and hearing organs.*

Myxine glutinosa - Atlantic hagfish

Upper*: Writhing hagfish photographed near Stavanger, southwest Norway. We placed bait to entice them from the muddy sediment. Depth 55 m.*
Left*: Mouthparts.*

Distribution: Found on both sides of the Atlantic; in the east from Morocco north to northern Norway. Western and northeastern coasts of the British Isles. Western Mediterranean.
Description: Eel-shaped, often pink in colour. Scales absent and skin very slimy. Limbs absent, but has a muscular fin on the posterior part of the back. Belly-side equipped with many slime-producing cells. Four tentacles arranged round the star-shaped mouth. Body length up to 60 cm.
Habitat: Dependent on soft bottoms, and does not thrive in water temperatures over 10°C. Also requires a high salinity > 32‰ . Depth 20 - 1300 m. Often buried, with only the head projecting from the sediment.
Biology: Mainly feeds on carrion, but will also take polychaete worms, crustaceans and squid. The hagfish uses its sense of smell to locate prey. Eyes rudimentary, skin-covered and with only a limited function. Light-sensitive organs in the skin. Because of its evolutionary history, the hagfish is an attractive research topic. However, its reproduction is still rather unclear. Before sexual maturity, their sex is indeterminate, because they have an immature set of both male and female organs. During sexual maturation, either the male or the female gonads develop. A copulatory organ is absent, so it is thought that fertilisation occurs externally. Females lay up to 30 eggs (14 - 25 mm diameter) per year. There is no larval stage, and the eggs hatch into 3 – 4 cm long juveniles.
Commercial value: The flesh is reputed to be tasty, but it is not exploited. ***Myxine*** attacks and damages fish catches, especially on bottom long-lines and nets.

SLIME-PRODUCING

In some languages, this fish is called the "slime-eel" due to its ability to produce large quantities of slime. This protects the skin during burrowing, and helps to maintain the body's salt balance. It also uses the slime to package its prey, which serves to keep it safe from other carrion-feeders.

Class Chondrichthyes - Cartilaginous fishes

Cartilaginous fish include sharks, skates and ratfishes, chimaeras. These all have a skeleton made of cartilage. Parts of this can be calcified, but proper bones are absent. The teeth are rooted in the gums, not in the jaw-bone, and are arranged in parallel rows. As the teeth are worn down, new ones develop to replace them. Cartilaginous fish lack a swim bladder, instead regulating their buoyancy using their outwardly-pointing fins and large oily liver. Most have several gill arches, but lack a gill cover (except the ratfish). The group includes approx. 800 species, including 339 sharks, 424 rays/skates and approx. 30 different types of ratfishes. A few species are common in coastal water, but the vast majority live in deep water. Here we present the most common species found along our coasts in relatively shallow water.

The ratfish (p. 482) is the only representative of cartilaginous fish within the order Chimaeriformes.

DID YOU KNOW...

Cartilaginous fish have internal fertilisation. Oviparous species lay eggs that develop after they are laid. In ovoviparous species, the eggs are hatched within the female's body. Viviparous species have a foetal development resembling that of mammals. Therefore, ovoviparous and viviparous species give birth to live young. Sexual maturation is slow, and the number of offspring low. These fish are therefore vulnerable to exploitation.

Super-order Selachimorpha - Sharks

*Sharks often are associated with warmer waters. However, there are around eight shark species that live permanently in our waters. Several more make shorter visits during summertime. The plankton-eating **basking shark,*** **Cetorhinus maximus** *is the largest of the permanent residents, with a length of up to 13.4 m and weighing 3 tonnes. Further, the **Greenland shark,*** **Somniosus microcephalus**, *and the **porbeagle**,* **Lamna nasus**, *are large sharks that generally live in deep water. The remaining five species are presented below.*

Family Squalidae - Dogfish sharks

Within this family are three species relatively common in north European waters. The two presented here are small, whereas the third, **Somniosus microcephalus**, *is a giant of up to 8 m in length. Members of this family lack the anal fin (located just behind the anus).*

Squalus acanthias - Spiny dogfish

The spiny dogfish is an active predator. Note the spines at the anterior of the dorsal fins.

Distribution: Both sides of the Atlantic, Black Sea, Mediterranean, west Africa to the Barents Sea. Very common.
Description: Two dorsal fins, both with stout spines at their anterior margins, equipped with poison glands that can cause painful wounds. Dorsal, or back, surface dark grey, with scattered paler patches, which disappear with age. Ventral, or belly, side pale grey. Blunt snout. Age up to 25 years. Females largest, up to 125 cm in length and weighing 9 kg.
Habitat: Found right up to the surface, down to several hundred metres depth. Prefers soft bottoms. Can occur in shoals of up to 20 000 fish.
Biology: Ovoviparous; eggs develop within the females. She gives birth to 4 - 8 young in open water. Gestation lasts between 18 - 22 months. The young stay in shallow water until they are around 50 cm long. Preys mostly on herring and codfish, but also invertebrates. Adults undertake large migrations. The 'norwegian' spiny dogfish migrates west of the Shetland and Orkney Islands in the autumn. Females move towards the coast to give birth between November - February.

SOUGHT-AFTER MEAT

The flesh of the spiny dogfish is sought-after and exported to western Europe. Since the 1960s, the catches have significantly decreased, and prices have increased. The stocks cannot withstand extensive fishing due to their slow maturation and low number of young produced. Norway has therefore introduced a minimum catch length of 70 cm.

Etmopterus spinax - Velvet belly

Distribution: Entire eastern Atlantic from Senegal, north to Iceland and northern Norway. Western Mediterranean.

Description: Our smallest shark, maximally 60 cm in length. Dorsal side blue-black, belly greyish black with light organs. Pale lateral stripes along the sides. As in the spiny dogfish, a spine at the anterior of each dorsal fin. Skin velvet-like (hence the common name), with a violet tinge. Snout long and rounded.

Habitat: Prefers soft bottoms in relatively deep water. Common on the continental shelf and in deep fjords from 200 - 900 m. Can occasionally be met at diveable depths.

Biology: In summer, females give birth to 6 - 20 young, 12 - 13 cm in length. Feeds on small fish, crustaceans and squid. Light organs present, in common with many deep-water fish. These are located on the belly and emit light in blinking series, probably in connection with mating. On the dorsal side under the skin is often a parasitic soft cirriped, ***Analesma squalicola***, a relative of the barnacle.

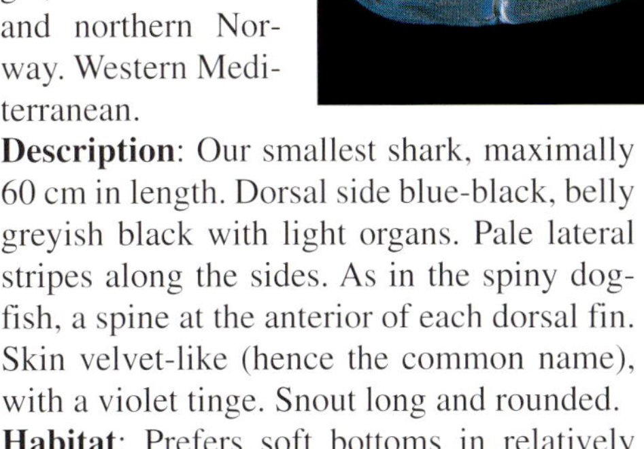

Parasitic amphipod on a velvet belly.

The large green eyes are typical for deep-living cartilaginous fish. Photographed at night at 45 m depth.

Family Carcharhinidae - Requiem sharks

Although this is the most diverse shark family, with more than 150 recorded species, only a few species are permanent residents in northern Europe. Most common is the tope shark. The up to 4 m long and 180 kg in weight ***blue shark****,* **Prionace glauca***, and the 1.6 m long* ***smooth-hound****,* **Mustelus mustelus** *occasionally appear off the Norwegian coast; both are common around the west coasts of the British Isles.*

Galeorhinus galeus - Tope shark

Distribution: Cosmopolitan; entire eastern Atlantic from northern Norway and Iceland to South Africa. Mediterranean, east coast of North and South America and Australia.

Description: Dorsal, or back side steel grey; belly whitish with mother-of-pearl sheen. As typical for the family, a torpedo-shaped body. Up to 2 m long and 30 - 40 kg in weight, but most individuals smaller. Length of the base of the 2nd dorsal fin around half that of the 1st dorsal fin. Tail fin with a characteristic incision in the lower half. As in other family members, eyes with a skin fold that can be drawn over the eyes.

Habitat: Generally in shallow coastal waters; larger individuals sporadically found down to 200-300 metres depth in wintertime. Usually keeps to the bottom, but also hunts pelagically. Undertakes seasonal migrations.

Biology: Feeds mainly on fish, especially whiting, and squid, but also takes various echinoderms and crustaceans. Ovoviviparous; females give birth to 25 - 35 live young, after around 10 months gestation. Young are 35 - 40 cm long at birth.

Commercial value: Not commercially exploited in Europe, but up to the 1950s, fished for its vitamin-rich liver. Popular as a sport fish, especially around the British Isles.

The tope shark has the classic 'torpedo-shaped' shark outline.

Family Scyliorhinidae - Cat sharks

*Most family members are small sharks, of which the **lesser spotted dogfish** and the **blackmouth catshark** are permant residents along our coasts. Another species, the **nursehound**,* **Scyliorhinus stellaris,** *is only occasionally recorded in southern Scandinavia, but common around the British Isles.*

Scyliorhinus canicula - Lesser spotted dogfish

Eggs from the lesser spotted dogfish often are attached to kelp stipes. Photographed at 40 m.

Distribution: From western Africa to northern Norway. Mediterranean.

Description: Up to 1 m in length and weighing around 2 kg. Small dark patches on the back. Basic colour sandy or brownish, also some pale patches. Tail fin oar-shaped.

Habitat: Common in shallow water, down to 100 metres. Prefers sandy, gravely and muddy bottoms, where it is found resting during daytime.

Biology: Internal fertilisation carried out throughout the year, but the eggs are laid in two capsules mainly between January to July. Egg capsules 5 - 6 cm in length, and attached to algae or soft corals, by means of metre-long threads extending from the corners. Eggs hatch 5 - 11 months after they are laid. Predator on small fish, but also various benthic animals. Nocturnal.

The skin has a sandpaper-like surface.

The lesser spotted dogfish (picture) can only be confused with the nursehound.

Galeus melastomus - Blackmouth catshark

The blackmouth catshark is seldom seen by divers. It usually occurs at depths below 100 m, but occasionally moves up to shallower water after dark.

Distribution: Found along most western European coasts from the Mediterranean north to northern Norway.

Description: Relatively small, females up to 90 cm long; males seldom exceeding 60 cm. Dorsal fins similar in size; anal fin long. Females with a pair of long copulatory organs extending beyond the pectoral fins. A characteristic when caught is the black inside of the mouth.

Habitat: Prefers relatively deep waters, usually between 50 - 200 metres depth, but also occurs down to 1200 metres. In fjords, we have seen it as shallow as 22 metres. Most common on banks, offshore. Primarily bottom-living, but occasionally makes excursions into the water masses. We have seen this shark twice in southwestern Norway during night dives. The shark was 'resting' on a steep slope and calmly disappeared into the depths as we approached.

Biology: Internal fertilisation, eggs laid in 5 - 7 brown horny capsules approx. 6 x 3 cm. Each capsule contains two eggs and has short 'teeth' at the corners. Capsules laid on sandy bottoms and probably hatch in winter. Mostly

Close contact with fish that usually live in deep water is an exciting experience. The blackmouth catshark has beautiful large eyes. This specimen was resting on a rocky ledge at around 22 m depth.

feeds on a variety of bottom-dwelling organisms, including various shrimps, bivalves and squids, but also takes fish such as the silver smelt, ***Argentina sphyraena*** and the bib, ***Trisopterus luscus***. Also some pelagic krill. Mainly hunts at night, resting on the bottom by day.

Commercial value: Very tasty meat, but not exploited commercially.

The blackmouth catshark typically has large reddish patches along the sides and a characteristic tail. Here photographed during a night dive in southwest Norway at 25 m depth.

Order Rajiformes - Skates and rays

In Northern Europe our permanent residents represent only one family, the Rajidae. However, the ***common stingray*****, Dasyatis pastinaca***, within the family Dasyatidae also occurs north to southern Scandinavia. Furthermore one species of the eagle rays (Myliobatidae), the* ***common eagle ray*****, Myliobatis aquila***, has been observed in southern Scandinavia. Skates and rays are recognised by their dorsally flattened body, and the large pectoral fins ('wings'), which are fused to the head. Unlike other flat fish, the gill openings and the mouth are on the underside. Most species have a clearly delineated tail region.*

Family Rajidae - Skates and rays

*Seven members of this family are commonly found along our coasts, but because most live in deep-water, we rarely see them. In addition to the two species presented here, the group includes the **common skate**,* **Raja batis,** *the **Norwegian**, or **black skate**,* **R. nidrosiensis**, *the **longnose skate**,* **R. oxyrhynchus**, *the **shagreen ray**,* **R. fullonica** *and the **round ray**,* **R. fyllae**. *The following are rare, or occur offshore; the* **sail ray**, **R. lintea**; *the **Arctic Skate**,* **R. hyperborea**; *the **spotted ray**,* **R. montagui** *(common south and west coasts of the British Isles) and the **sandy ray**,* **R. circularis**.

Raja clavata - Thornback ray/ Roker

Distribution: Entire eastern Atlantic coast north to northern Norway. Mediterranean and Black Sea. Iceland. One of our most common skates.
Description: Recognisable by its blunt snout and parts of its belly being rough with small skin-teeth. Numerous rows of spines on the body disc and three rows of spines on the tail. Brown, with small dark-brown and large pale patches. Males up to 90 cm in length; females up to 120 cm, weighing up to 18 kg.
Habitat: Prefers sandy bottoms between 20 - 300 metres depth. Unlike most of our other skates, can be found right up to the shore. Often partially buried in the sediment.
Biology: Like most skates, it lives on the bottom, feeding on various benthic organisms. Occurs offshore in winter. Spawning migrations to the coast begin in springtime, and mating occurs *en route*. About 15 - 20 black, rectangular egg capsules laid in summer; these often found washed up on the shore. Can give an electric shock if the tail is touched. Can reach at least 20 years of age.
Commercial value: Very tasty flesh. Considerable quantities are fished by trawl and longlining around the British Isles, but not to any great extent in Norway.

The thornback ray can occur in several colour variations.

Raja radiata - Starry ray

The starry ray is often seen in very shallow water. The upper photograph shows the mouth and exhalent gill openings.

***Below**: Detail of the eye, showing an attached leech sucking body fluids.*

Distribution: Eastern parts of the British Isles and north to Spitsbergen. Also west coast of north America. Entire Norwegian coast. Probably our most common ray.
Description: Short snout, greyish brown with a rough surface. Between 13 - 17 large, strong spines mid-dorsally, extending back to the 1st dorsal fin (located on the tail). Belly smooth and white. Up to 90 cm long and weighing 17 kg.
Habitat: Sandy and muddy bottoms from 20 - 400 metres, at temperatures between 1 - 10 °C. Often moves to shallow waters, 5 - 10 metres depth, at nights.

Biology: Like the thornback ray, undertakes regular spawning migrations from offshore to coastal waters. The sexually mature females move towards the shores in springtime; males around a month later. Egg capsules laid all year round, but mostly during March - August. These capsules, often seen washed up on the beach after a storm, are oblong with perpendicular stripes and elongate horns at each corner. Feeds mainly on various benthic fish and crustaceans.

Order Chimaeriformes - Chimaeras/ Ratfishes

Within this group, only one family, the ***Chimaeridae****, and only one species, is represented in northern coastal waters. Unlike all the other cartilaginous fish, members of this order have an external gill cover and the body shape is more reminiscent of bony fish. The head is large, adults lack scales and the body ends in a long, whip-like tail. At the anterior of the first dorsal fin is a spine connected to a poison gland. They live mostly in deep water.*

Chimaera monstrosa - Ratfish/ Rabbitfish

Distribution: Western Africa, north to the Kola Peninnsula. Mediterranean, Iceland, entire Norwegian coast.

Description: Characteristic appearance with large head. Body narrowing, ending in a

DEADLY POISON

The triangular, high dorsal fin of the ratfish has a stout spine connected to a poison gland that produces a deadly poison.

At night, divers may be fortunate enough to encounter the ratfish, which is a tremendous experience. ***Inset****: Detail of the eye.*

At 15 m depth in the Trondheimsfjord, mid-Norway.

er, this bony fish is distinguished by its large prominent scales and blunt snout. Further, it is black inside the mouth and gill cavity.

Habitat: Soft bottoms, mostly in deep water, between 50 - 1000 metres depth, shallowest in spring and summer. However, in certain fjords, it can be seen as shallow as 15 metres at night.

Biology: Fertilisation is internal; eggs spawned in spring and summer, in club-shaped egg capsules, 17 cm x 3 cm, with a long thread at each corner. Juveniles around 11 cm at hatching. Feeds on bivalves, brittle-stars, crustaceans and small fish. Commonly found on shrimp beds.

whip-like tail. Eyes large, with green lenses. Mouth located under a short snout. Skin smooth and silvery. Up to 1.5 m long (including tail), weighing up to 2.5 kg. Not easily confused with other fish in our waters, although there is another species with a whip-like tail; ***Coryphaenoides rupestris***. Howev-

Commercial value: The liver oil previously was used as a lubricant and as a domestic remedy against ailments such as stings from the **greater weever**, ***Trachinus draco*** and a festering finger.

Class Osteichthyes - Bony fishes

*This class is usually divided into the **lobe-finned fishes** (Sarcopterygii), which include **Coelacanths** and the **lung fishes**, and the **ray-finned fishes** (Actinopterygii) which include all the other groups. Bony fish are the most numerous and diverse of all the fish. The skeleton is made of true bone and the gill arches are covered with a gill covering that has a single opening. Unlike sharks and rays, bony fish can flatten or fold their fins. Most have a swim bladder to regulate buoyancy. However, those bony fish that are very well-developed swimmers or live on the bottom, lack swim bladders. Most use external fertilisation, and spawn large quantities of either eggs or sperm. In most species, the larvae more or less drift freely as plankton during the first weeks after hatching. Most bony fish are exclusively marine, although some live wholly, or partly, in fresh water.*

The topknot, **Zeugopterus punctatus**, *can be seen at only a few metres depth, usually firmly clinging to the substrate.*

Order Clupeiformes - Anchovies and herrings

This group comprises relatively early-evolved fish. The pectoral fins are positioned low on the body, and the pelvic fins are relatively far back. Further, the fins have soft, jointed ribs. The body is covered in gleaming silver scales, and the tail is cleft. Herring fish often form large shoals, and feed on plankton, such as copepods and planktonic larvae. Because of the huge shoals that can occur, these fish are important for fisheries. They are exploited world-wide, either directly by fishing, or indirectly, in that they are an important food source for other commercially fished species. In addition to being eaten by humans and fish, herring are important in the diet of marine mammals and seabirds. The group includes two families represented in our waters; ***herrings–Clupeidae*** *and* ***anchovies–Engraulidae.***

Clupea harengus - Herring

Distribution: Northern distribution. In the north Atlantic from the Bay of Biscay in the south to the Barents Sea, the White Sea and Novaja Semlja in the north.

Description: Gleaming silvery, somewhat laterally compressed, up to 50 cm long. Dorsum, or back, a deep blue. Large scales forming a rounded keel along the belly. Large, incised tail fin. Prominent overbite, and large eyes. Single dorsal fin lacking rays; pectoral fins usually with 9 rays. Based on their habitat, four sub-species have been identified and these also have characteristic appearances. ***Clupea harengus marisalbi*** and ***C. h. suworowi*** are found in the White Sea and the Kara Sea. The **Baltic herring**, ***C. h. membras***, lives in the Baltic and the **Atlantic herring *C. h. harengus*** is the main form throughout the Atlantic and Norwegian Sea. The latter is further sub-divided into various tribes. In general, herring that live in the open sea tend to be larger than those spending most of their lives close to the coast and inside fjords or inlets. Some water bodies have their own herring populations, such as in Trondheimsfjorden, Sognefjorden and Lysefjorden in Norway. Three other members of Clupeiformes can be confused with the herring in our waters. The **sprat**, ***Sprattus sprattus***, the **sardine**, ***Sardina pilchardus*** and the **anchovy**, ***Engraulus encrasicholus***. The latter is considerably more slender, more laterally compressed (from the sides), with a prominent underbite. The up to 20 cm long sprat is distinguished by the pectoral fin being located in line with the anterior edge of the dorsal fin, whereas it is positioned behind it in herring. In contrast to the herring and sprat's dark blue back, the sardine has a dark green back, and the gill coverings have radial stripes. Sardines reach up to 25 cm in length, have a southern distribution and are only occasional visitors in our waters.

Habitat: The herring is pelagic, occurring in large shoals near the coast, on the continental shelf or in fjords and inlets, from the surface to around 200 m. Spawning is carried out on hard

HERRING IN THE CITY'S COAT OF ARMS

Already in the Viking period there where considerable herring fisheries in eastern England, along the channel coasts and the Dutch coast. Cities such as Great Yarmouth in east England and Enkhuizen in Noord-Holland developed a strong economy due to these fisheries, and expressed their gratitude in their cities' coat of arms, both of which display three herrings.

bottoms, usually between 40 - 70 metres (10 - 200 metres) depth. Herring in the Norwegian Sea belong to three different stocks, comprising the Atlanto-Scandinavian herring; Norwegian spring-spawning herring, Icelandic spring-spawning herring and Icelandic summer-spawning herring. All have a common feeding ground between Iceland and Jan Mayen, but spawning occurs either around Iceland or the Norwegian coast. The Norwegian spring-spawning herring stocks live along the Norwegian coast, in the Norwegian Sea and in the southern parts of the Barents Sea. These spawn along most of the Norwegian coast. In the North Sea, there are at least two additional herring stocks; one spring-spawning and another that spawns in autumn.

Biology: Largely feeds on planktonic crustaceans, particularly copepods, but also various planktonic larvae. Sexually mature at 2 - 5 years, depending on the habitat. Each population spawns in a particular area at a particular time, at salinities between 33 - 34 ‰ and temperatures between 4 - 6 °C. Eggs not pelagic; usually lie in thick layers on the bottom. The Norwegian spring-spawning herring spawns between February and March. Eggs hatch after around 3 weeks, depending on the temperature. Larvae drift northwards with the coastal currents, and disperse along the coast. Some have their nursery grounds in the southern Barents Sea. Growth is rapid and by August, the juveniles are around 4 cm long, depending on the area. Around 30 cm long at sexual maturity. Herring can spawn several times, but growth declines after the first spawning. Can reach up to 25 years of age. Much energy is 'lost' during spawning, and herring are preferentially fished before they spawn. After hatching, the larvae bear a yolk sac that provides nutrition for the first 8 - 10 days. After this, they must feed for themselves, on zooplankton, especially copepods. Should there be few copepods in the water, most of that herring year-class will die. Several poor year-classes in a row leads to reduced stocks and fluctuations in herring stocks are a natural phenomenon. The difference between the weakest and strongest year classes can be as much as 1:100. However, overexploitation exaggerates the problem, and can lead to an unnaturally long time before the stocks have recovered.

Commercial value: Until the end of the 1960s, herring was the most economically important fish in Europe. During the 1960s, herring were over-fished and this, combined with poor recruitment in several year classes, led to herring being a rare commodity until the 1990s. History shows that herring have 'disappeared' every 100 years or so, and the previous period of 'empty sea' lasted from 1874 to 1896. However in post 2nd world war times, large quantities of herring were caught in the mid fifties and mid sixties. In 1965 - 1967, approx. 1.2 million tonnes were brought ashore; the greatest catches of all times. However, in the late sixties there was a dramatic drop in the population. The herring catches have been ground into fishmeal at coastal herring oil factories, or salted and made into marinated fillets.

Herring form spectacular schools.

Sprattus sprattus - Sprat

Like herrings, the sprat occurs in pelagic schools, feeding on zooplankton.

Distribution: Southern distribution. In Europe from the Black Sea, Mediterranean and along west-European coasts, North Sea, Kattegat, Skagerrak and the Baltic. Southern part of Norwegian coast.

Description: Resembles herring, but much smaller; usually between 2 and 12 cm long (but up to 20 cm), and even more laterally compressed. Underbite and large eyes. Large silver glittering scales. Most characteristic are the angled scales forming a sharp keel along the belly. This keel feels like a saw blade to the touch. Pectoral fin attached in line with the anterior of the dorsal fin, unlike the herring, where the pectoral fin is situated further back. Two sub-species are recognised; one unique to the Baltic and the other occurring throughout the remaining distribution area.

Habitat: Pelagic, occurs in large shoals from the surface down to approx. 150 metres depth. Undertakes vertical migrations. On clear days, tends to stay in deeper water but at nights and on cloudy days, it often can be seen as far up as the surface. In summertime it seldom occurs deeper than 50 metres.

Biology: Feeds on zooplankton, especially copepods. Larvae also eat phytoplankton. Spawning is carried out during spring, either near the coast or in open water. Different populations have different spawning grounds and spawning times. The Skagerrak and Kattegat are important spawning grounds. Sprat are ready to spawn at 2 years of age, when they are around 12 - 13 cm long. Eggs are pelagic and hatch after around one week. Can reach up to 7 years of age.

Commercial value: Sprat fisheries are carried out in summer and autumn, using purse seine, bottom- and floating trawls, and they are kept alive for at least three days so that the gut can empty, before being delivered for canning. They are sold either as sprat; but also are (incorrectly) marketed as anchovies. Sprat compete with herring for food, and weak herring stocks often lead to increased sprat stocks, and vice versa.

Order Anguilliformes - Eels

Eels have a long, snake-like body, a long anal fin and they lack pectoral fins. Most of the 720 member species live in deep water in the sea. In our waters, three families are represented, each represented by one species; **Conger conger**, **Anguilla anguilla** *and* **Nemichthys scolopaceus**. *The latter has a very thin body ending in a whip-like tail.*

Conger conger - Conger eel

Conger eels can be up to 3 m long, and 110 kg in weight.

Distribution: From north-west Africa to the British Isles, mid-Norway and Iceland. Also Mediterranean and Black Sea.
Description: The conger eel is distinguished from the common eel by its overbite and by the dorsal fin situated anterior to the pectoral fins. Also larger; up to 3 m. long and 110 kg. Females seldom over 1 m.
Habitat: Found from very shallow water down to around 250 metres depth. Occurs down to 4000 metres in connection with spawning. Prefers cliffs and overhangs or rocky bottoms, with available hiding places. Some become permanent residents on shipwrecks.
Biology: Sexually mature at 5 - 15 years of age. At sexual maturation, an irreversible and fatal change occurs in the body. The digestive canal and other vital organs degenerate in favour of the production of gonads, which take over up to half the body weight. The

The conger eel can be distinguished from the common eel by its large eyes and overbite.

skeleton becomes fragile and the teeth are lost. Reproduction occurs at 3000-4000 metres depth in the Mediterranean and in the Atlantic west of Portugal. After spawning, the individuals die. Larvae live in the plankton for one to two years, until they settle to the bottom. Growth is very rapid–individuals in aquaria have been known to increase in weight from 1 to 40 kg within five years. The conger eel is a voracious predator that eats almost any type of larger prey, but most commonly fish, squid and large crustaceans.
Commercial value: Conger eel are caught by rod or line, and some appear as by-catch. In Europe, 10.000 - 15.000 tonnes are landed annually. The flesh is relatively coarse and it is most often smoked. Due to its size, the conger eel is a much-desired catch by sea fishermen. Commonly, a whole herring or a half mackerel or similar is used as bait.

Anguilla anguilla - Common eel

Distribution: Found as far south as the Canary Isles, Mediterranean, Black sea and north to Iceland. Entire Norwegian coast and north-east to Murmansk.
Description: Snake-like body form; females up to 1.5 m long and weighing up to 13 kg. Males considerably smaller. Larvae change shape as they approach the coast, from transparent, leaf-shaped larvae to a more adult-like form known as glass eels by the time they are 65 mm in length. After drifting for a year, they migrate into brackish or freshwater areas and the back turns brownish-black and the belly is yellowish-brown. This stage is called yellow eel.

DID YOU KNOW...

Fish species that hatch in fresh water, but spend much of their lives in the sea are called **anadromic** fish. Eels, which grow up in freshwater, but hatch in the sea, are **catadromic**.

Divers most often encounter the common eel in brackish water areas at a few metres depth. It swims along the bottom almost exclusively after dark.

The common eel has a keen sense of smell. By day it remains hidden, only emerging at night.

After 6 - 25 years in brackish water or freshwater, the eels are ready to migrate again. The back becomes completely black and the belly turns silver-white; a stage known as silver eel. There is no growth during this stage. The snout becomes more pointed and the eyes larger. The scales are concealed within the skin and the dorsal fin is very long, beginning posterior to the pectoral fins. The dorsal fin, anal fin and tail are continuous. Underbite. Can be confused with the **conger eel**, ***Conger conger***, but the latter is considerably larger and has an overbite.

Habitat: Catadromic, developing in freshwater, where it spends most of its life, and then migrating to the sea to spawn. The eels remain in hiding during the day, such that divers usually only see them after dark.

Biology: The life cycle of the eel has long been a mystery, most of which was solved only during the last century. Eel larvae hatch in the Sargasso Sea, and the newly-hatched larvae follow the Gulf Stream and the 'North-Atlantic current to Europe. Researchers are still not sure how the silver eel navigates back to the Sargasso Sea to spawn. The yellow eel probes into the sediment to find food, usually feeding on fish, eggs, crustaceans, snails, bivalves and insect larvae. Eels can move across land for short distances, provided the conditions are moist enough.

Commercial value: In some countries, eels are commonly fished, and are popular among sports fishermen. In brackish areas and in the sea, they are caught in traps, but in rivers and other freshwater areas, they are caught using earthworms as bait or with lead lures. This type of fishing is effective in the evening and at night.

WHICH SEX?

The type of nursery grounds determine which sex the juvenile eels become. It appears that most eels developing in freshwater become females, whereas those growing up in brackish water become males.

Glass eels are common in shallow brackish water, where they hide under stones by day.

Order Salmoniformes - Salmons

Species within the salmonid family are recognised by an elongate, streamlined body with a centrally located dorsal fin and a small rib-less adipose fin between the dorsal fin and the tail fin. Members of this family are widespread in the northern hemisphere. Some are stationary freshwater fish, whereas others are anadromic and migrate between the sea and freshwater.

Salmo salar - Atlantic salmon

The streamlined shape and strong tail fin make the Atlantic salmon a powerful swimmer.

Distribution: Along the European coast from Portugal in the south to the southern parts of the Barents Sea in the north, around Iceland and the southern coast of Greenland. To the east as far as the Pechora River in Russia. Baltic salmon seldom migrate out of the Baltic. Also found along the north American Atlantic coast from Connecticut in the USA to the outer Ungava Bay in Canada.

Description: The salmon is a muscular, streamlined predator fish with a large mouth. The upper jaw extends back to the posterior margin of the eyes. The back and upper sides have black x-shaped flecks against the glittering silver background. There are few, if any, such patches below the lateral line. This, together with the colour of the adipose fin (the rudimentary fin between the dorsal fin and the tail fin), distinguishes it from the trout. In the salmon, the adipose fin is greyish, whereas that in the trout has an orange-yellow tip. Juvenile salmon are known as parr, which have a brown back and 10 - 12 large grey-green 'fingerprints' on the sides. A process known as smoltification occurs when the parr become ready to migrate to sea. The juveniles become completely silvery in colour and are called smolt. The silvery colour is retained until the salmon is sexually mature and migrates back up into its native river. These mature fish, lose their silvery colour and become yellowish brown, covered in red and black spots. The Atlantic salmon can reach up to 1.5 m in length, up to 40 kg in weight and can become up to 10 years of age.

Habitat: The salmon is an anadromic fish. Reproduction occurs in freshwater, and young fish migrate to sea to develop until they are ready to spawn (with two exceptions in Norway). In freshwater, the salmon favours cold water flowing over stony substrates, but it lives pelagically in the sea. The duration in fresh water varies from one to six years, whereas the duration in the sea can be up to five years. Salmon always try to return to the same river in which they were spawned in order themselves to reproduce. Most succeed. However, the occasional migration to

At sexual maturity, Atlantic salmon becomes brownish in colour. Here a male.

the wrong river is an important source of genetic exchange between populations and prevents inbreeding.

Biology: Feeds on aquatic insects in fresh water. In the sea, it feeds mainly on herring, sandeels, capelin, krill, large pelagic amphipods and squid. Growth rate in the sea can be rapid. After one year, salmon generally weigh 1-4 kg, after two years, 3-12 kg and after three years, 6-20 kg. The very largest salmon are usually males.

Commercial value: There is little commercial salmon fishing in the sea; in recent times the value is in sports-fishing of wild salmon in rivers, particularly in Norway and Scotland. The salmon aquaculture industry has escalated from the mid 1980s to the present day, and has become the most important industry in rural areas, as well as a national export product.

In northern European rivers, the popularity of the salmon as a sports fish has escalated into a passion for many fishing enthusiasts. Rod-fishing became firmly established in the 1800s, when the 'salmon-lords' moved in on Scottish and Norwegian rivers. The world's biggest catch record in freshwater is 35.8 kg.

Salmo trutta - Trout

A sea trout photographed at night in southwest Norway.

Distribution: Along the European Atlantic coast from the White Sea in the north to Portugal in the south. Inland trout are naturally occurring in Europe, North Africa and northwest Asia, but it has also been introduced in other areas lacking in natural stocks.

Description: Being a member of the same genus, it resembles the salmon. The sea-trout is the same species as the regular brown trout; the different habitats cause changes in growth and appearance. Sea trout can be distinguished from salmon as follows: in trout, the upper jaw extends beyond the posterior margin of the eye, whereas in salmon, it is shorter. Further, the trout has a greater number of almost circular black spots both above and below the lateral line. Such spots in the salmon are almost x-shaped and almost absent below the lateral line. Young trout usually have an orange patch on the adipose fin.

Habitat: Favours oxygen-rich lakes and rivers that are not too warm. Trout that developed in rivers near the coast will usually migrate out to sea, and, like salmon, will return to their native river to spawn.

Biology: In freshwater, mainly feeds on zooplankton and various aquatic insects. Diet more varied after migration, including various crustaceans and small fish such as sprat and herring. Migration from the rivers occurs after the trout has developed in the river for between two to five years (exceptionally one year). Migration usually occurs from early May (southern Norway and Scotland) to June/July (northern Norway). At this stage, trout smolt are 10-25 cm long. Unlike the salmon, only a minority of trout migrate over large distances. Sea-trout usually stay in the near vicinity of their native river and usually will migrate back up the river before winter. Large, sexually mature trout migrate upriver first, often in July/August, whereas the smaller trout often wait until October before they migrate. Trout that migrate into fjords usually double their weight after a year, usually weighing around 1 kg after three seasons. Sea trout can reach up to 1.4 m in length and up to 20 kg in weight. Can become more than 20 years old.

Trout have round spots distributed also below the lateral line.

TIPS FOR SEA-TROUT FISHING:

The best time for trout fishing in rivers is from the middle of July, when sea-trout migrate up-river, until the end of the season, usually in early to mid September. Most effective fishing is at twilight or at night. Wet flies, nymphs, worms, smal lures and spinners (where permitted) give good results. Good fishing spots often are where the water breaks in the outlet of the river pools. Sea-trout can be fished using lures, flies and bait, and will bite most of the year, but less so during the coldest periods. When using a lure, preferably a herring lure or a thin spoon-lure, best results are achieved by reeling it in relatively quickly near the surface. For fly-fishing in the sea, streamers, imitations of small fish, crustacea or bristleworms, are best. As with lures, these should be reeled in quickly. When using bait, a narrow bob or float should be used, with thin slivers of mackerel or herring, but whole, fresh small herring, sprat or sandeels are even better. In springtime, sea-trout favour shallow, warmed inlets. In such places, the best fishing times are early morning or at sunset.

Order Osmeriformes

This group includes the ***argentine*** *family, the Argentinidae, represented by the* ***lesser silver smelt*****,** **Argentina sphyraena,** *and the* ***capelin*** *family, represented by the* ***pelagic capelin*****,** **Mallotus villosus**. *Within the Argentinidae is also the* ***greater argentine****, which occurs in relatively deep water. These two families relatively recently were removed from the order Salmoniformes, the* ***salmonid fish****.*

Argentina sphyraena - Lesser silver smelt

The lesser silver smelt is a deep-water fish that occasionally occurs in shallow water. Here photographed at 25 m depth during a night dive in southwest Norway.

Distribution: From the west coast of Italy north to northern Norway.

Description: Up to 25 cm long; can become up to 15 years old. Relatively small, with large eyes. The diameter of the eye is of a similar length to the distance from the tip of the snout to the anterior margin of the eye. Scales large, with small spines on their posterior margins. Approx. 50 - 54 scales along the lateral line. Glittering silver with a greenish back, often with glittering blue stripes along the sides. Can be confused with the **greater argentine,** ***Argentinidae silus*****,** but the latter occurs in 100 metres depth or below, has a brownish back and stout spines at the posterior margin of the scales (66-69 scales along the lateral line). The diameter of the eyes in the **greater argentine** is greater than the distance from the snout to the anterior margin of the eye.

Habitat: For the most part a pelagic fish; but periodically retreats to the bottom. Usually deeper than 50 metres, but occasionally observed as shallow as 20 metres on soft bottoms in Norwegian fjords.

Biology: Spawning is carried out in deep water during spring and summer. Both the eggs and larvae are pelagic. Mostly feeds on zooplankton; also takes polychaete worms.

A beautiful fish, with large eyes typical of deep-water fish.

Order Gadiformes - Codfishes

In northern European waters, 28 codfish species occur, within three families; the ***hakes, Merluccidae; grenadiers, Macrouridae,*** *and the* **cod** *family,* **Gadidae***. All codfish lack proper fin rays, although some have a stiff first ray in the first dorsal fin. Further, the pelvic fins, where present, are located in front of the pectoral fins. Of these 28 species, only the commonest are presented here. The grenadiers comprise 4 species; the* ***onion-eye grenadier,*** **Macrourus berglax***; the* ***rat-tail,*** **Coryphaenoides rupestris***; the* ***hollowsnout grenadier,*** **Coelorhynchus coelorhynchus** *and the* ***softhead grenadier,*** **Malacocephalus laevis***. All these are deep-water species. Within the whiting family, only one species is represented in northern Europe and the remaining 23 species of codfishes belong to the cod family. The following are not presented here: the* ***blue whiting,*** **Micromesistius poutassou***; the* ***silvery cod,*** **Gadiculus argenteus***; the* ***Norwegian pollock,*** **Theragra finnmarchica***; the* ***greater forkbeard,*** **Phycis blennoides***; the* ***blue ling,*** **Molva dypterygia***; the* ***shore rockling,*** **Gaidropsarus mediterranus***; the* ***three-bearded rockling,*** **Gaidropsarus vulgaris***, the* ***four-bearded rockling,*** **Enchelyopus cimbrius***; the* ***threespot eelpout,*** **Ciliata septentrionalis** *and the* ***Arctic rockling,*** **Onogadus argentatus**.

Polar cod, **Boreogadus saida,** *only reaches 46 cm. Its an arctic fish. From Svalbard.*

The European hake; a streamlined codfish. Here from southwest Norway at 25 m depth.

Merluccius merluccius - European hake

The European hake is a deep-water fish that occasionally occurs in shallow water at night.

We have observed these fish digging using their pectoral and pelvic fins. They throw sand onto their backs, perhaps as a form for camouflage. Here photographed in southwest Norway during a winter night dive.

Distribution: Eastern Atlanteren from Senegal, Mediterranean and north to Iceland; common up to mid-Norway; becoming rarer farther north. All around the British Isles.

Description: Up to 135 cm in length and 15 kg in weight. Elongate body, with one short and one long dorsal fin. Head narrow, with a long underbite mouth, equipped with strong curved teeth. Inside of mouth and gill covers black. Back greyish black; sides and belly silver grey. Small scales. Lateral line pale, and almost straight.

Habitat: From the surface down to 1000 metres; commonest between 150 - 500 metres. Hunts along the bottom, or in the water masses; in the latter case usually in shoals.

Biology: A ferocious predator mainly taking pelagic fish such as mackerel, herring and sprat at night. Bottom-living by day. Spawning occurs mostly during spring and summer; northern stocks spawn later than southern ones. Known spawning grounds in the Kattegat are as shallow as 30 - 40 m.

Commercial value: Tasty white flesh popular as a restaurant dish in Europe. Not fished to any great extent.

VORACIOUS PREDATOR

The European hake is a ferocious predator that will continue to attack prey in nets, even when hauled on-board.

Trisopterus luscus - Bib/ Pouting

The bib often has pale transverse stripes along the body, but not this specimen.

Distribution: Northeastern Mediterranean, north to the south of Norway. All around the British Isles.
Description: Closely resembles the **poor cod**, ***Trisopterus minutus***, making observations somewhat uncertain. Both are small codfish, rarely exceeding 35 cm in length and up to 3 kg in weight. Slight overbite and a marked barbel, equal to eye diameter. Large eyes and a high back. A vertical line drawn from the anus to the dorsum (back) will cross the mid-point of the dorsal fin, whereas in the poor cod, this line crosses its posterior edge. The bib, or pouting, usually has a dark patch at the basis of the pectoral fins, which is lacking in the poor cod. Further, the bib often has broad transverse stripes, alternating pale and dark.
Habitat: Shallow water from 10 - 100 metres; rarer down to 300 m. Occurs in small schools around cliffs and reefs. Seldom seen individually. Moves to deeper, warmer water during winter. Nocturnal, often hidden during the day. Favours shipwrecks.
Biology: Preys on small fish, crustaceans and occasionally octopus or squid. Spawning occurs during March to April at around 50 - 70 metres depth. Sexually mature at 1 year of age, at which point it has a length of just over 20 cm.

Trisopterus esmarkii - Norway pout

REDUCED TO FISHMEAL

The Norway pout is one of the most important commercial fish species in the North Sea. Catches are ground up into fishmeal, and the oil extracted.

A Norway pout, photographed at 40 m depth during a night dive.

Distribution: Northeast Atlantic from the Bay of Biscay north to Iceland, Norway and Svalbard.
Description: Small codfish, up to 25 cm long. Very large eyes; diameter greater than the length of the snout. Small barbel and pronounced underbite. Dark lateral line. Small dark patch at the root of the pectoral fin. Resembles the **poor cod**, ***Trisopterus minutus***, but the latter has an overbite.
Habitat: Usually occurs in relatively deep water on muddy bottoms, 80 - 300 metres; occasionally shallower.
Biology: Mainly feeds on crustaceans, but also small fish such as gobies. Occurs in large schools. Sexually mature after 1 - 3 years, depending on temperature. Spawning occurs from January to July; earliest in southern areas. Seldom exceeds 6 years of age.

The large eyes are typical of deep-water fish.

Trisopterus minutus - Poor cod

Distribution: From Morocco in the south, northern Mediterranean and north up to southwest Norway. Faeroes.
Description: Up to 30 cm in length, but rarely exceeds 20 cm in northern waters. Resembles the **Norway pout**, ***Trisopterus esmarkii***, but has an overbite. Eyes large with a diameter approximately that of the distance from the front of the eye to the tip of the snout. Pale brown, with a white belly. Can also be confused with the **bib**, ***Trisopterus luscus***, but distinguished, by, amongst others, the following characteristic: a vertical line drawn from the midpoint between the 1st and 2nd dorsal fin down to the belly, this will end at the anal opening in the poor cod. In the bib, the anal opening lies further forward.
Habitat: Occurs from 10 metres down to 300 metres depth. Highest densities occur a little offshore, but young individuals can be seen in fjords and around the coast. Very common, and often seen by divers at night. The poor cod is a curious fish and often approaches right up to divers' masks.
Biology: Preferentially feeds on small crustaceans such as krill, copepods, amphipods and mysids, but also will take small fish. Spawning occurs between March and July in open waters, from 50 - 100 metres.

Poor cod can be found even in the inner parts of harbours. It is particularly fond of boulder reefs.

Juvenile poor cod.

Merlangius merlangus - Whiting

At night, whiting often migrate to shallow water. They have been known to take advantage of the light from our diving torches to capture illuminated prey.

Distribution: Black Sea and north-eastern Mediterranean. Atlantic waters from Gibraltar to the Barents Sea. Iceland, Norway (most abundant up to southwestern parts).
Description: Slender codfish with overbite and pointed snout. Back faintly brownish; otherwise glittering silver colour with white belly. Dark patch at the base of the pectoral fin. Outer parts of the pelvic and dorsal fins form a white edge. Lateral line only faintly curved over the pectoral fin. Adults lacking barbel. Up to 55 cm in length.
Habitat: Bottom-living; prefers sandy and muddy bottoms from 10 - 200 metres depth.
Biology: Spawning occurs from January to July, most intensively between March and April. During this time, whiting aggregate in large schools, including in northern parts of the North Sea and the Skagerrak. Both eggs and larvae are pelagic.

A whiting larva takes refuge within the tentacles of the stinging jellyfish, **Cyanea capillata**. *Juvenile whiting, haddock and the atlantic horse mackerel often make use of this sheltered habitat until they settle to the bottom.*

A young whiting among the protective tentacles of the stinging jellyfish, **Cyanea capillata**.

WHITE DELICACY

The whiting is named after its white flesh, which is considered a delicacy. It is moderately fished commercially in the North Sea, but highly sought-after by sports fishermen. The flesh tends to spoil quickly.

Melanogrammus aeglefinus - Haddock

Distribution: Both sides of the north Atlantic. In the eastern part from Portugal to Spitsbergen, Iceland and Novaja Semlja. Entire Norwegian coast, but most abundant in the north.

Description: Easily recognisable by the prominent dark patch below the first forsal fin. Marked overbite, with a small mouth and short barbel. First dorsal fin high and pointed. Greyish in colour, with a pink tinge. Lateral line prominent and dark. Up to 110 cm in length and 19 kg in weight.

Habitat: Benthic dwelling, prefers sandy clay and gravel bottoms, but also found on sand substrates. Occurs between 10 and 300 metres depth, shallowest in the summer and at night.

Biology: Spawning occurs at 100 - 150 metres depth during March to June. Eggs and larvae pelagic. In common with the whiting, larvae seek refuge among stinging jellyfish. Feeds on various benthic organisms, octopus and squid as well as other fish.

Commercial value: An important commercial species caught by line, or by "fly-shooting", Danish seine nets and trawls. Norwegians mainly use haddock to make fish cakes and puddings, whereas in Britain and Ireland, it appears battered and fried in fish and chip shops. Icelanders prefer haddock to cod. In Norway alone (2002), 55 153 tonnes of haddock were landed, with a first hand value of approx. 60 million Euro.

It is possible to come close to this otherwise shy fish during night dives.

Pollachius pollachius - Pollack (Pollock)/ Lythe

Distribution: Mediterranean from west coast of Italy and further west. Atlantic waters from Morocco and north to Iceland and northern Norway. Main distribution around Bay of Biscay and southern and western parts of the British Isles.

Description: A codfish with a prominent underbite. Lateral line bends downwards below the midpoint of the 1st dorsal fin. Resembles the **saithe**, ***P. virens***, but the latter has a straight lateral line. Up to 130 cm in length and over 10 kg in weight.

Young specimens of pollack have a very different appearance to adults. Note the prominent underbite.

Habitat: From the surface down to approx. 200 metres; pelagic or on the bottom. Often

The pollack is a popular sports fish. It prefers rocky bottoms and often swims just above the algae.

observed by divers.
Biology: Usually seen singly by divers, seldom in schools, although adults form schools at spawning time in spring or early summer. Eggs pelagic. Mainly feeds on sand lances or various herrings or codfish. Also takes various crabs and shrimps. Up to 15 cm in length by its first autumn. A specimen measuring 120 cm will be more than 15 years old.
Commercial value: A good food fish, but not widely exploited. Popular with hobby fishermen, because large individuals can be caught in shallow coastal waters.

Pollachius virens - Saithe/ Coalfish

Distribution: Both sides of the Atlantic. Eastern Atlantic from the Bay of Biscay north to Novaja Semlja. Very common.
Description: A muscular codfish; barbel inconspicuous or absent. Lateral line straight. Weak underbite. Back dark green, almost black, progressively becoming silver along the sides. Silvery white belly. Darkens with age. Pelvic fins reddish white, whereas the others are a dark olive-grey, except the anal fin, which has the same white colour of the belly at its base. Young individuals have a lighter greenish back. Resembles the **pollack**, or **lythe**, ***Pollachius pollachius***, but distinguised by the straight lateral line.
Habitat: Probably the codfish that to the greatest degree occurs near the surface, especially during summer. Large saithe prefer deeper water, and often stay close to the bottom.
Biology: Typically occurs in schools, and carry out large migrations. Small saithe mostly feed on pelagic crustaceans and fish larvae filtered from the water by means of the gill arches. Larger saithe additionally take fish such as sand lances, capelin and sprat. Its hunting methods have been widely stud-

Saithe often can be seen resting on the bottom at night.

The fish divers see swimming around kelp often are saithe.

Below: *Like the pollack, saithe have an underbite.*

ie; G. O. Sars described how a school of saithe encircles aggregations of fish larvae, closing in, thereby chasing them to the surface, such that the sea "boils". Sexually mature at 5 - 6 years of age. Reproduction occurs from January to April at around 100 - 200 metres depth. Eggs pelagic, hatching after 6 - 15 days. In August, small saithe hatched the previous spring gather around the coast, measuring between 14 - 22 cm in length. The following year, they reach around 35 cm; 40 cm by the third year and 60 cm by the fifth.
Commercial value: One of the most important commercial fish species, caught by purse seine, trawl, nets and long-lines. Popular sports fish, widely fished by rod. The flesh is a little fattier than in cod or haddock. In 2002, in Norway alone, 203 922 tonnes of saithe were landed, with a first hand value of over 107 million Euro.

30 YEARS OLD
Saithe, or coalfish, can reach up to 120 cm in length and weigh up to 21 kg. It is thought that some can live for up to 30 years.

Gadus morhua - Cod

Below: *Cod; probably one of our most well-known fish.*

Distribution: Widespread in the North Atlantic, on the eastern side from the Bay of Biscay north to Iceland, Svalbard, Greenland and Novaja Semlja.
Description: Elongate codfish, with large head and often swollen belly. Overbite; prominent barbel. Colour brown-flecked, against an ash-grey background; belly silver–grey. Colour varies somewhat according to habitat-coastal cod living amongst kelp is reddish. Lateral line white and curving. Can reach up to 1.8 m long, weighing an impressive 55.6 kg.
Habitat: Coastal cod stocks remain in shallow water and do not migrate to any extent. The more pelagic Norwegian Arctic cod has its nursery grounds in the Barents Sea and moves in to the Norwegian coast to spawn. Found from the shore down to 600 metres.

CAPELIN IMPORTANT FOR COD STOCKS

In the Barents Sea, the rich Norwegian-Arctic cod stocks are heavily reliant on capelin. This was not taken into account during the 1980s, when capelin was severely over-fished. The 'disappearance' of capelin in 1986 marked the beginning of several years of crisis for the Barents Sea cod fisheries. This is unlikely to will happen again, because ecologists are more actively involved in monitoring fish stocks and setting fishing quotas.

Biology: Along our coasts there are several local cod populations. Norwegians have chosen to call those local cod populations that do not undertake large spawning migrations "coastal cod" and the migrating stocks "oceanic cod", such as the North Sea cod and the Norwegian Arctic cod. The latter moves in to the Norwegian coast from its nursery grounds in the Barents Sea when it is sexually mature. The Norwegians have a special word, "skrei" for this stage, and the largest fish reach their spawning grounds near Lofoten off the coast of northern Norway at the end of January. At this time, the flesh, together with the vitamin-rich roe and liver, are eaten with great relish. Young cod developing in the Barents Sea are known in Norway as "capelin cod", after their staple diet. The cod in British waters migrate to spawning grounds, such as in the central North sea, between February - April. Females spawn up to 5 million eggs in several portions over a period that can extend over two months. Cod is a generalist feeder and feeds on almost anything available.

Commercial value: Probably our most important commercial fish species, with large national and export markets. Depending on nation, cod is exported in dried, salted, frozen, fresh and smoked forms. Because the various cod stocks fluctuate, there are often strong national and international disputes about fishing rights and quotas. In 2002 in Norway alone, 228 750 tonnes of cod were landed, with a first hand value of approx. 330 million Euro.

The cod can vary its appearance according to habitat. Upper specimen from northern Norway; lower from mid-Norway.

Brosme brosme - Tusk / Cusk/ Torsk

Distribution: Both sides of the Atlantic. Eastern side from northern parts of the British Isles to Iceland, Norway, Svalbard and the Barents Sea.
Description: Elongate, but robust shape, distinguished from other codfish by having a single, extended dorsal fin, continuous with the caudal, or tail fin, as is also the case with the long anal fin. These fins are all dark in colour, with a white edge. Lateral line dark, curved near the anus. Colour usually an even brown, but young fish often have 5 - 6 transverse stripes. Barbel present; weak overbite. Scales deeply embedded, giving the skin a scale-less appearance. Can become over 1 m long, weighing around 16 kg.
Habitat: Prefers depths between 50 - 1000 metres; occasionally occurs shallower, but usually found on offshore banks, amongst the coral ***Paragorgia arborea***, between 200 - 400 m.
Biology: At the depths mentioned above, spawning occurs between April and August. Eggs and larvae pelagic. Larvae settle to the bottom at around 50 mm length. Sexually mature between 6 - 10 years of age, then with a length of 40 - 50 cm. Feeds on a variety of benthic organisms including prawns, shrimps, crabs and bristleworms.

DID YOU KNOW...

The Swedish researcher Euphrasén, already in 1794, described the tusk, as having firm, white, tasty flesh, relatively fatty, with a flavour reminiscent of crab or lobster, which undoubtably accounts for it being considered a delicacy. Many still regard it as the prize of all the codfishes. It is mostly fished by long-lines or trawl, but also has become popular amongst sport fishermen, who use bait such as mackerel.

This tusk was found resting at around 40 m depth near a steep rock face.

This specimen was photographed at the marine aquarium in Bergen, Norway.

Molva molva - Ling

Small individuals are camouflaged.

Distribution: From Gibraltar in the south, north to Iceland, Norway and the Barents Sea.
Description: Slender, elongate shape; first dorsal fin short, the second long and continuous with the tail fin. Head small and narrow. Long barbel; longer than the eye diameter. Upper side brownish-grey; belly white. Fins with white outer edge. Up to 180 cm long; can weigh over 37 kg.

The ling has a long barbel.

Habitat: Adults prefer deeper water from 100 - 1000 metres. Young fish often observed by divers at 15 - 30 metres. Usually occurs singly on hard bottom or sandy substrates with large stones. By day usually hidden under stones or rocky overhangs; by night often lying openly on the bottom.
Biology: Spawning takes place between April - June at 100 - 300 metres depth. Females can spawn as many as 20 - 60 million eggs. Larvae pelagic; deeper than 50 metres. Reaches 18 cm after the first year; sexually mature at 6 - 8 years. Mainly feeds on herring, mackerel, cod, flounders, as well as crabs and other benthic organisms.
Commercial value: Ling is fished at the edge of the continental shelf, mainly using long-lines. Tasty, somewhat coarse flesh. Like the **torsk**, has become a popular sports fish, caught using bait.

The body is long and thin.

The ling is a typical cave-dweller, often found in rocky crevices or shipwrecks, especially those lying deeper than 40 m. In our experience, it is a curious fish, which is easy to lure out of its hiding place. These pictures were taken at around 20 m depth.

Raniceps raninus - Tadpole fish/ Little torsk/ Lesser forkbeard

The tadpole fish usually is black, but some individuals can appear greyish.

Distribution: From the Bay of Biscay to northern Norway.
Description: Gleaming black in colour in its natural habitat, with a broad head and body tapering towards the tail. First dorsal fin consists of three short free rays; second dorsal fin and anal fin long. Small barbel. Up to 30 cm in length. Can hardly be confused with other fish.
Habitat: Reputed to be relatively rare throughout its distribution area, probably because it can be difficult to see. It hides under stones, in crevices and between algae on hard substrates, often right up to the shore. Rarely down to 100 metres. However, once divers are accustomed to spotting it, they are surprised by how commonly it occurs. It is very shy, and almost impossible to photograph. Can be observed swimming freely at night.
Biology: Spawns from March to September, at temperatures above 10°C. Eggs and larvae pelagic. Feeds on most kinds of benthic organisms, but prefers starfish and brittle stars.

This fish is an expert at hiding, and good photographs are rare.

Ciliata mustela - Five-bearded rockling

The five-bearded rockling usually is seen only after dark. The five 'feelers' are used to locate food.

Distribution: From Portugal north to Iceland and Norway.

Description: As the name suggests, this rockling species has five barbels; four located on the snout and one on the 'chin'. Two dorsal fins; posterior one longest. Anterior dorsal fin consisting of a single ray, often hidden in a ridge along the back. Pelvic fin long. Upper side reddish to dark brown; paler belly. Up to 45 cm in length. Several other rockling species occur in our waters; the three-bearded rockling, Gaidropsarus vulgaris and the **four-bearded rockling**, ***Enchelyopus cimbrius***. The **shore rockling**, ***Gaidropsarus mediterraneus***, which also has three barbels, is more rare. Two species are found in deeper water; the **arctic rockling**, ***Onogadus argentatus*** and the **northern rockling**, ***Ciliata septentrionalis***.

Habitat: Shallow water down to 20 metres; moves to deeper water in summer during spawning. Occasionally found in rock pools. Occurs on both hard and soft bottoms.

Biology: Spawning takes place from January to July. Eggs and larvae pelagic; larvae settle to the bottom at around 45 mm length. Remains on the bottom throughout the rest of its life, hunting various bottom-living organisms, which it seeks out using the barbels. The groove within which the 1st dorsal fin is situated is densely provided with sensory cells, which most likely detect olfactory (smell) substances.

Left*: The* **shore rockling Gaidaropsarus mediterraneus***, can reach up to 35 cm in length. The photographed specimen was found hidden under a stone in southwest Norway.*
Right*:* ***Three-bearded rockling****,* **Gaidaropsis vulgaris***; our largest rockling, reaching up to 53 cm in length and weighing up to 0.75 kg. Photographed at the marine aquarium in Bergen, Norway.*

Order Lophiiformes - Anglerfishes

The anglerfish is a favourite amongst divers. Adults are fearless, have few enemies, and seem to enjoy being stroked on the back.

Members of this family are largely found in warm waters and in the deep sea. Most are associated with the bottom, and have relatively low activity levels, spending most of their time waiting for prey to swim past their mouths. In most anglerfish, the first ray of the first dorsal fin is modified into a 'fishing rod', or lure. Only anglerfish within the family Lophiidae are permanent residents in northern European waters. However, the ***Sargassum anglerfish*****, Histrio histrio**, *occasionally appears, having drifted all the way from the Sargasso Sea.*

Lophius piscatorius - Anglerfish/ Monkfish/

Distribution: Widespread across the entire eastern North Atlantic, but also found along the east coast of America. From western Africa, Mediterranean, Black Sea and north to Iceland, Norway and the Barents Sea.
Description: Due to its characteristic appearance, unlikely to be confused with any other

Anglerfish spend much time lying in wait of prey. When startled, the 'fishing rod' is laid flat.

The first fin ray is modified into a 'fishing rod'.

fish. The head is enormous compared to the rest of the body; in fact this fish looks as if it consists mainly of head and tail. Mouth extremely broad, and a number of fringes dangle from below the 'chin'. Pectoral fins very large. Light brown in colour, with darker patches. Belly side flattened, almost white. Eye iris attractively marbled. Records exist of up to 60 kg anglerfish being caught.

Habitat: Exclusively bottom-living, preferring soft bottoms, where it often lies partially buried awaiting prey. However, can also be found within kelp forests. Occurs from the shore down to around 1800 metres. Most frequently observed by divers in June/July and througthout autumn, on slopes that extend to deep waters.

Prey (here a lumpsucker) is swallowed whole.

Biology: Spawning occurs in spring in deep water. Newly-hatched larvae aggregate in up to 11 m long gelatinous bands. These float up to the surface, and are only found above deep water. Juveniles settle to the bottom at around 6 cm length. Diet varied, but mostly fish are on its menu. However, large anglerfish have been known to take diving ducks and we have caught one with a 3 - 4 kg salmon in its stomach.

SYMBOL OF BAD LUCK

Because of its shaggy and rough appearance, the anglerfish for a long time was regarded as a trash fish and a bad omen. Stories tell of fishermen in the 1800s being so afraid of this fish that they never took it on board. It was reputed to mean impending disaster, and many a fisherman hurried ashore after meeting this dreaded symbol of bad luck.

DID YOU KNOW...

The anglerfish has a sophisticated method of catching its food. The modified first ray of the first dorsal fin is long and moveable, with a meaty tip. This is lowered down in front of the mouth to attract curious fish. When a fish is close enough, the anglerfish quickly opens its huge mouth, creating a suction effect that propels the prey to its doom. The extent to which the Anglerfish makes the effort to lunge forward depends on the size of the prey.

The monkfish has an amazing ability to blend into the background by camouflage.

Another we saw was in the process of swallowing a fully-grown lumpsucker.

Commercial value: Extremely tasty firm white flesh, sold in the UK masquerading as deep-fried 'scampi', because of its shellfish-like flavour and texture. Caught in nets or by long-line. Prized by divers, who spear them with knives. The anglerfish has played its part in medical history–insulin, important in the treatment of diabetes, was first isolated from the pancreas of this fish.

Order Beloniformes - Needle fishes

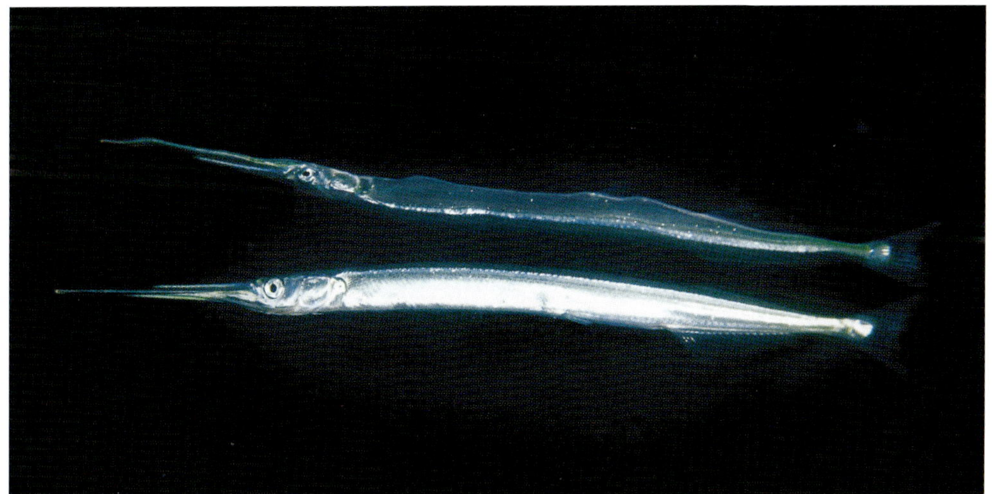

*This order comprises five families, all of which live freely in the water masses. In addition to the **garpike** described below, the **Atlantic flyingfish**, **Cheilopogon heterurus** and **saury pike**, **Scomberesox saurus**, also occur in our waters. Only one species occurs within the garpike family, **Belonidae**, which are characterised by their long thin shape, elongate beak-like snout and small scales. They live in the upper water masses.*

Belone belone - Garpike/ Garfish

Distribution: Widespread in the north-eastern Atlantic from the Canary Islands to the Norwegian coast; becoming rarer in the northern parts. Mediterranean and Black Sea.
Description: Highly distinctive appearance; long and thin with a beak-like snout. Glittering silver underside with small scales. Dark green back. Fins somewhat backwardly displaced; pelvic fin situated at around mid-body level. Can reach up to almost 1 m in length, but still not exceeding 1.3 kg in weight. Unlikely to be confused with any other fish in our waters.
Habitat: Occurs pelagically in schools, preying on other fish in the upper water layers. Remains around the Norwegian coast from the summer until September or October.
Biology: Spawning takes place in shallow water during early summer. Eggs spawned in portions, approx. 3.5 mm in diameter, equipped with attachment threads such that they attach to vegetation and other similar objects. Mostly feeds on small pelagic fish such as sprat, small saithe, sandeels and larvae, including those of codfish.

HIGHLY PRIZED

The garpike is highly prized in Mediterranean and Black Sea areas. Those who have eaten it will have noticed that the bones turn green when cooked.

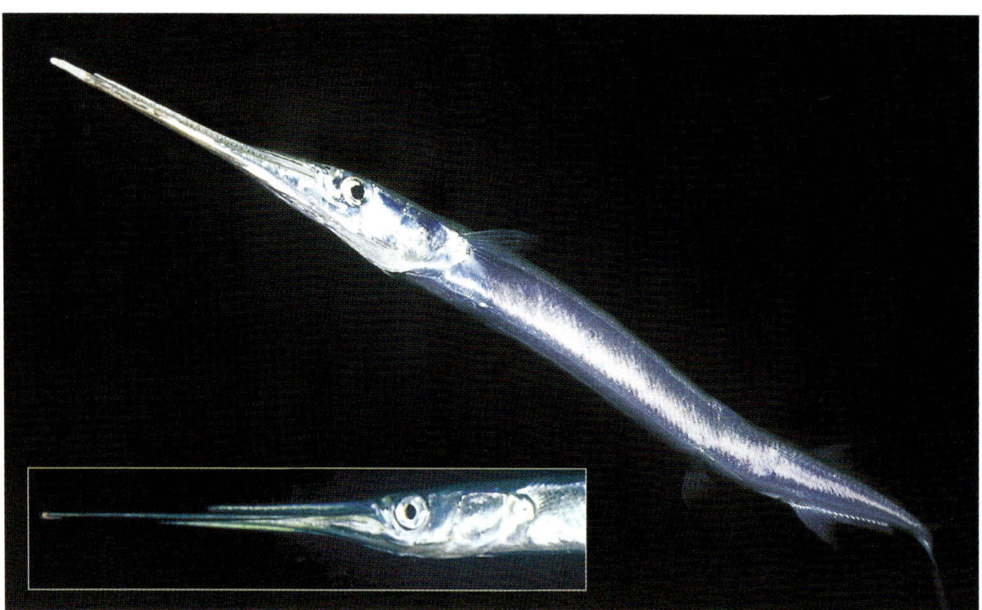

The garpike is a summer visitor in northern Europe.

DID YOU KNOW...

The garpike is one of the few fish that will bite even when the line is towed behind a fast-moving boat. This shows it to be a fast and effective hunter.

Order Gobiesociformes - Clingfishes

This order comprises three families, one of which, the ***clingfish*** *family* ***Gobiesocidae****, occurs in our waters. The body is small, drop-shaped, with a short dorsal fin located directly above the short anal fin. All have a well developed sucker disc on the ventral, or belly, side just behind the head, and all lack a swim bladder. Only a single species occurs in northern waters (three more species around the western and south coasts of the British Isles. There are more than 100 species world-wide.*

Diplecogaster bimaculatus - Two-spotted clingfish

This two-spotted clingfish was only 3 cm in length. Photographed at 20 m depth in southwest Norway.

Distribution: Eastern Atlantic from Gibraltar north to the western British Isles to the Faeroes, northern part of Swedish west coast and to mid-Norway.
Description: Brownish red; only 4.5 cm long. Broad anterior end; somewhat dorsally (from the top) flattened. Dorsal and anal fins short, situated in line with each other relatively far back on the body. Eyes markedly convex, with large pupils. Males with two red patches ringed with yellow on the 'neck', and one behind the pectoral fins.
Habitat: Hard bottoms, preferably current-exposed sites. Attaches itself by suction to objects, such as a stone or bivalve shell. Common between 5 - 10 metres, but also occurs down to 100 m.
Biology: Spawning occurs during late summer. The around 200 small, golden eggs often are laid in an empty bivalve shell and are guarded by the parents, usually the male. Mainly feeds on small crustaceans, such as ostracods and copepods.

This specimen was found curled up under a horse mussel shell at around 12 m depth. From Hoy in the Orkney Islands.

Order Syngnathiformes - Pipefishes and sea horses

The broad-nosed pipefish is a master of disguise.

In northern Europe, this order is represented by eight species within two families. The majority of pipefishes and seahorses are found in warmer waters. All have a small mouth, located at the end of a narrow, elongate snout. The ***longspine snipefish*, Macroramphosus scolopax**, *is a rare guest from southern waters. More commonly seen are the seven other species within the* ***pipefish*** *family,* ***Syngnathidae***. *Except for the* ***long-snouted seahorse*, Hippocampus ramulosus**, *distributed along the western and southern coasts of the British Isles. These all have a long, thin body with an external armour of bone rings. In some, such as* ***Nilsson's pipefish*** *and the* ***greater pipefish***, *the bony plates give the fish an angular appearance. These species are stiff, and have a reduced motility. In the* ***snale pipefish*** *and* ***worm pipefish***, *the skin is thicker, smooth and conceals the bony plates. A peculiarity of these fish is their ability to move their heads, and the eyes also can be moved independent of the other. The prey, usually small crustaceans, is sucked into the mouth of the fish. Many pipefish are totally reliant on their ability to camouflage themselves - they have little chance of escaping danger due to their reduced ability to swim. Divers can therefore experience close contact with these fish.*

Syngnathus typhle - Broad-nosed pipefish; also deep- or high-snouted pipefish

Distribution: From Gibraltar north to northern Norway. Black Sea, Mediterranean, Baltic.

Description: Like other pipefishes, a thin, almost thread-like shape. The shape of the head and snout is used in identification. In the broad-nosed pipefish, the length of the snout comprises more than half the length of the whole head. Snout markedly flattened from the sides, with a longitudinal ridge along the upper edge. Colour variable, but often brownish with paler patches. Females up to 35 cm long; males 24 cm. Distinguished from the **greater pipefish**, ***S. acus***, and **Nilsson's pipefish**, ***S. rostellatus***, amongst others, by the lack of prominent longitudinal ridges.

Habitat: From the shore down to around 20 m. Usually 'hanging' vertically among eelgrass and other algae, making it difficult to spot. Also brackish water.

Biology: Feeds mostly on small crustaceans, but also takes juveniles of other fish.

The snout makes up more than half of the length of the head, and is flattened from the sides.

BROODY MALES

The broad-nosed pipefish, in common with other pipefishes, has an unusual division of sex roles. Mating time is usually mid-summer. Males receive eggs from females during a mating dance, where the tails are coiled around each other. Females transfer 10 - 20 eggs in each batch, and the ritual is repeated several times. Males often receive eggs from several females. The eggs are kept in specialised brood pouches on the males' belly. Males are 'pregnant' for around four weeks before the around 50 - 250 eggs hatch. The juveniles may stay in the males' brood pouch for several weeks after hatching. After they leave the brood pouch, they stay near the male for some time, and he actively looks after the young. In times of danger, the juveniles hide in the brood pouch.

Syngnathus acus - Greater pipefish

Distribution: From Morocco north to the Faeroes and mid-Norway. Black Sea and Mediterranean.

Description: Resembles the **broad-nosed pipefish**, ***S. typhle***, but the snout is tube-shaped and lacks the upper keel. Body robust and angular in cross section; not oval. Snout longer than half the length of the entire head, with an inflated tip. Head prominently raised behind the eyes. Colour variable. Back, brownish-grey or dark greyish yellow; sides paler. Approximately 20 relatively broad, pale transverse stripes. Usually a dark stripe from the snout tip to just below the eye. Dorsal fin also with dark stripes. Dorsal, pectoral and caudal (tail) fins well developed. Up to 50 cm in length.

The greater pipefish resembles the broad-nosed pipefish, but the snout is not flattened.

Pipefish have mouths shaped such that their diet is limited to small crustaceans, molluscs and fish larvae. Here the attractively coloured greater pipefish.

Habitat: From the shore down to approx. 20 m. Often seen 'hanging' vertically and well hidden amongst various algae and eelgrass. Also seen lying on the bottom. Thrives in brackish water.
Biology: Mating rituals generally similar to the broad-nosed pipefish. Spawning occurs between May and July and the eggs hatch after around 5 weeks in the male's brood pouch. Feeds on small crustaceans, molluscs and fish larvae.

Syngnathus rostelatus - Nilsson's pipefish

At only 17 cm in length, Nilsson's pipefish is the smallest of our pipefish.

Distribution: From the Bay of Biscay, north to western Norway.
Description: Resembles the somewhat larger **greater pipefish**, ***S. acus***, but with a shorter snout; less than half the total length of the head. Snout upwardly curving. Back and sides dark grey or brown; belly pale grey. Body with numerous small yellowish-white patches and dark transverse stripes. Dorsal fins without patches. Our smallest pipefish; reaching only up to 17 cm in lenght.
Habitat: Shallow water down to 20 m. Found on clay or sandy bottoms; particularly

Note the typically angular outline.

common amongst eel-grass. Thrives in brackish water; often found right up to river mouths.
Biology: Lifestyle similar to the **broad-nosed pipefish, *S. typhle***. The around 100 eggs remain in the male's brood pouch for around 3 weeks. Larvae 13 - 14 mm long at hatching. Short planktonic phase before settling to the bottom and adopting an adult lifestyle.

Entelurus aequorus - Snake pipefish

The characteristic colouring makes the snake pipefish easy to recognise.

Distribution: From Portugal to Iceland and northern Norway (rarer in the north).
Description: Our largest pipefish. Females up to 61 cm long; males up to 40 cm. This species cannot change colour as readily as the others, and can be identified by its characteristic transverse stripes along the sides. Skin thick, covering the bony structures, giving a smooth, oval appearance. Eye iris is copper-yellow coloured. Pectoral, pelvic and caudal (tail) fins absent.
Habitat: Usually occurs in deeper water than its relatives, and has been recorded as deep as 100 m. However, in northern waters it is commonest in coastal waters shallower than 10 metres. Retreats down to warmer water layers

Male snake pipefish can have up to 1000 eggs attached to their bellies. They lack the brood pouch seen in Nilsson's pipefish.

in winter. Often seen hanging vertically in the water with its head up, or with its tail coiled around a stem.
Biology: Spawning takes place in mid-summer, when males and females form pairs. Females attach around 1000 eggs to the males' bellies. Males lack brood pouches–eggs are attached by a mucus layer extending from the head of the male and back to the anal opening. Eggs stay attached for four weeks before hatching. In common with pipefishes in general, the eggs receive blood containing oxygen and nutrition from the male. Feeds on small crustaceans, fish larvae and some small molluscs.

As with other pipefish, the snake pipefish cannot flee to any great extent if discovered.

The ***straight-nosed pipefish***, **Nerophis ophidion**, *has but one fin; the dorsal fin. Large females usually have a bluish stripe along the belly. Females can reach 30 cm and males 25 cm in length. Distributed from west Morocco to mid-Norway, Mediterranean and Black Sea; between 2 - 25 metres depth, particularly among eelgrass.*

Nerophis lumbriciformis - Worm pipefish

The worm pipefish

Distribution: In the Atlantic from west coast of northern Africa north to the British Isles, Kattegat, Norwegian coast (except the very north).
Description: Snout relatively short, approx. 1/3rd of the total head length, and upwardly curving. Also distinguished by having 17 - 19 bony rings between the anus and the head; whereas other species have more than 25. Anal opening directly below the anterior edge of the dorsal fin. Females up to 17 cm long; males up to 15 cm. Colour variable according to habitat, but often yellowish-brown or greenish.
Habitat: Occurs from the tidal zone down to approx. 30 metres, most commonly on mixed hard bottoms with a dense algal vegetation.
Biology: As in other pipefish, feeds mostly on small crustaceans and fish larvae. Males carry up to 150 eggs (usually 60 - 90) in four rows along the belly. These usually hatch in September, at which point the larvae are 9 - 10 mm long. Larvae settle at the bottom after one to two months; at around 30 mm length. Sexually mature after two years.

Order Gasterosteiformes - Sticklebacks

The sea stickleback seems to thrive under boats and hanging algae.

This order comprises two families, of which the stickleback family Gasterosteidae is represented in our waters by the ***three-spined stickleback*****, Gasterosteus aculeatus** *and the* ***sea stickleback*** *(next page).*
The three-spined stickleback is found in fresh, brackish and seawater. The third species in our waters, the ***nine-spined stickleback,*** **Pungitius pungitius**, *is a freshwater fish only occasionally found in seawater.*

Spinachia spinachia - Sea stickleback

The dark stripe through the eye is typical for the species.

The sea stickleback is common amongst the brown alga **Halidrys siliquosa.**

Distribution: From the Bay of Biscay to Norway; common north to mid-Norway, rarer in the north. Baltic. All around the British Isles.

Description: Slender, with long snout and tail; looking at first glance like a mixture between a 'regular' fish and a pipefish. Seen from the side, the dorsal and anal fins form a fan. In front of the dorsal fin are 14 - 17 upright spines. Brownish-black stripe extending from the corners of the mouth, across the eye socket and cheek to the root of the pectoral fin. Otherwise, colour varies according to background, but sides often pale yellow with dark green patches. Belly white. Males distinguishable from females by their pectoral fins which reach all the way back to the small pelvic fins. Up to 20 cm in length, but smaller in the northern parts of its distribution area.

Habitat: Amongst eelgrass and other algae down to approx. 10 metres. Often brackish water. Very difficult to spot when motionless amongst vegetation.

Biology: Males start preparations for spawning by building a nest within seaweed. Nest constructed by 'glueing' pieces of algae together using a sticky thread secreted by the kidneys. Spawning carried out between April and July. Monogamous; only takes one partner. The ripe female is lured into the nest after a specialised mating dance, and lays 150 - 200 eggs. The female dies shortly after spawning and the male takes care of the eggs, providing them with protection and moving water. After 4 - 5 weeks of such paternal care, the male also dies, at which point the larvae have used up their yolk sacs and are ready to take care of themselves. Can reach sexual maturity already the same autumn, and are ready for mating the following spring/summer. Maximally reaches 1.5 years of age.

WEATHER FORECASTER

Together with pipefishes, the **sea stickleback** has long been used in weather forecasting. People used to dry one of these fish and suspend it from a piece of thin twine at its centre of balance. The direction in which the fish rotates on its thread varies with changing humidity. In damp weather, the dried fish absorbs water, becoming heavier and therefore stretching the twined thread, causing it to rotate. And so rain was forecast. When the air becomes drier, water evaporates from the fish, causing it to become lighter, and therefore rotating the opposite way. In this case, fine weather could be expected.

Gasterosteus aculeatus - Three-spined stickleback

The three-spined stickleback occurs in fresh-, brackish and seawater.

To see the three-spined stickleback, divers should look at 0 - 2 m depth.

Distribution: Widely distributed, both in freshwater and in the sea. Entire Europe north to the Barents Sea.

Description: Small, usually 5 - 7 cm long; maximally 8 cm in freshwater. Can reach up to 20 cm in seawater. Characterised by usually three (2 - 5) spines located in front of the dorsal fin. Body glittering silver, with a blueish back. More or less covered in bony plates; scales absent. At spawning time, males develop an attractive red belly and blue eyes. Females develop colour nuances of brown and copper-yellow.

Habitat: Lives in schools in shallow water, often in areas with much vegetation. Occurs in seawater, brackish and freshwater. Some individuals migrate between freshwater and seawater. Active at dusk.

Biology: Diet varied, including insect larvae, small crustaceans, segmented worms, molluscs and larvae. Reproduction occurs in May - June, and ripe males make nests in hollows on the substrate and defend a small territory. Other males are chased away and females are enticed to the nests through a mating dance during which the male displays the red belly, signalling his readiness to spawn. The female swims into the nest and lays her eggs, rapidly followed by the male who fertilises them with sperm. Several females may be enticed into the same nest. The male guards and broods the eggs after spawning, and they hatch after 8 - 15 days, depending on the temperature. The 2 - 4.5 mm long larvae remain in the nest, or in the immediate vicinity for around 4 - 6 days, actively protected by the male. Thereafter the larvae enter a pelagic stage, until they reach a lenght of 25 mm, after which they move to sandy or rocky beaches. Sexually mature at 1 - 2 years. Most die after their first spawning, and only exceptionally reach three years of age.

Commercial value: Has been exploited to a certain degree in Denmark, the Baltic and the White Sea, and used for fishmeal, liver oil, fertiliser and bait.

DID YOU KNOW...

The three-spined stickleback has been extensively used in ethology research (the study of animal behaviour). One amusing anecdote tells of a ripe male with an established territory, held in a laboratory with controlled conditions, who displayed a curious behavioural change at a certain time each day. After much head-scratching and hypothesis testing, the researchers realised that the male reacted aggressively each time the red post-office van drove past. The male, whose tank had a window position, obviously thought he was being threatened by a rival.

Order Scorpaniformes - Scorpionfishes and flatheads

Myoxocephalus scorpius.

This order comprises a large number of species. In northern European waters, around more than 20 species within six families occur, but the systematic relationships are somewhat unclear. The ***rockfish*** *family* ***Sebastidae*** *is represented in northern Europe by the* ***redfish***, **Sebastes marinus**, *the* ***deepwater redfish***, **Sebastes mentella**, *and the* ***bluemouth***, **Helicolenus dactylopterus**, *in addition to the* ***Norway redfish*** *presented here. Within the sea robin family* ***Triglidae***, *the two species described here are permanent residents along North Sea coasts, but further three or four species are occasional visitors. Within the* ***bullheads*** *-* ***Cottidae***, *there are more than 10 marine species in northern Europe. Many have an Arctic distribution:* ***two-horned sculpin***, - **Icelus bicornis**; ***Arctic staghorn sculpin***, - **Gymnocanthus tricuspis**; ***Atlantic hook-eared sculpin***, - **Artediellus atlanticus**; ***ribbed sculpin*** - **Triglops pingelii** *and the* ***fourhorned sculpin***, - **Triglopsis quadricornis**. *These are small species that may be difficult identify. Of the* ***tadpole sculpins***, - ***Psychrolutidae***, *we find the northern* ***polar sculpin***, - **Cottunculus microps** *and the* ***pallid sculpin***, - **Cottunculus thompsoni**. *The* ***poachers***, - ***Agonidae***, *are represented by the* ***hooknose***, - **Agonus cataphractus**, *and the* ***Atlantic poacher***, - **Leptagonus decaganos**, *in Arctic waters. The* ***lumpfish*** *family*, - ***Cyclopteridae***, *apart from the three species represented here, includes another 10 species in northern European waters. Most of these have northern and Arctic distributions.*

Taurulus bubalis.

Sebastes viviparus - Norway redfish

The Norway redfish is a trusting fish and will allow divers to approach it.

Distribution: North-western parts of the British Isles, Iceland and north to northern Norway.

Description: Resembles the more familiar, **redfish**, ***Sebastes marinus***, but the norway redfish is as the name indicates, considerably smaller; up to 25 - 30 cm in length. Back and sides yellowish red to red; somewhat paler than the redfish. Belly white. Gill coverings (operculae) often with a dark patch at their posterior margins. Above the lateral line are 5 - 6 more or less visible dark, irregular transverse patches. Distinguished from other scorpionfishes by the fact that all 5 spines on the gill operculae are backward-pointing and have sharp tips. The redfish has blunt spines and the lower one points downwards. The Norway redfish has 70 - 80 scales along the lateral line, whereas the redfish has 80 - 90. A third species, the **bluemouth**, ***Helicolenus dactylopterus***, can be distinguished from the other two by the insides of the mouth and gill coverings being black. The latter occurs in 200 - 800 metres depth. One more species, the **deepwater redfish**, ***Sebastes mentella***, occurs between 300 - 1000 metres depth.

Habitat: Found from 10 - 200 metres depth. Remains close to the coast, and enters further into fjords and inlets than does the redfish. Divers often observe it swimming over stony bottoms in small groups. Most commonly found in areas where the bottom slopes down to greater depths.

Biology: In common with other scorpionfish, the Norway redfish gives birth to live young. Sexual maturation occurs at around 18 cm length. Between June and July, each female gives birth to up to 30 000 young. Larvae pelagic until they reach around 6 cm in length. Extremely slow growth rate and can reach up to 30 years of age. Mainly feeds on bottom-dwelling organisms such as crustaceans, bivalves and small fish.

Commercial value: Tasty fish popular among sports fishermen. Appears as a by-catch during shrimp trawling and sold as an industrial fish.

FRIENDLY FISH

The Norway redfish is friendly towards divers, often letting itself be touched. Often it will even follow divers.

Eutrigla gurnardus - Grey gurnard

Distribution: From Madeira north to Iceland and northern Norway. Mediterranean and Black Sea. All around the British Isles.

Description: Characteristic appearance; head relatively large, with the eyes as the highest points. Pronounced overbite. Body tapers evenly, narrowing towards the large caudal, or tail fin. Upper side grey or greyish brown. First dorsal fin bears dark patch. Can be confused with the **tub gurnard**, ***Trigla lucerna***, but the latter lacks the grey gurnard's enlarged scales with a spiny keel along the lateral line. However, the most obvious difference is in the length of the pectoral fins. In the grey gurnard, these do not extend as far back as the anterior edge of the anal fin, as is the case in the tub gurnard. Further, in our waters, the tub gurnard is more red in colour than the grey gurnard, although the Mediterranean form also is reddish. Can reach up to 45 cm in length and 1.2 kg in weight.

Habitat: Usually sandy bottoms, but also found on other substrates from the shore down to around 150 metres depth. Can come right up to the surface at nights.

Biology: Spawning takes place from April to August. Females spawn up to 300 000 pelagic eggs, which hatch after 8 - 10 days. Larvae 3 - 4 mm long at hatching and live pelagically until they are 3 cm in length, at which point they settle to the bottom. Feeds mainly on various bottom-dwelling organisms such as crustaceans and bristle-worms. However, other small fish also are on its menu.

Commercial value: No commercial value in our waters, despite the fact that the white flesh is firm and tasty.

BUZZING SOUND

The family Triglidae are known as sea robins, perhaps because they make a trilling or buzzing noise when taken out of the water. This noise comes from vibrations in the swim bladder.

The grey gurnard is shy and a difficult fish to photograph. Note the characteristic fin rays used as 'walking legs'. Here from 15 m depth. Inset: The colours can vary.

Trigla lucerna - Tub gurnard

Young individuals are common in the intertidal zone.

Distribution: From Africa's west coast, Mediterranean and north along Europe's Atlantic coast to northern Norway. Main distribution is from the English Channel and south. Not as common as the **grey gurnard**, ***Eutrigla gurnardus***.

Description: Resembles the grey gurnard, but can be distinguished by the following characteristics: pectoral fins extending back beyond the anterior margin of the anal fins. The scales along the lateral line are only weakly enlarged and lack the spiny keel. In our waters, it has a reddish colour. Becomes larger than the grey gurnard; females can reach up to 75 cm long and weigh up to 5.2 kg.

Habitat: As for the grey gurnard, closely associated with the bottom; preferentially sands or mud. In our waters, found from the shore down to around 150 m depth, but occasionally occurs at the surface.

Biology: Reproduction is carried out between March and October. Breeds May - July in the Channel. Feeds on various bottom-living organisms and also fish, to a greater extent than the grey gurnard. The tub gurnard is reputed to have been observed leaping across the water surface with the large pectoral fins spread out like wings! In common with the grey gurnard, makes a buzzing sound.

Commercial value: The flesh is white, firm and tasty. No commercial value in our waters, but popular among hobby fishermen due to its size.

The tub gurnard is markedly larger than the grey gurnard. The pectoral fins are impressive.

Myoxocephalus scorpius - Shorthorn sculpin

The shorthorn sculpin is found on all kinds of substrates. It hunts by surprising its prey.

***Left**: Here a goby is being swallowed whole. Photographed during a night dive.*

Distribution: From the Bay of Biscay to Svalbard, Iceland and the Baltic. Very common.
Description: Head large in relation to the rest of the body, with a variety of spines and bony plates. On the gill covering are two spines that do not extend beyond the margin of the operculum. Lateral line straight; lacks spines. Colour variable. Back usually dark brown or greenish brown; sides with numerous dark spots or stripes. Belly pale. Around spawning time, females develop an orange belly, whereas the male belly becomes reddish with white spots. At 60 cm, this is the largest of the Cottidae. In our waters, more than 10 marine species occur, including the **fourhorned sculpin**, ***Triglopsis quadricornis***, which also occurs in freshwater. However, all these are considerably smaller than the shorthorn sculpin, and can be difficult to identify.
Habitat: Bottom-living; common in shallow coastal waters. Often seen from the shore and down to 30 metres depth, but has also been caught at 250 m. The larger specimens often lie unprotected on sandy bottoms, whereas smaller individuals are camouflaged amongst algae.
Biology: Spawns during winter between December to March. Eggs laid in clusters, approximately the size of a hen's egg, placed between algae and stones. Each cluster contains up to 2700 eggs. Eggs guarded by the males until they hatch, around 4 - 12 weeks after spawning. Larvae hatch at around 6 - 8 mm length and live pelagically the first months. The shorthorn sculpin lies camouflaged on the bottom whilst waiting for prey; usually small fish or crustaceans. It attacks by shooting out of its place of refuge and swallowing the prey whole.
Commercial value: None, although the meat is tasty.

Taurulus bubalis - Longspined bullhead

The long-spined bullhead usually occurs in kelp forests, often among the stipes.

Distribution: Eastern Atlantic from Portugal, along the Norwegian coast to the Kola peninsula. Scattered stocks in the Mediterranean. Baltic.

Description: Smaller than its relatives; only 17 cm in length. In other respects, resembles the **shorthorn sculpin**, ***Myoxocephalus scorpius***, but distinguished by having four spines on the gill covering, of which one extends beyond the edge of the operculum. Lateral line with small thorns (lacking in the shorthorn sculpin). Colour varies according to surroundings, but back often brown or greenish, with four darker transverse stripes along the sides. Often reddish in kelp forests. Resembles the norway bullhead, but the latter is much smaller; maximally up to 7.5 cm long. Three soft rays and one spiny ray in the pelvic fin. Further, a flap of skin hangs from the upper corners of the mouth.

The species is distinguished by the flap of skin at the mouth corners and by the single spine on the gill covers.

Habitat: Common on hard bottoms down to approx. 30 metres. Often well camouflaged amongst algae.

Biology: Spawning takes place during the transition between winter and summer, i.e. between February and April. Eggs yellowish-brown; usually laid in rocky crevices or under stones. Eggs 1.5 - 1.8 mm in diameter; larvae 6 mm long at hatching. Sexual maturation occurs at 2 years of age. Feeds on small crustaceans, bristle-worms, bivalves, brittle-stars and small fish.

The long-spined bullhead is a camouflage expert and appears in a wide variety of colours.

Microenophrys lilljeborgi - Norway bullhead

The Norway bullhead is rather rare and lacks a spine on the gill cover, distinguishing it from the long-spined bullhead.

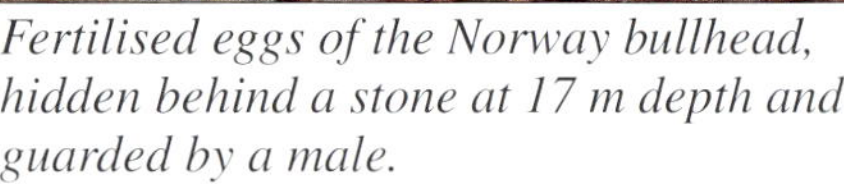

Fertilised eggs of the Norway bullhead, hidden behind a stone at 17 m depth and guarded by a male.

Distribution: Limited distribution from the west coast of the British Isles, Faeroes, northwest coast of Sweden and north to mid-Norway.
Description: The smallest member of the group in our waters. Females up to 7 - 8 cm in length, but rarely over 5.5 cm. Resembles the **longspined bullhead**, ***Taurulus bubalis***, but can, in addition to size, be distinguished by the following characteristics: the Norway bullhead has two (not three) soft rays and one spiny ray in each pelvic fin, and the flap of skin at the mouth corners is absent.

The flap of skin at the mouth corners is absent.

Habitat: Hard bottoms, from the shore down to approx. 80 m. Often among algae or under or below stones, particularly in areas of dense calcareous algae such as ***Corallina***.
Biology: Spawning carried out in batches; eggs usually hidden behind stones. Males guard the eggs continuously, and will allow themselves to be petted on the back. Larvae 4 mm in length at hatching. Live pelagically until they settle, at around 10 - 14 mm length.

The eye often is greenish.

The ***ribbed sculpin***, **Triglops pingelii**, *is a northern, circumpolar species, up to 20 cm long; can be confused with the* ***moustache sculpin*** *(below). Note the four small spines on the gill covering. Here photographed at Svalbard.*

The ***twohorn sculpin***, **Icelus bicornis**, *is an Arctic form, but also found along the entire Norwegian coast. It can reach up to 10.5 cm in length. Typically with two rows of thorns along the midline of the body, and thorns along the dorsal fin.*

Triglops murrayi - Moustache sculpin

Distribution: A northern form, found along the northwestern coasts of the British Isles, around Denmark, Swedish west coast, Norwegian coast and north to the Barents Sea. Spitsbergen, White Sea and Greenland.
Description: Elongate body, more slender than most flatheads. Head relatively small; gill covering with four small spines. Resembles the **ribbed sculpin**, ***Triglops pingelii***, which is only found north of Vestfjorden, northern Norway. The moustache sculpin has a black patch at the rear margin of the first dorsal fin, and dark transverse bands on the tail fin. Females reach up to 17 cm lange; males somewhat smaller.
Habitat: Bottom-living in deep water, usually between 50 - 250 m, but occasionally occurs shallower–we have observed it as shallow as 10 meters in southwest Norway. With the exception of the White Sea stocks, it requires a water salinity of 33 - 34 ‰ and temperatures around 2 - 3 °C.
Biology: Feeds on various bottom-living organisms; reproduction probably late autumn or winter.

The head of the moustache sculpin resembles the Shorthorn sculpin whereas the body resembles the grey gurnard. Here pictured at 30 m depth.

The moustache sculpin is an elegant fish found in relatively shallow water.

Agonus cataphractus - Hooknose / Pogge

The hooknose is a strange-looking fish, usually found on soft bottoms in shallow water. The numerous barbels are used to seek small prey organisms.

This fish is not shy, trusting in its bony projections to deter predators.

Distribution: From the English channel to Iceland, Norway and the White Sea. Also southern Baltic.
Description: Body covered with rows of bony plates, making it angular in outline. Head large and broad, with stout thorns on the beak-like snout. Two barbels on the snout above the mouth, and numerous barbels under it. Colour brown to brownish grey. Males' pectoral fins take on an orange hue around spawning time. Up to 20 cm in length. Highly characteristic; unlikely to be confused with any other fish in our waters. In northern Norway and the Arctic, another member of the poacher family occurs, the **Atlantic poacher**, ***Leptagonus decaganos***. However, the latter is smaller than the hooknose, has a longer tail, considerably greater distance between the dorsal fins and fewer barbels.
Habitat: Lives on the bottom; prefers sandy or gravelly substrates. Usually in shallow water (5 - 20 m), but may migrate to deeper water in winter (500 m).
Biology: Spawning takes place in shallow water between January to April. Eggs yellowish eggs, 2 mm in diameter, and laid in clusters on the bottom. These hatch in spring. Newly hatched larvae are 6 - 8 mm in length and live pelagically until they settle in midsummer, around 2 cm in length. Feeds on a variety of bottom-living organisms, including crustaceans, bristleworms, snails and bivalves.

DID YOU KNOW...

The hooknose is an unusually trusting fish. Divers can pick it up and put it back down again, without it seeming particularly disturbed by the experience. It most likely relies on its bony armour plating for protection against predators.

Cyclopterus lumpus - Lumpsucker

The lumpsucker is a trusting creature that lets divers come close.

CAVIAR

The Danish "Limfjord" caviar is in fact lumpsucker roe, dyed black to resemble 'real' sturgeon caviar.

Distribution: Eastern Atlantic from the coast of Portugal north to the White Sea and the Baltic. Common in the sub-littoral zone from January to September.

Description: Highly characteristic appearance, unlikely to be confused with any other fish. Shape rather round and squat, with seven rows of bony projections along the sides and along the belly. First dorsal fin conceals a high comb of bony projections. Skin thick. Most characteristic is the strong suction disc located on the belly between the pectoral fins. Males have a larger head and pectoral fins than females. During spawning, the male's belly turns an attractive red, whereas females are greenish. Females reach up to 63 cm in length and 5.5 kg in weight; males up to 55 cm. Juveniles resemble the adult form, often greenish in colour.

Habitat: Occurs right up to the shore during spawning time. In winter, it is found swimming freely in the water masses over greater depths. Small juveniles are often observed attached by suction to seaweed and kelp in shallow water (1 - 5 m).

Biology: Migrates to the coast in springtime to spawn, which is carried out in shallow water amongst algae. Females spawn up to 400 000 pale yellow eggs in batches over a two week period. After some time, the eggs turn green. Larvae well-developed upon hatching, and already after four days can attach themselves to the substrate by means of the ventral disc. Larvae remind of small, green tadpoles with large eyes.

Commercial value: Lumpfish fisheries are known from hundreds of years back. The

GUARDING THE EGGS

Male lumpsuckers guard the eggs with their lives, until they hatch around 60 days after spawning. If a female has unwisely laid her eggs in so shallow water that they are exposed at low tide, the male will remain with them, even if that means drying out. These males usually become the victims of sea-birds.

Female lumpsuckers are less commonly seen by divers and are distinguished by their blue-green colour tones.

This 2 cm long young lumpsucker resembles the adult form. Below: 3 cm long specimen.

flesh of the male is regarded as a delicacy in many countries, although in Norway and Britain and Ireland, it usually is considered a trash fish. Around Iceland, Greenland, Russia, Denmark and northern Norway, a regulated lumpsucker fishery is carried out for the roe. Each female can yield up to 0.7 kg roe.

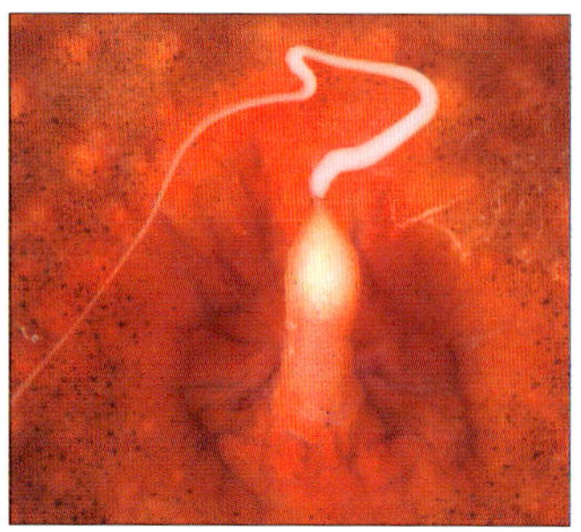

The male reproductive organ is located behind the sucker disc.

DID YOU KNOW...

Adult lumpsuckers eat a lot of plankton. The young keep in close contact with seaweed and kelp until they are around two years old. They then migrate away from the coast and do not return until they are sexually mature. During this period, lumpsuckers live freely in the water masses, hunting pelagic crustaceans and jellyfish. Adult lumpsuckers do not eat when they are in shallow water.

Liparis liparis - Striped seasnail

Because of its small size, the striped seasnail is seldom seen by divers. It usually is attached to kelp blades, but occasionally lies on the bottom.

Distribution: Distributed in the eastern Atlantic, within a limited area from the British Isles, Iceland, Norway and the Baltic.
Description: Outline tadpole-like, with a large head and a body tapering backwards towards the tail. A number of seasnails occur in northern European waters, but only the **striped seasnail** and **Montagu's seasnail**, ***Liparis montagui***, are common near the shore. Striped seasnail distinguished from Montagu's seasnail by the dorsal and anal fins being continuous with the caudal, or tail fin.

In Montagu's seasnail, the dorsal and tail fins are clearly separated. Colour varies, but those hiding amongst algae are usually brownish. Up to 18 cm in length; larger than Montagu's seasnail with its modest length of 6.5 cm. Sucker disk present on belly, in common with its relative, the lumpsucker.

Habitat: Bottom-living, usually only observed by divers in wintertime among algae in shallow water. Often attached to a kelp stipe (***Laminaria*** sp.) with the tail curved back along the body. Lives in deeper water during summer; recorded down to 300 m.

Biology: Spawning carried out during midwinter. Eggs 1.5 mm in diameter; laid in small clusters amongst algae and bottom-living organisms. Larvae hatch after 6 - 8 weeks and live pelagically until around 1.2 cm length. Feeds on various crustaceans, small fish, jellyfish and algae.

Liparis montagui - Montagu's seasnail

Montagu's seasnail is distinguished from the striped seasnail by the dorsal fin being separate from the tail fin.

Distribution: From the north coast of France, around most of the British Isles, Iceland, Svalbard, Norwegian coast to the White Sea.

Description: Can be confused with the **striped seasnail**, ***Liparis liparis***, but distinguished by the dorsal fin separate, not continuous with, the tail fin. Also of a smaller size; reaches only up to 6.5 cm in length. Colour variable, often pale brown, but can be greenish or yellowish.

Habitat: Closely associated with the bottom; mostly seen by divers in shallow water during winter. Prefers hard bottom; often attached to algae or the undersides of stones. Rarely occurs deeper than 30 metres.

Biology: Spawning takes place between January and May. Around 700 eggs per female, laid amongst various attached bottom organisms. Eggs hatch after around 6 weeks, and the 3.5 - 4 mm long larvae live pelagically until they settle at around 1.2 cm length. Mostly feeds on a variety of small crustaceans.

Note the clear division between the dorsal and tail fins.

Order Perciformes - Perch-likes

This is the largest group amongst the vertebrates, comprising close to 8000 species within 150 families. Not surprisingly for such a large group, there is much variation between its members, such that there are few common characteristics. However, all perch-likes have fins with rays. In most, the pelvic fins have one spiny ray and five or less soft rays. Most have two dorsal fins. The scales usually bear spines. Members of 13 families are presented here. The families are described below.

Family Carangidae - Jacks and pompanos

Members of this diverse family have two free-standing spiny rays in front of the anal fin. Appears together with the ***pilot fish*****, Naucrates ductor***, north to the British Isles*

During summer, young Atlantic horse mackerel may seek protection within the tentacles of stinging jellyfish.

Trachurus trachurus - Atlantic horse mackerel

Distribution: Warm water species generally found throughout the eastern Atlantic from South Africa to Iceland and Norway. Mediterranean and Black Sea. Migrates into northern waters during springtime, and moves south again in autumn.
Description: Body slender; underbite. Around 70 - 80 spiny scales along lateral line. Back blueish; sides silvery. Dark patch at the posterior margin of the gill cover. First dorsal fin short, with eight rays. Up to 60 cm in length and 1.5 kg in weight.
Habitat: Occurs in schools from the surface down to around 100 metres. Often swims together with mackerel, sprat and herring. Divers often observe juveniles seeking protection amongst the tentacles of the stinging jellyfish, ***Cyanea capillata***. This also the case for whiting and haddock young. The young leave the jellyfish when they reach around 4 cm in length. Often seen in small schools after reaching 4 - 15 cm in length. Some individuals swim in close contact with the goldsinny wrasse.

During autumn and winter, Atlantic horse mackerel often are seen in schools. Here from 40 m depth in October.

Biology: Spawning occurs pelagically in summer. Up to 140 000 eggs per female; 1 mm in diameter. Eggs and larvae pelagic. Reaches 15 cm in length after the first year. In winter, mainly feeds on bottom fauna, but feeds pelagically in summer. Larger fish also take squid and octopus.
Commercial value: Widely fished commercially in southern Europe. Little commercial value in northern European waters, other than being an important food source for other fish.

Family Mullidae - Goatfishes

Two species are more or less common in northern Europe. The ***plain red mullet*, Mullus barbatus** *is an offshore species. All goatfish live closely associated with the bottom. They have a pair of long feelers on the head, which seek out food by touch and smell. The head is relatively large, with a steep snout and eyes located high on the head. Forked tail fin and large, loose scales. Most goatfish have beautiful colours.*

The striped red mullet changes colour throughout the day. During daytime it has a prominent red stripe along the head and body.

Mullus surmuletus - Striped red mullet

Distribution: Belongs in the Mediterranean, but occurs south to the Canary Islands and north to southern Norway.

Description: Characteristic appearance with two long forked barbels, or feelers. Eyes situated high and anterior on the head. Short snout. Upper law lacking teeth. Forked tail and large, loose scales. Beautiful red marbled appearance; particularly prominent after dark. During daytime, pigmentation in the form of up to five longitudinal red stripes, one dark stripe and three yellow stripes. Up to 40 cm in length and 1 kg in weight. Unlikely to be confused with other fish in inshore waters.

Habitat: Soft bottoms; occurs from the shore down to 100 m depth; most commonly between 20 - 60 metres. Usually occurs in relatively large schools, but often occurs singly at the extremes of its distribution area.

Biology: The long sensory feelers register both touch and smell and are used to locate prey buried in the sediment. As soon as the fish finds a prey organism, usually a bristle-worm, small crustacean or mollusc, it stops and digs it up. In the Mediterranean, spawning occurs in shallow water during May - June, later in northern water. Eggs pelagic and the young also live pelagically until reaching 5 cm in length, feeding on zooplankton.

Commercial value: Highly valued as a food fish, especially in the Mediterranean, where it is fished commercially.

By night, it often has strong red colouration.

SKY HIGH PRICES

The striped red mullet was the most highly prized fish amongst the Romans. On ceremonious occasions, the fish were brought live to the table and the diners selected their fish. The fish was slaughtered at the table, so that the guests could observe the beautiful colour changes that the fish underwent during its slow death. The price of such a meal was 'sky high'.

Chelon labrosus - Thick lip grey mullet

The thicklip grey mullet is a strong fish, popular amongst sports fishermen.

Distribution: From Senegal in the south to Iceland and northern Norway. Mediterranean and western Black Sea.

Description: Torpedo-shaped outline, robust, with a somewhat blunt front end and a dominating upper lip. Back covered with glittering silver scales, which are dark on their upper sides and pale in the lower parts. Two short dorsal fins, widely spaced; the first with four spiny rays (three long and one short), and the second with 9 - 10 soft rays. Up to 75 cm in length and 5 kg in weight. Can reach 20-25 years old. The **thinlip grey mullet**, ***Liza ramada,*** also occurs in northern waters and can be distinguished by the distance between the eyes being equal to the width of the mouth. In the thinlip grey mullet, the mouth is narrower than the head at eye-level. Further, the upper lip in the thicklip grey mullet is thicker or equal than half the eye diameter, and additionally is covered with thin skin warts. In the thinlip grey mullet, the upper lip is thinner than the eye pupil diameter, and lacks warts. A third mullet, the **golden grey mullet**, ***Liza aurata***, also sporadically occurs in northern waters. The latter resembles the thinlip grey mullet, but can be distinguished by its longer pectoral fins and the lower part of the eye socket being pointed at the rear part. Further, the golden grey mullet has a golden patch on the gill covers, and lacks the black patch at the basis of the pectoral fins.

Habitat: Occurs in small groups, and attracted to bays with warmed water. Also thrives close to river mouths, and young individuals may move up into the river. Keeps to shallow water. At dusk, may be seen leaping across the water surface, and in calm water, individuals can be seen swimming right up to the surface.

Biology: Feeds mostly on various algae, but also takes small invertebrates and detritus (organic deposits). Sexual maturation occurs at four years of age, at around 30 - 35 cm in length. Spawns between June-August in the English Channel and around Ireland; earlier farther south.

Commercial value: Popular amongst sports fishermen; normally caught using small floats with various types of bait, including bread and maize as well as shrimps, maggots and earthworms. Fished with nets in southern Europe. Excellent food fish.

Family Labridae - Wrasses

Most wrasses display fantastic colours. Here the corkwing wrasse, **Symphodus melops**.

Most wrasses occur in warm waters, and those occurring in northern European waters have a generally more southern distribution. Included, but not presented in this book is the ***rainbow wrasse***, - **Coris julis**, *sporadically a southern guest along the North Sea coasts and western Norway. They all have prominent lips. The mouth has stout teeth and can be everted. The body is rather elongate and flattened from the sides. Many wrasses are extremely colourful. Three of 'our' wrasses are known as cleanerfish, because they eat parasites on other fish.*

Symphodus melops - Corkwing wrasse

Distribution: From the coast of Morocco to mid-Norway. Faeroes. Mediterranean and western parts of the Baltic.

Description: Characteristic dark, kidney-shaped patch behind the eyes, and a dark patch at the tail root, although these not always prominent. Males attractively marbled in green, blue and brown colours. Females more uniformly brown. Dorsal fin with 15 - 18 spiny rays and 8 soft rays. Anal fin with 8 - 10 soft rays and 3 spiny rays. Around 32 - 36 scales along the lateral line. Up to 30 cm

EATS PARASITES

Corkwing wrasse, rock cook and cuckoo wrasse are highly effective in removing sea lice from farmed salmon. Because these have different depth preferences, best results are achieved by using a mixture of these species. This is an efficient and environmentally friendly alternative to chemical de-lousing and a good example of biological pest control. One wrasse per 50 salmon is considered an effective stocking density.

in length but rarely more than 15 - 20 cm. Reaches up to 10 years of age.

Habitat: As for other wrasses, lives amongst algae in shallow water. Young individuals also occur on sandy bottoms amongst eelgrass. Moves to deeper and warmer water in winter, but seldom below 30 metres.

Biology: Males build nests and often entice several females into them. Mating rituals with an introductory mating dance occur during May - July. Males defend their territories around the nests until the eggs hatch. Larvae live pelagically until they settle to the bottom around autumn. As for other wrasses, large females can change sex and become males. Mostly feeds on various small crustaceans and bivalves, especially blue mussels.

The corkwing wrasse has blue-greenish marbled pigmentation. Note the kidney-shaped patch behind the eye and the dark patch at the tail base.

Centrolabrus exoletus - Rock cook

The rock cook typically has bluish marbling, most prominent around the head.

Distribution: European distribution; from Portugal, south and west coasts of the British Isles, as well as east coast of Scotland. Sparsely distributed in southern parts of Scandinavia.

Description: Resembles the **corkwing wrasse**, ***Symphodus melops***, but distinguished by the following: rock cook lacks the black patch behind the eye and at the tail root. Finer marbling; coloured stripes more blueish and

less green. Lateral line markedly curving below the posterior part of the dorsal fin. Anal fin with 4 - 6 spiny rays; in contrast with the 3 spiny rays in the corkwing wrasse. Small, seldom exceeds 12 cm in length, but records up to 18 cm reported.
Habitat: Shallow water, often right up to the shore, on hard bottoms amongst algae or soft bottoms amongst eelgrass. Down to 25 metres; deeper in winter than summer.
Biology: Spawns in spring. Males build nests among algae into which they entice females. Males guard the eggs until hatching. Mainly feeds on small crustaceans.

The rock cook displaying its attractive colours.

Labrus bimaculatus - Cuckoo wrasse

Male cuckoo wrasse have characteristic waving bluish stripes.

Distribution: From Senegal, western Africa north to the western and northeastern British Isles. Mediterranean. Scandinavia from Swedish west coast to mid-Norway.
Description: Adults highly characteristic. Body slender, with elongated head and pointed mouth. Sexually dimorphic; i.e. males and females strongly differing in appearance; in fact, for a long time, these were thought to belong to different species. Females reddish to orange, with three black patches at the transition between the rear part of the dorsal fin and the body sides. Males blue, with attractive darker marbling. Females seldom over 30 cm in length, whereas males can reach 35 cm. Can become 17 years old or more.
Habitat: Bottom-living, prefers hard bottom. In summer, often seen amongst algae right up to the water surface. Moves to deeper water in winter, usually from 15 to 200 metres depth. As

***Above:** Female cuckoo wrasse*
***Below/ right**: Female in the process of becoming male.*

One of our most beautiful fish. It is easiest to photograph at night, when it is resting.

with other wrasses, active during daytime; can be seen resting in crevices or under stones at night.

Biology: Spawning occurs between May - July. Males build nests of algae between stones or in crevices, into which the females are enticed. During the mating dance, the male undertakes a fantastic display of colour, flashing from dark to pale and dark again in a matter of seconds. Females lay around 1000 eggs, which hatch after 10 days. Feeds on various crustaceans, including barnacles, as well as bivalves and snails.

Commercial value: The flesh is sweet to the taste and suitable for frying.

CHANGES SEX

In common with many other perch-like fish, the cuckoo wrasse is a hermaphrodite. The capacity to be both male and female is present from birth; the sex of the fish often is determined by external factors. Most newly-hatched larvae develop into females, but some become males, with the same external characteristics as the females. These are known as primary males and live as males for all their lives. However, some individuals are what is termed secondary males - these have all changed sex. Colour also changes in connection with the sex change, but not necessarily at the same time as the internal changes. In some, colour changes before the internal 'real' sex change occurs, whereas in other individuals, this occurs first afterwards. These sex changes occur at the earliest at 7 years of age and are socially regulated. Males keep a harem of females. In turn, the females have a hierarchical ranking system. When a secondary male disappears, the dominant female will begin to change sex and take over his role. Primary males are a minority and do not appear to take part in reproduction; females seem to prefer to mate with secondary males.

Ctenolabrus rupestris - Goldsinny wrasse

Children fishing from piers and quays may be familiar with the bait-thieving activities of the goldsinny wrasse. Note the dark patch at the base of the tail.

Distribution: From Morocco northwards along the west coast of the British Isles. Mediterranean. Abundant in Scandinavia from Swedish west coast to mid-Norway; scarcer further north.

Description: Relatively slender and, like the **rock cook**, ***Centrolabrus exoletus***, one of our smallest wrasses. Distinguished by the dark patch on the dorsal fin between the first and fifth spine, and another just above the root of the tail. Colour orange; darkest along the back. In spawning colours, males have more dark red spots under the anterior part of the lateral line. Females have more dark stripes on the belly side. Usually 10 - 12 cm in length, but can reach up to 18 cm. Up to 10 years of age.

Habitat: Bottom-dwelling; common in shallow water, from the shore between stones and algae down to approx. 20 metres. Deepest in winter.

Biology: Spawning occurs in summer. Males establish territories before mating. Unlike most wrasses, eggs pelagic. Feeds on small crustaceans, bivalves and snails, or anything else available. Known as the 'bait-thief' among fishermen.

The strong teeth are used to grasp small molluscs.

Acantholabrus palloni - Scale-rayed wrasse

The scale-rayed wrasse is occasionally seen as shallow as 10 m depth.

SO-CALLED RARE

The scale-rayed wrasse is considered to be relatively rare throughout its recorded distribution area. However, we believe this to be because it is difficult to catch using traditional fishing methods. These fish usually live well-hidden, often in groups, under rocks and cliffs. In Norway, our observations suggest that the scale-rayed wrasse is abundant in fjords deeper than 20 metres and in outer coastal areas below 40 metres.

Distribution: From the Guinea-bay to western Norway. Southern coasts of the British Isles. Previously believed to be rare, but probably simply because it is seldom caught.

Description: Brownish; belly pale yellow. Characteristic dark patch on the upper part of the tail root, and a long dark patch on the dorsal fin at the border between the spiny and rays. These are particularly prominent in adults, as seen on the photographed specimen. Also typical are the prominent teeth on the gill covering. Younger individuals lacking these features can be confused with the **goldsinny wrasse**, ***Ctenolabrus rupestris***. However, the scale-rayed wrasse can be distinguished by having 14 - 21 spiny rays and 7 - 10 soft rays in the dorsal fin, 4 - 6 spiny rays and 5 - 8 soft rays in the anal fin and 40 - 45 scales along the lateral line. Sexual dimorphism absent. Recorded up to 27 cm in length.

Habitat: Biology and habitat poorly known, but thought to prefer deeper water than our other wrasses. The specimen shown here was photographed at 20 metres depth, but divers have also noted it as shallow as 5 metres. Has been caught at 270 metres depth.

Biology: Reproduction and feeding habits almost unknown. The few existing records suggest that the diet is varied, consisting of fish, bristle-worms and bivalves.

Labrus bergylta - Ballan wrasse

Distribution: From Morocco to mid-Norway.
Description: The largest of our wrasses. Stout shape; high in relation to its length. Largest individuals reaching 60 cm in length and 3.5 kg in weight; cannot be confused with any of our other fish. Small specimens can, however, be mistaken for other wrasses, but can be distinguished by having 3 spiny rays in the anal fin; whereas the others have 5 - 6 spiny rays. Dorsal fin with 19 - 20 spiny rays and 9 - 11 soft rays. Along the lateral line are 41 - 47 scales. Large individuals with a raised portion on the upper side of the snout. Colour variable; prominent brown marbling. Up to 25 years old.
Habitat: Common in kelp forests in summer; retreats to deeper water in winter. Down to 200 metres.
Biology: Spawning takes place in summer. As in other wrasses, males establish territories and build nests of algal fragments bound together by secreted mucus. Females spawn up to 1300 egg, which hatch after 8 - 12 days depending on temperature. Like the **cuckoo wrasse**, ***Labrus bimaculatus***, the ballan wrasse is a hermaphrodite. All larvae develop first into females and become sexually mature after 5 - 6 years. A few turn into males at this point. When the mature females are 10 - 14 years old, they change sex and become males. Almost all fish over 40 cm in length are males. Prey includes bivalves, snails and crustaceans.
Commercial value: The flesh is highly prized by connoisseurs.

All large ballan wrasse are males. Females change sex at around 10 - 14 years of age.

ANGLING

Angling for ballan wrasse is popular. Ragworms tend to be the best bait, with lugworm, fish strip, mussels, limpets and pieces of crab also proving successful. A sliding float rig is a very good way to fish for wrasse from the rocks. All wrasses display a knack for sucking worms off the hooks; to stop them, a tiny square of squid or bits of mackerel behind the hook barb is effective.

Family Trachinidae - Slimeheads

A small family with 4 species. The ***lesser weever****,* **Echiichthys vipera** *is common from Denmark and southwards.*

The greater weever prefers warm shallow water and sandy bottoms. It can be a hazard for bathers.

Trachinus draco - Greater weever

POISONOUS

Immediate medical attention is required if stung by a greater weever. The sting should be bathed in warm water until reaching the doctor. The poison is particularly dangerous for children. Treat this fish with extreme caution should it appear on your hook or in your net.

Distribution: From Morocco north to mid-Norway. Mediterranean.

Description: Elongate, up to 10 cm in length. Brown and blue-green metallic marbling. Ventral, belly side pale, with light brown markings. Evenly tapering towards the tail. Prominent underbite. Eyes situated high and anterior. One short and one long dorsal fin; the short one with 5-7 spiny rays. The first of these, like the spine on the gill covering, is connected to a poison gland.

Habitat: Lies on the bottom during daytime, often partially covered in sand, at night it often leaves the bottom to hunt. Common around many sandy beaches in summer at between 4-10 metres depth, but recorded as deep as 150 metres. Seldom seen so shallow as to be a danger to bathers.

Biology: Spawning takes place in shallow water during June to August. Eggs pelagic and hatch already after 4 - 5 days. Mostly feeds on various small fish and crustaceans.

Commercial value: Not commonly eaten in northern Europe, but used as a food fish in southern Europe. When preparing the fish–take care to remove the poisonous tissue at its anterior end.

The greater weever usually lies partially buried in sand. Here from 10 m depth on a warm summers day.

Family Blenniidae - Combtooth blennies

Combtooth blennies have an elongate body, with a large rounded head, a long, continuous dorsal fin and all lack scales. Most are temperate water species, but three occur north of the English Channel. In addition to the shanny, the ***butterfly blenny***, **Blennius ocellaris** *and the* ***tompot blenny***, **Parablennius gattorugine**, *occur around the British Isles.*

Lipophrys pholis - Shanny

An attractive fish with effective camouflage.

Distribution: From Morocco north to the British Isles and southwest Norway. Can occur in large numbers at certain localities.
Description: Small and elongate; up to 20 cm in length. Large head and long dorsal fin with a prominent incision at the level of the anal opening. Skin lacking scales and colour varies according to habitat.
Habitat: Extremely shallow water; tidal zone–photographers should look around at 1-2 metres depth. Hides in small crevices or under stones. Tolerates periodic air exposure and some aquarium-inhabitants have been seen to 'crawl' onto land; an ability unique amongst European fish.
Biology: Feeds on any available invertebrates in the tidal zone; specialist in snatching barnacles with their feeding legs out. Spawning takes place between April-August. Females stick their eggs under stones or shell remains and the males guard them. During this time, the males do not feed, and are almost coal-black in colour, with the exception of the almost completely white lips. Fast growth rate; reaches 8 cm in length within one year.

Divers seldom observe the shanny, because it occurs so shallow, often from the surface to only 20 cm depth.

Family Pholididae - Gunnels

Unlike the Blenniidae, most gunnels are found in cold waters, mostly in Arctic parts of the Pacific. Only one species is found in our waters.

The rock gunnel occasionally is seen partially buried in sand, but it usually remains close to a suitable hiding place.

Pholis gunnellus - Rock gunnel

Distribution: Eastern Atlantic from west coast of France north to the White Sea. Large parts of the Baltic, Iceland, Norway.
Description: Eel-like body, with small head. Brown with 9 - 13 characteristic black spots along the back at the base of the single, continuous dorsal fin. Spots surrounded by a narrow white ring. Up to 25 cm long.
Habitat: Bottom-living, found amongst algae right up to the intertidal zone. At low tide often lies amongst moist algae, under stones or in rock pools. Usually shallower than 20 metres, but has been caught as deep as 110 metres.
Biology: Spawning takes place in late autumn/ early winter in the northern part of its distribution area. Around the British Isles spawning occurs first in January to February. The 80 - 200 eggs are attached to the substrate in 2 - 3 cm diameter clusters, often under stones or in crevices. Eggs guarded by the parents, usually the female. Eggs hatch after 40 - 60 days into 9 mm long larvae. The

DID YOU KNOW...

There is a good chance of finding rock gunnels under algae and, as the name suggests, under rocks and stones. It wriggles vigorously if discovered.

The characteristic dark patches are surrounded by a yellow ring.

brooding parent does not feed during this time. Larvae live pelagically until approximately 3-3.5 cm in length. Diet varied, including crustaceans, bristle-worms, carrion and fish eggs. The rock gunnel is an important prey item for many seabirds, particularly the black guillemot and the common heron.

Eggs of the rock gunnel can be seen from the shore and below.

Family Zoarcidae - Eelpouts

As is the case for gunnels, eelpout family members belong in cold waters and 15 species are represented in our area. Most of these occur offshore in deep water, or around Arctic Svalbard and Jan Mayen. Four species occur along the Norwegian coast. In addition to the ***viviparous blenny****, these include the following:* ***Sars' wolfeel*****, Lycenchelus sarsii**, *and* ***Vahl's eelpout*****, Lycodes vahlii**, *at depths greater than 150 metres. The* ***three-spot eelpout*****, Lycodes rossi** *is found along the very north coast of Norway, recorded up to 10 metres depth.*

The viviparous blenny usually appears at night.

Zoarces viviparus - Viviparous blenny

Distribution: From the English Channel north to the White Sea. Norwegian coast. Large parts of the Baltic. Also north and northwest coasts of the British Isles.

Description: Eel-like body. Up to 60 cm in length. Head flat on the underside. Lips thick. Dorsal fin long; continuous with the anal fin, as is also the case with the anal fin. Dorsal fin with a pronounced indentation at its posterior part. Several colour nuances; commonly with brown back and yellow-brown belly. A row of dark spots along the top of the back, extending on to the dorsal fin. Scales small, deep-sitting and scarcely visible on the smooth, slimy skin.

Habitat: Bottom-living; prefers hard bottoms amongst algae or hidden under stones. Found in the intertidal zone and down to approx. 40 metres. Also found in brackish water.

Biology: Gives birth to up to 400 live young. Ripe females found all year round. Feeds on various bottom-living, or benthic, organisms, including crustaceans, snails, bivalves or eggs and larvae of other species.

The viviparous blenny has an eel-like body, a relatively large head with prominent lips and large pectoral fins.

DID YOU KNOW...

The viviparous blenny has internal fertilisation and is **viviparous**, i.e. females give birth to live young (hence the latin species epithet <u>viviparus</u>). Mating usually occurs in the autumn, and the fertilised eggs develop inside the mother until they hatch around a month later. At this point the larvae are 14 mm long. These further develop within the ovaries and are born three to four months later, at around 4-5 cm in length. If seen with a swollen belly around late autumn or winter, there is a good chance it is a pregnant female.

The viviparous blenny thrive in muddy harbour sediments, often amongst the eel-grass, **Zostera marina.**

Family Stichaeidae - Pricklebacks

Most members of this family occur in the northern Pacific, but three species occur in northern Europe.

Chirolophis ascanii - Yarrell's blenny

Yarrell's blenny is a 'cave-dweller', often seen in pairs, well hidden in rocky crevices.

Distribution: Northern distribution from the British Isles north to Iceland and the Murmansk coast. Scandinavia from the Swedish west coast and entire Norwegian coast.

Description: Eel-like body with an erect bushy head appendage between the eyes. Blunt snout. Can remind of a small wolffish, but only reaches 25 cm in length. Colour variably brown with pale flecks. Adult males also bear an appendage on the first fin ray. Eyes protruding and surrounding tissue ringed with dark pigment.

Habitat: Often well-hidden in rocky crevices or deep under rocks, from a few metres depth down to approx. 40 metres. Has been caught down to 200 metres. Commonly seen in pairs.

Biology: Spawning takes place in late autumn or early winter. Egg clusters small; guarded and tended by the male until hatching after 5 - 6 weeks. Diet includes bivalves, bristleworms and hydroids.

WELL-HIDDEN

Yarrel's blenny is a challenging fish to photograph in its natural environment. It usually occurs deep within a narrow crevice or well-hidden under large rocks.

Leptoclinus maculatus - Daubed shanny

The daubed shanny, photographed at Svalbard. Most frequently occurs on soft bottoms.

Distribution: Arctic distribution; found on both sides of the north Atlantic. Western side from the northern part of the Swedish coast and north along Norwegian coast, Svalbard, Iceland and Greenland. East to the Kara Sea. Northeast coast of North America.

Description: Body markedly elongate. Up to 15-20 cm in length. Single dorsal fin extending from the above the basis of the pectoral fins almost to the caudal, or tail fin. First fin rays very short and free (not connected by fin membrane). Pelvic fin extending from the anal opening approximately at the middle of the fish, and back almost to the tail fin. Mouth relatively large, with the corners in line with the mid-eye point. Colour yellowish-brown with numerous brown patches. Often five dark patches on the dorsal side. Tail fin with marked transverse dark brown stripes and with an almost straight edge. This distinguishes it from the similar **snake blenny**, ***Lumpenus lampraetaformis***, which has a pointed, elliptical tail edge.

Habitat: Occurs from 15 metres depth down to approx. 400 m; shallowest in winter. Bottom living on both hard and soft substrates.

Biology: Spawning occurs in shallow water in winter. Females spawn around 1000 eggs; larvae pelagic. Prey items include bristle-worms and various crustaceans.

Lumpenus lampretaeformis - Snake blenny

Distribution: Somewhat more southern distribution than the **daubed shanny**, ***Lumpenus maculatus***. Found from the southern parts of the Baltic (probably a relict form from the last ice age), North Sea basin from the English Channel and northern parts of the British Isles and north to Iceland, Norway and Spitsbergen. Greenland's west coast and east to Novaja Semlja. Northeast coast of North America.

Description: Resembles the daubed shanny, but distinguished by the pointed, eliptical tail edge and the anal fin beginning considerably further forward than the middle of the fish. Also becomes somewhat longer; up to 49 cm. Characteristic luminous patches on the gill coverings (only seen on live fish).

Habitat: Prefers soft bottoms from around 30 metres depth down to approx. 200 m. Builds a Y-shaped tunnel system.

Biology: Spawning carried out during mid-winter. Sexually mature after three years; then a little less than 20 cm long. Lays between 600 - 1100 eggs. Feeds on various benthic invertebrates, including bristle-worms, small crustaceans, brittle stars, snails and bivalves.

Left: *Juvenile; 5 cm long.* ***Right and above***: *The snake blenny can become almost ½ m long.*

In certain fjords with loose muddy sediments suitable for tunnelling, the snake blenny can occur in large numbers. During night dives, it will allow divers to approach.

Occasionally, a fully-grown wolffish can be seen in the kelp forest, although generally it prefers soft sediments.

Family Anarhichadidae - Wolffishes

Of the six species within this family, three are present in northern European waters. All are bottom-living, with an elongate body and a large head with strong jaws and teeth. The ***northern wolffish*****, Anarhichas denticulatus***, is distinguished from the two species presented below by its dark, blue-black colour, a thicker body and a longer snout. The latter also occurs in deeper waters; between 60 - 900 metres.*

The wolffish **Anarhichas lupus** *has an intimidating appearance, but it is a peaceful fish if not provoked.*

Anarhichas minor - Spotted wolfish

Distribution: Arctic distribution. Eastern Atlantic from Novaja Semlja, Svalbard, northern to mid-Norway, Iceland and the southern coast of Greenland.

Description: Elongate body, with strong head and jaws. Jaws densely provided with teeth, also in the roof of the mouth and the 'throat'. Pectoral fins large, with 21-22 rays. Single long dorsal and anal fin. Resembles the **wolffish**, ***Anarhichas lupus***, and the **northern wolffish**, ***Anarhichas denticulatus***, but distinguished by being paler in colour and, from a length of 10 cm, prominent dark round spots on the body and dorsal fin. Also known as the leopard-fish in some languages. The largest of our wolffish; up to 190 cm in length and weighing more than 25 kg.

Habitat: Usually soft bottoms, often sand and muddy substrates with scattered boulders. In northern Norwegian fjords, can occasionally be seen in shallow waters around 25 metres; but most commonly further offshore, between 100-550 m.

Biology: Internal fertilisation, as in other wolffish. Moves in to the Norwegian banks from the Barents Sea in spring to spawn, which occurs in around 100 - 250 metres depth. Large females spawn up to 50 000 eggs, laid in spherical clumps on the bottom. Males guard the eggs. Larvae large and well-developed at hatching; rapid growth rate. Measuring 15-20 cm in length at one year and 45 - 60 cm after five years; sexually mature at 7 - 10 years. Feeds on bottom-living organisms including sea urchins, bivalves, snails and brittle stars. Larger individuals also take other fish such as cod and flatfish.

Commercial value: Fished to a considerable degree in northern Norway. Prized as a fillet fish due to the firm and tasty white flesh. As with the northern wolffish, the skin often is cured and the attractive leather used for novelty items and even clothing. Attempts have been made to farm this species.

The spotted wolfish is a northern form rarely seen by divers. This specimen from the marine aquarium in Bergen, Norway.

Anarhichas lupus - Wolffish

Large head with strong pectoral fins and a fearsome array of teeth.

Distribution: Eastern Atlantic from Novaja Semlja, Svalbard, Norway, Iceland and eastern Greenland. Common around northern parts of the British Isles; becoming rarer towards the Bay of Biscay.
Description: Elongate body, with a strong head and jaws. Jaws densely provided with teeth, also in the roof of the mouth and 'throat'. Pectoral fins large, with 18 - 20 rays. Single long dorsal and anal fin. Pelvic fins absent. Resembles its close relative the **spotted wolffish**, ***Anarhichas minor*** and the **northern wolffish**, ***Anarhichas denticulatus***, but distinguishable by the colour. The Spotted wolffish has dark flecks or spots against a pale background, whereas the northern wolffish is considerably darker; also thicker and with a longer snout. *Anarhichas lupus* usually is grey, often with a number of vertical stripes along the sides, a feature typical for adults on sandy or clayey substrates. In coastal areas amongst algae, younger individuals of a more brownish colour often are seen. Can reach around 120 cm in length and weighing over 20 kg. Thought to become more than 20 years of age.
Habitat: Generally bottom-living, mainly seen on sandy bottoms; optimal conditions appear to be sandy or clayey bottoms with scattered rocks. Less commonly seen on hard bottom, amongst algae. Can occur in large numbers in certain fjords-it is not uncommon to encounter 3-4 specimens during a single dive. Can be found in fjords in shallow water all year round, but in more open coastal areas it tends to move to deeper, warmer water layers in winter. Found from the shore down to 450 metres depth. Fully grown individuals often remain in the same place and can be observed there year after year.
Biology: Sexual maturation occurs around 6-7 years of age, at which point the fish are 50-60 cm in length. Spawning occurs in winter, from November to December. Eggs yellow, up to 6 mm in diameter. Laid in spherical clumps about the size of a football; often under stones or

PEACEFUL FISH

The wolffish is known amongst fishermen as an agressive beast. When it is hauled onboard, it will bite anything within reach. In fact, a fully-grown wolffish has no problems in shredding an oar paddle, and so fingers should be guarded well. However, in the water the wolffish is by contrast a very peaceful and friendly animal. It has few, if any, natural enemies and seldom feels threatened. We have several times stroked their backs without mishap. It is said that wolffish will attack divers, but then usually after the diver has first attacked and stuck a knife into its body.

rocky overhangs on sandy bottoms. Males guard the eggs for a period of around 2-3 months until hatching. Males puff themselves up and click their teeth if divers approach too close during this time. If the diver does not go away, the male will flee. Brooding males have been recorded from 20 metres depth and deeper; some even down to 200 m depth. Larvae 12 mm in length at hatching and remain close to the bottom until the large yolk sac is absorbed around 3 - 4 months later. At this point they move into the water masses at depths between 100 - 200 metres, before settling to the bottom again after some months. Feeds on various bottom-living organisms, particularly sea urchins and scallops. Also takes other echinoderms such as starfish and brittlestars, as well as various bivalves and snails. In the wolffish, the first set of 'milk teeth' are replaced by adult teeth; young fish in the process of changing teeth will not feed during this time.

Young wolffish less than 5 cm long are rarely seen. Here from a sandy bottom at 15 m depth.

Commercial value: Much sought-after as a fillet fish, but not fished to the same extent as the spotted wolffish. Popular with sports fishermen and divers.

Family Ammodytidae - Sandeels/ Sand lances

This family comprises 12 species that live in shallow water around the coasts of the Atlantic, Pacific and Indian oceans. Four species are found in northern Europe. However, two of these, the ***smooth sandeel*****, Gymnammodytes semisquamatus** *and the* ***lesser sandeel*****, Ammodytes marinus**, *are not presented here, because they are extremely difficult to separate.*

DID YOU KNOW...

January 2004: EU ministers have extended the ban (initiated in 2000) on sandeel fishing for a further 12 months in an effort to protect the UK's seabird population.

The great sandeel is common on sandy bottoms during autumn and winter.

Ammodytes tobianus - Small sandeel

Distribution: From the coast of Portugal to Norway, Iceland and the northwest Kola peninsula. Baltic.

Description: Small, slender, up to 20 cm in length. Glittering silver. Single long dorsal and anal fin. Scales located in skin folds; those on belly in regular groups. Scales present on base of tail fin. Palate lacking teeth in midline. Pectoral fins extending beyond anterior of dorsal fin. The latter two characters distinguish it from the **great sandeel**, ***Hyperoplus lanceolatus***. Resembles the **lesser sandeel**, ***A. marinus***, but the latter has 55 - 63 fin rays, whereas the Small sandeel has 51-57 rays. The lesser sandeel lacks scales at the tail fin base. Another species, the **smooth sandeel,** ***Gymnammodytes semisquamatus***, can be distinguished by the lack of skin folds at the upper margin of the belly, and by its lack of scales on the anterior body.

Habitat: Closely associated with sandy bottoms, within which it frequently buries. Common from the shore to approx. 30 metres depth. In contrast to the nocturnal great sandeel, active by day; usually buried within sand by night.

Biology: Spawning occurs at different times within the distribution area; mid-winter in the North Sea, late winter at Iceland, autumn in the English Channel and both spring and autumn in Baltic. Eggs pear-shaped, attached to sand grains and usually hatch after 10 days, but longer if the eggs are buried within the sand. Feeds on plankton in the water masses by day; can aggregate in large schools where water currents lead to an accumulation of plankton. Feeds by everting the jaws.

INDUSTRIAL FISH

Large quantities of sandeels are fished industrially in the North Sea. Although the several sanddeel species are involved, recent studies from the fishing banks off southwest Norway show the catches mostly to consist of the lesser sandeel, not the small sandeel. The fisheries have increased over the past years, and there is concern for overfishing. A reduction in sandeel stocks has serious implications for other fish species, puffins and other seabirds.

A small sandeel peeps from its burrow.

Hyperoplus lanceolatus - Great sandeel

Not many fish are capable of such spectacular disappearing acts as the great sandeel. The fish dive straight down into the sand and 'disappear' in a fraction of a second.

Distribution: From the coast of Portugal north to Iceland, mid-Norway and northwest Kola Peninsula. Baltic, except for the northernmost parts.

Description: Elongate, slender, silver-glittering. Up to 32 cm; larger than other sandeels. Occurs singly or in schools. Distinguished from the **small sandeel**, ***Ammodytes tobianus***, and the **lesser sandeel**, ***A. marinus***, by having two teeth in the upper jaw, and scales covering the entire body. Further, pectoral fin extending just up to the anterior margin of the dorsal fin. Dark patch on snout immediately in front of the eye. Up to 5 years of age. In contrast to the small sandeel, active by night; especially immediately after sundown and just before sunrise.

Habitat: Sandy bottoms into which it buries. Occurs most commonly between 20 - 40 metres depth; moves down to deeper water in winter. Despite this, divers still frequently observe it in shallow water (1 - 10 m), even during winter.

Biology: Spawning occurs between April to August, in 20 - 100 metres depth; latest farthest north. Eggs spawned pelagically and sink to the bottom; hatch after approx. 3 weeks. Sexually mature at two years of age.

Commercial value: Occurs in modest amounts in catches, together with the small sandeel. In the North Sea, the great sandeel comprises only 1 % of the total sandeel catches.

Both the small sandeel and the great sandeel can be seen by divers on sandy bottoms. These often school together and can be difficult to separate. The great sandeel is distinguished by a dark patch in front of the eye.

Family Gobiidae - Gobies

Jeffrey's goby, **Buenia jeffreysii,** *at the west-coast of Norway, Iceland and the British Isles. It is up to 6 cm in length and recognisable by the long second ray in the first dorsal fin.*

This is a very large family, comprising around 800 species, 18 of which occur in northern Europe. Most are small and live in shallow water, often right up to the surface. In most gobies, the head is relatively large and the eyes are situated high on the head. Most have large fleshy lips. The body is round in section and tapers evenly towards the tail. All European gobies have two distinct dorsal fins, and most are difficult to separate from each other. The 12 most common northern European gobies are presented here.

Lesueur's goby, **Lesuerigobius frisii,** *at the west-coast of the British Isles, southern Norway and the west-coast of Sweden. This goby has a characteristically pointed tail fin, and the specimens we have seen have attractive yellow spots. The species makes a protective tunnel within the sediment.*

Crystallogobius linearis - Crystal goby

The crystal goby is the most transparent of our fish. Note the eyes and the 'metallic' stomach.

Distribution: Mediterranean and further north in Atlantic waters along the west coasts of the British Isles. Scandinavia from the Kattegat and north to mid-Norway. Faeroes.
Description: Small form, up to 5 cm in length. Almost transparent, except for the eyes and some pigment spots along the lower jaw and at the bases of the ventral fins. This distinguishes it from other northern European fishes. An exception is the very similar **transparent goby**, ***Aphya minuta***, but this is distinguished by having larger scales. The crystal goby lacks scales altogether. Prominent underbite. First dorsal fin with 2 spiny rays; second dorsal fin with 1 spiny ray and 18 - 20 soft rays. Pelvic fins present in males but absent in females. At spawning time, males develop two large 'fangs' in the lower jaw. Reaches up to 1 year of age.
Habitat: Transparency suited to its pelagic lifestyle. The crystal goby and the transparent goby are the only pelagic gobies. Usually occurs somewhat offshore, most frequently between 20 - 80 metres depth.
Biology: Spawning occurs between May to June. Eggs often laid in empty polychaete worm tubes, often those belonging to *Chaetopterus*, at between 20 - 40 metres depth. Eggs guarded by the males, which often themselves also hide in the tubes. The parents die after the eggs hatch. Feeds mainly on plankton.

Transparency is an effective form of camouflage for a pelagic fish.

Aphya minuta - Transparent goby

The transparent goby resembles the crystal goby, but is distinguished by having large scales.

Distribution: From the coast of Morocco north to the British Isles. Scandinavia from western Denmark, Kattegat, north to southwestern Norway.

Description: Almost transparent; up to 5 cm in length. Resembles the **crystal goby**, ***Crystallogobius linearis***, but distinguished by its larger scales, which are lacking in the latter. Internal organs visible through the transparent scales.

Habitat: Occurs in pelagic schools, occasionally in thousands. May also occur in brackish water amongst eelgrass. From the surface and down to approx. 60 metres depth.

Biology: Spawning occurs on the bottom between May to August. Females lay around 3000 eggs, placed in an empty bivalve shell or similar object. Parents die shortly after the eggs hatch. Feeds on various zooplankton.

HE MINDS THE KIDS

As in several other fish, in the gobies, it is the male who is responsible for parental care. After the male goby has found a suitable 'nest', often an empty shell, or an abandoned bristle-worm tube, he entices the female towards it. Male gobies are not monogamous, and often entice several females to their nests. The females lay their eggs in his nest, which are then looked after by the male. The eggs hatch after 1 - 2 weeks, depending on temperature, and the larvae move up into the water masses. For most gobies, this signals the end of a short life. Having secured the propagation of their genes, they die.

The ***diminutive goby,*** **Lebetus scorpioides,** *as the name suggests, only reaches up to 4 cm in length and occurs between 15 - 375 metres depth. It lives on sand, mud or gravel substrates. The specimen in the picture was found under a bivalve shell at 15 metres depth in southwest Norway.*

Pomatoschistus minutus - Sand goby

Distribution: Black Sea, Mediterranean, Eastern Atlantic from Gibraltar north to Northern Norway, Faeroes and Baltic.

Description: Small, up to 10 cm length. Elongate; eyes located high on the head. First dorsal fin bearing 6 - 7 rays and a dark patch at the posterior margin. Resembles the **common goby**, ***Pomatoschistus microps***, but lacks the latter's black, diagonal bands at the base of the pectoral fins. Colour uniform grey, with scattered diffuse dark patches. Scales present on the anterior of the body; lacking in the **Norway goby**, ***P. norvegicus***, with which the sand goby often is confused. Reaches up to 2.5 years of age.

Habitat: Common from the shore down to approx. 20 metres; occasionally deeper. Usually on sandy bottoms.

Biology: Spawning occurs from March - August, dependent on the water temperature. Males entice females into their nests, which often comprise empty bivalve shells or similar objects. Females lay eggs one at a time, which are attached together within the shell. Males guard and tend the eggs until they hatch after around 9 days. Several females may be enticed into the same nest. Feeds on various small crustaceans such as copepods, amphipods, isopods and various larvae. Most die after their second summer.

Divers often see the sand goby, darting out of reach if approached too closely.

Like other gobies, the colouration of the sand goby matches its surroundings.

Pomatoschistus pictus - Painted goby

The painted goby resembles the sand goby, but is distinguished by the shape and distribution of the dark pigment patches.

The painted goby is a ferocious predator for its size. Here it has caught a crystal goby, **Crystallogobius linearis.**

Up to 95 mm in length and reaching up to 2 years of age.

Habitat: Favours shell-sand or gravely substrates, but also found on sandy bottoms among eelgrass. Occurs in shallow water down to approx. 50 metres depth; deepest in winter.

Biology: Reproduction takes place after the first over-wintering; spawning occurs between April - July. As in most other gobies, eggs laid in empty bivalve shells or similar objects, and guarded by the male. Parents die after the eggs hatch. Larvae live as plankton for 1 - 2 months, before settling to the bottom at around 10 mm in length. Feeds mainly on small crustaceans.

Distribution: Eastern Atlantic from northern Spain to the Kattegat and western Norway.

Description: Resembles the **common goby**, ***Pomatoschistus microps***, but distinguished by its prominent black spots arranged in a single row, or occasionally double, row on the dorsal fin. First dorsal fin with 6 spiny rays; second dorsal fin with one spiny ray and 8 - 9 soft rays. Scales medium large; 35 - 41 scales along the lateral line. Dorsal side brown, with a row of large, dark patches along the sides.

Pomatoschistus microps - Common goby

The common goby shares with many other gobies a preference for soft bottoms.

Distribution: Eastern Atlantic from Gibraltar to mid-Norway. Mediterranean and Baltic.
Description: Resembles the **painted goby**, ***P. pictus***, but distinguished by diffuse reddish-brown flecks on the dorsal fin, a single dark diagonal fleck at the base of the pectoral fin and one at the rear of the first dorsal fin. Back greyish brown; sides paler and belly white. Up to 65 mm in length and reaching up to 3 years of age.
Habitat: In contrast to the painted goby, it prefers river mouths and brackish water. Occurs in shallow water down to approx. 30 metres dept. Prefers stony sandy bottoms or clay substrates.
Biology: Spawning takes place in summer; eggs guarded by the male. In Norway, the parents die after their first spawning, whereas around the British Isles they usually spawn twice. Feeds largely on small crustaceans, but also takes other bottom-living animals.

If careful, gobies will allow divers to come close. Here a common goby peeps from its cave.

Pomatoschistus norvegicus - Norway goby

The Norway goby closely resembles the sand goby.

Distribution: Scattered distribution in the Mediterranean; extending around the western and northern coasts of the British Isles. Norway, Faeroes. Not recorded from Iceland or Denmark. Because the Norway goby has been confused with the **sand goby**, ***Pomatoschistus minutus,*** the distribution records are somewhat uncertain.

Description: Resembles the sand goby, *P. minutus,* but distinguished, amongst others, by lacking scales on top of the head and 'throat', as well as by fewer scales along the lateral line (55-60 compared to 59-75). Also closely resembles the **common goby**, ***P. microps***, but the latter can be distinguished by a diagonal band at the base of the pectoral fins. Males usually with 5 - 6 narrow, dark transverse stripes on the body sides.

Habitat: Lives somewhat deeper than the sand goby; from approx. 30 to 300 metres depth.

Biology: In southern Norway, known to become sexually mature at around one year of age.

This Norway goby has a leech attached to its left eye and a parasitic copepod on the first dorsal fin. Photographed at 30 m depth in southwest Norway.

Gobius niger - Black goby

The black goby often lies partially hidden under rocks or overhangs.

Distribution: From western Africa to mid-Norway. Mediterranean and Black Sea. Baltic.

Description: The largest of our gobies; up to 17 cm in length, almost twice as large as other gobies. In addition, it can be recognised by the 4th fin ray in the first dorsal fin being longer than the others. Robust tail fin. Generally darker than other gobies, but colour varies according to habitat. Old males generally blue-black, whereas females and younger males brown. Scales large. Reaches at least 3 years of age.

At 17 cm in length, the black goby is the largest of our gobies.

Habitat: Usually occurs on sandy, clay or muddy bottoms, but can also occur on hard bottoms amongst algae. From the shore down to approx. 65 metres depth. Tolerant of low salinities and commonly found around river mouths. Individuals often remain in the vicinity of more or less permanent hiding places, often rock crevices, or under stones and similar objects. Rather shy; rapidly retreats into its hiding place if approached too closely.

Biology: Spawning occurs between May and August. Sexually mature at two years of age. Males develop colourful fins at mating time. The up to 6000 eggs are spawned on hard substrates and guarded by the male. After the spawning period, many of the parents die. Feeds mainly on small bottom-dwelling crustaceans, bristle-worms and bivalves.

Here a black goby seeks protection behind two sea anemones.

Commercial value: Used to a certain extent as bait, especially in Denmark.

Thorogobius ephippiatus - Leopard-spotted goby

The attractive leopard-spotted goby is a common sight at certain hard-bottom localities.

Distribution: Atlantic from Madeira north to the British Isles and mid-Norway.
Description: Up to 13 cm in length. Characteristic colours. Greyish-brown to steel-blue base colour with numerous dark, reddish patches on the head and body. Fins with blue edges. Unlikely to be confused with other gobies.
Habitat: usually found in relatively shallow water, on substrates with rocky overhangs and large rocks. Most frequently occurs in partially sheltered localities at between 10 - 20 metres depth.
Biology: Around the English Channel, spawns between May - June. Prey organisms include gammarid amphipods and bristleworms. Other aspects of its biology are poorly known.

DID YOU KNOW...

The leopard-spotted goby is a good example of a species' ability to change along with a change in habitat. Around 15 - 20 years ago, this species was almost unknown along the Scandinavian coasts. During the last 15 years this goby has become very common in Norway. It looks like it has found its place in the norwegian fauna.

This goby is easiest to observe by night- by day it hides under stones or in rocky crevices.

Gobiusculus flavescens - Two-spotted goby

*A female two-spotted goby resting on a kelp blade. Near the tail is a young starfish (*Asterias rubens*), which recently has settled to the bottom.*

Distribution: Eastern Atlantic from Gibraltar and north to northern Norway. Faeroes, Baltic, western Mediterranean.
Description: Small form; up to 6 cm in length; colour brown. Distinguished by a round, dark spot surrounded by a thin yellow ring at the base of the tail fin. Males also have a similar spot below the first dorsal fin. Further, the first dorsal fin has 7 rays, in contrast to other gobies, which have 5-6 rays.
Habitat: Mainly found amongst brown algae or eelgrass in shallow water; most common between 1 - 5 metres depth. Occurs in small schools. Often moves to deeper water in winter; down to 20 metres. Fish caught on rocky shores with hand nets often are this species.
Biology: Spawning occurs between May and August. Females lay eggs on the bottom, or often attached to kelp holdfasts. Eggs guarded by the male. Sexually mature at one year of age in the southern parts of its distribution area, but can take two years in more northern waters. Parents die after spawning. Mostly feeds on various copepods, but also takes larvae of larger crustaceans.

The two-spotted goby has a dark patch at the base of the tail fin. Males can be distinguished from females by an additional dark patch just below the first dorsal fin.

Family Scombridae - Mackerels, tunas and bonitos

Mackerels, tunas and bonitos are streamlined fish always on the move. All lack a swim bladder. Another feature common to the group is a row of small, free-standing fins between the dorsal fin and the tail and on the lower side of the fish between the anal fin and the tail base. The first dorsal fin has stout fin rays. The unpaired fins can be folded into a groove. Six of the 48 species within the family are recorded from northern European waters. These are as follows: ***Atlantic mackerel*****, Scomber scombrus***;* ***bullet tuna*****, Auxis rochei***;* ***little tunny*****, Euthynnus quadripunctatus***;* ***plain bonito*****, Orcynopsis unicolor***;* ***Atlantic bonito*****, Sarda sarda***, and the* ***northern bluefin tuna*****, Thunnus thynnus**. *The latter can reach a large size. A tunafish measuring 2.6 m in length weighs around 300 kg, although older stories tell of tunas weighing up to 600 kg. The sports fishing record is 444 kg.*

Mackerel are typical school fish.

Scomber scombrus - Atlantic mackerel

Distribution: Eastern Atlantic from the Canary Islands north to the Kola coast, Iceland and Norway. Baltic, Mediterranean and Black sea.

Description: Muscular, streamlined body suited for rapid swimming. Scales small; back typically patterned with dark blue on a greenish background. Belly white or silvery. Two dorsal fins and one anal fin, the latter situated directly below the rearmost dorsal fin. A row of small, free-standing fins between the dorsal fin and the tail, and the anal fin and the tail base. Up to 70 cm in length and 3.5 kg in weight; mostly not exceeding 40 cm and 700 g.

Habitat: Occurs in constantly moving schools; usually in the upper layers, but in winter can occur down to 200 - 250 metres depth.

Biology: Feeds largely on planktonic crustaceans and fish larvae. Moves into coastal waters in summer to spawn. Temperature sensitive and feeds only sparsely during the cold season; those caught in early spring are therefore thin. Reproduction between May and June, usually in open water, but occasionally in coastal areas, at water temperatures between 12 - 13°C. Females spawn in batches at night, close to the surface; each female can in total spawn more than 1 million eggs. Rapid initial growth rate; around 20 cm long after one year 1, but rate decreasing thereafter.

Commercial value: Caught by a variety of methods (spoon hooks, bait, hand-held longlines), but the largest amounts taken by purse seining. Mackerel fisheries escalated at the end of the 1960s, with around 840 000 tonnes landed in from the North Sea 1967, but in the 1970, North Sea stocks became almost decimated. Mackerel is extremely popular poached, fried, smoked and canned in various sauces. Considered a healthy food due to its high content of polyunsaturated fats (omega-3 fatty acids).

MANY NAMES

Italians have 39 common names for mackerel.

Family Callionymidae - Dragonets

This family includes around 40 species in total; three of which occur in northern Europe. The dragonet, **Callionymus lyra** *and* **C. maculatus** *are represented here. The third species, the* ***reticulated dragonet*****, C. reticulatus** *is less common.*

The eyes are situated high and far back on the head.

Callionymus lyra - Dragonet

Distribution: Eastern Atlantic from the Azores to Morocco, north to Iceland and northern Norway. Mediterranean, Black Sea.

Description: In common with other family members, an elongate body, a large, broad, dorsally (from the top) flattened head, with a relatively pointed snout and eyes located high on the head. Body tapering evenly towards the tail. Fins large, especially in males. Gill openings high; allows respiration even when partially buried in sand. Gill covers (operculae) with a four-pointed spine. Two large dorsal fins, of which the first has 4 spiny rays. Second dorsal fins each with 9 soft rays. Colour speckled yellow-brown. Males with curved blue stripes which, combined with the large blue and yellow striped fins make them one of our most beautiful fish. Males up to 30 cm in length; females around 20 cm. Reaches at least 6 years of age. ***Callionymus maculatus*** is distinguished by a shorter snout (about equal to the eye diameter) and second dorsal fin with four horizontal rows of round black spots. In this species, males reach only 15 cm in length; females 11 cm. The more rare **reticulated dragonet**, ***Callionymus reticulatus***, resembles *C. maculatus*, but is distinguished by having only three tips on the operculum spines. Further, the reticulated dragonet has blue flecks along the back (absent in *C.*

Males have spectacular blue marbling.
Inset*: Detail of eye.*

DAZZLING COLOURS

Reproduction in the dragonet occurs early in summer and begins with an exotic mating dance. The male courts the female by spreading his colourful fins and blinking his eyes. The eyes can even produce bursts of green or yellow light. During the actual mating process, the pair swims belly-to-belly vertically towards the surface. Their anal fins form a kind of tube within which the eggs and sperm are mixed. The mating dance is often repeated several time–and males often are not satisfied with just one partner.

***Above**: This female dragonet is eating a bristle-worm.*

***Left**: Female. Male dragonets are more colourful than females.*

maculatus), and those on the 2nd dorsal fin are arranged in vertical bands. This species reaches only up to 10 cm in length. *C. maculatus* usually occurs deeper than 30 metres, whereas the reticulated dragonet occurs shallower than 40 metres.

Habitat: Often partially buried in sand. Most abundant between 20 and 100 metres depth, but also found at only a few metres depth.

Biology: Spawning occurs in early summer. Eggs float to the surface; larvae live pelagically until late autumn, when they settle to the bottom, at approx. 10 cm length, and adopt the adult lifestyle. Feeds on various bottom-dwelling organisms, including bivalves, bristle-worms, small crustaceans and snails.

Commercial value: Although the flesh is white, firm and tasty, there are no dragonet fisheries.

Callionymus maculatus

Callionymus maculatus *is smaller than the dragonet. Males have attractive blue pigmentation.*

This fish is most often seen on loose sandy or muddy bottoms.

Distribution: In Europe from the Mediterranean and Black Sea. Atlantic coast north to mid-Norway.

Description: Resembles the **dragonet**, but distinguished by a shorter snout (approx. the eye diameter) and rows of black spots on the second dorsal fin. In males, the second dorsal fin has four rows of dark spots. Blue spots absent. First fin ray of the first dorsal fin very long in males. Males reach up to 15 cm in length; females 11 cm. For comparisons with the reticulated dragonet: see under the dragonet, ***Callionymus lyra***.

Habitat: Lives deeper than the dragonet, from 30 metres down 650 metres. Like the dragonet, prefers sandy bottoms, where it partially buries. Occasionally caught in shrimp trawls.

Biology: feeds on various bottom-dwelling animals including snails, bivalves, small crustaceans and bristle-worms. Reproduction in northern Europe in spring; believed to be sexually mature by its second summer. Males reach 11 - 12 cm in length in 3 years.

The ***reticulated dragonet,*** **Callionymus reticulata**, *has a three-tipped spine on the gill cover. It can be confused with* C. maculatus *(above) but has different colours along the back.*

Order Pleuronectiformes - Flatfishes, flounders and soles

This characteristic group of fish comprises more than 600 species world-wide. In northern European waters, there are 18 species, within 4 families.

The topknot, **Zeugopterus punctatus** *is extremely well camouflaged and will 'cling' firmly to hard substrates, often rocky overhangs.*

WANDERING EYES

Flatfish are either **right-sided** or **left-sided**. Flatfish larvae look more or less the same as any other fish larvae. However, during development, a peculiar phenomenon occurs. The eyes 'wander' towards one body side, such that the other side lacks eyes and is then known as the '**blind side**'. The side bearing the eyes is called the '**eyed side**'. The flat shape has been selected through evolution to suit their lifestyle, which in most cases is strongly associated with the sea floor. In a right-sided flatfish, the eyes are located on the right side, when the fish is held up with the mouth pointing downwards. Similarly, a left-sided fish has the eyes located on the left side when held in this way. In many flatfish, all members of the same species are either left or right sided. However, some species have both left and right-sided members.

Family Scophthalmidae

Members of this flatfish family have a southern distribution. As a general rule of thumb, left-sided flatfish found in northern European waters belong to the Scophthalmidae, although there are occasional exceptions. There are around 210 marine species within this family and 6 occurring in northern Europe. Of these, the **scaldfish, Arnoglossus laterna**, *is not presented here.*

SOUGHT-AFTER

The flesh of the turbot is white, firm, fine-grained and much sought-after. In Roman times, it was as highly-regarded as pheasant. The turbot therefore became known as the "water pheasant", an expression still used in France to this day. The flesh is of the best quality between September to April.

Psetta maxima - Turbot

Distribution: From Gibraltar north to northern Norway. Black Sea, Mediterranean, Baltic.

Description: In common with the other members of the family, it can be distinguished from flounders because it is left-sided-with both eyes located on the left side. Characterised by its almost round outline. Skin of eyed side lacking scales, but provided with numerous bony plates or spines. Colour varies according to bottom type; often grey or brownish, with dark brown spots. Blind side white, often with dispersed dark patches. Eye with narrow, copper-yellow ring around the pupil. First dorsal fin ray branched. Anal fin absent. Weighing up to 25 kg; the second largest of the European flatfish. Females largest.

Habitat: Usually occurring on sandy bottoms, but also other substrates from the shore down to around 100 metres. In France, the best quality specimens are caught on hard bottoms. Youngest individuals found in shallow water; larger ones prefer greater depths and deeper in winter. Sexually mature fish move to shallow water in summer to spawn.

Biology: Spawning occurs usually between 10 - 20 metres depth during April to August. Females sexually mature at 4 years of age; males one year earlier. Females each spawn from a few million up to 15 million pelagic eggs. Larvae pelagic until reaching around 25 mm in length. Settling occurs in shallow water and youngest individuals remain in only a few metres depth until reaching 8 - 10 cm in length.

The turbot typically is sandy-coloured, with black patches.

A ferocious predator; feeding on various mostly bottom-dwelling organisms including decapods, echinoderms, molluscs and small fish such as sandeels, herring, sprat and whiting. Largest individuals feed preferentially on bottom-living fish such as flounders, dragonets and gobies.

Scophthalmus rhombus - Brill

The brill is left-sided, with a greyish-brown or greenish upper side. ***Inset****: detail of eye.*

In common with many other flatfish, the brill is a master of disguise.

Distribution: Eastern Atlantic from Gibraltar north to mid-Norway.
Black Sea, Mediterranean, western Baltic.
Description: Left-sided. Eyed side usually greyish-brown with darker patches. A row of yellowish-white patches along the lateral line and at the base of the dorsal and anal fins. Unlike the **turbot**, ***Psetta maxima***, scales present but lacking bony spines. Lateral line markedly curving over the pectoral fin basis. First fin ray of dorsal fin branched. Up to 80 cm in length and 7.5 kg in weight; females largest.
Habitat: Prefers sandy bottoms, but also occurs on other substrates. Shallow water, from the shore down to approx. 70 m; deepest in winter. Particularly young fish also found in brackish water.
Biology: In northern Europe, spawning between May and June. Females sexually mature at around 25 - 30 cm in length and spawn just under one million pelagic eggs. Larvae live pelagically until settling to the bottom at 12 - 30 mm in length. Feeds on various fish, including sandeels, herring, sprat, Norway pout, whiting and haddock, but also takes crustaceans, particularly shrimps, prawns and small crabs.
Commercial value: Tasty flesh, but not as high quality as that of the turbot. Occurs as by-catch in Danish seine. Low commercial value but popular amongst sport-fishermen.

Zeugopterus punctatus - Topknot

Distribution: From the Bay of Biscay north to northern Norway.
Description: Left-sided. Almost rectangular outline; height of body around half its length. Dorsal fin beginning already at the snout and extending well past the base of the tail fin, as also is the case for the anal fin. Scales on eyed side hirsute, or hairy, giving a velvet-like appearance. Colour varies with habitat; commonly mottled brown and yellow. Black patch immediately behind bend of lateral line. Dark line through the eye. Up to 25 cm in length.
Habitat: Typically hard bottoms; often found 'glued' to rocky substrates, well-camouflaged on encrusting fauna such as the sponge, ***Halichondria panicea***.

The topknot has a somewhat rectangular outline.

Topknots are most common in shallow waters, often in rock crevices.

Especially common in shallow water in exposed areas. Found down to approx. 40 m.
Biology: Spawning generally occurs from May to June; earliest farthest south. Eggs and larvae pelagic; larvae settle to the bottom during the autumn, at a length of approx. 2,5 cm.
Commercial value: Tasty flesh, but of no commercial value due to its small size.

Phrynorhombus norvegicus - Norwegian topknot

The pigmentation is spectacular.

Distribution: From the eastern part of the Bay of Biscay north to Iceland and northern Norway. Absent from the northern part of the English Channel and north to western Denmark. Occurs in the Kattegat and to the eastern Baltic.
Description: Left-sided. Up to 13 cm in length; the smallest of the European flatfishes. Females largest. Length around 2.5 times that of the height. Snout length approximately that of the eye diameter. Scales large and spiny; rough to the touch. Colour can change according to bottom type, but usually brownish, with mottled darker brown and pink pigmentation, making it an attractive and colourful fish.

The camouflage colours are extremely effective on hard substrates covered in red calcareous algae.

Habitat: As for the **topknot**, ***Zeugopterus punctatus***, the Norwegian topknot is a typical hard-bottom fish; can cling to the substrate by suction. Usually found in deeper water than the topknot; prefers somewhat more protected localities, from 10 - 200 metres depth.
Biology: Spawning occurs from April to August. Females sexually mature at around 8 - 9 cm in length. Eggs and larvae both pelagic; larvae settle to the bottom at around 1 - 1.5 cm in length. Mainly feeds on small crustaceans including amphipods and isopods, as well as bristle-worms and fish larvae.

BEAUTIFUL FISH

According to Erling, the book's photographer, with almost 3000 dives under his belt, the Norwegian topknot is the most beautiful fish in northern waters.

Lepidorhombus whiffiagonis - Megrim

Distribution: In the eastern Atlantic from Gibraltar north to Iceland and southwest Norway. Western Mediterranean.
Description: Skin almost transparent. Up to 60 cm in length. An unusually elongate flatfish; length around 2.3 - 3 times the height. Colour yellowish-brown or greyish-yellow, with more or less prominent dark patches. Mouth large, with underbite; at first glance reminiscent of the American plaice, ***Hippoglossoides platessoides***, but the latter is right-sided. Dorsal fin in the megrim twice as large on the eyed side compa-

Left-sided flatfish with a large mouth.

red to the blind side. Anterior margin of lower eye situated further forward than that of the lower eye located anterior to the upper eye.
Habitat: Found on mixed bottoms, often hard substrates. Often found attached by 'suction' to the underside of rocks. Usually found below 50 metres depth.
Biology: Reproduction occurs during March to May. Eggs and larvae pelagic. Feeds on various bottom-dwelling organisms.
Commercial value: Dry flesh; much sought-after. Appears as by-catch in trawls, south of the English Channel.

TRANSPARENT

The internal organs of the megrim are visible through the body wall when held up to the light.

The megrim is usually found deeper than 50 m.

Family Pleuronectidae - Flounders

Members of this family are those most would associate with the term 'flatfish'. Of the around 100 species within the family, eight occur in northern Europe. The deep-water ***Greenland halibut***, **Reinhardtitus hippoglossoides**, *is not represented here. Most are right-sided, but left-sided individuals do occasionally occur.*

The European plaice is a willing photo model.

Limanda limanda - Dab

Distribution: From the northern part of the Bay of Biscay north to Iceland, Norway and the White Sea. Southern Baltic.
Description: Right-sided. Upper side with coarse, spiny scales. Colour often uniform greyish-brown, or with rust-brown patches. Blind side white and smooth. Can be confused with the **European plaice**, ***Pleuronectes platessa***, but distinguished by the lateral line, which in the dab describes a semicircular curve over the pectoral fin. Anal spine present at the anal opening. Can become over 40 cm long, although not usually above 25 cm, and weighing up to 1.5 kg.
Habitat: Sandy or other soft substrates; often partially buried. Prefers seawater, but tolerates brackish water-sometimes also found in freshwater. From 2 - 150 metres depth.
Biology: Time of spawning varies across the distribution area; in the North Sea around April to June. Eggs and larvae pelagic. Larvae settle to the bottom at a size of 13 - 16 mm. Competes with the European plaice for food. Feeds on various bottom-dwelling invertebrates, such as crabs, shrimps, bristleworms and small echinoderms, but also takes fish such as gobies and capelin.
Commercial value: The dab is a fish which is not specifically trawled for because its commercial demand is lower than other flatfish. Marketed fresh, dried or salted, smoked and frozen. Tasty flesh, but not as highly regarded as that of the European plaice.

Flounders trust in their camouflage and let divers approach close. Note the large eyes.

TOLERATES FRESHWATER

Several flounders can tolerate freshwater. Both the dab and the flounder can be found in streams and rivers, but these are usually young individuals.

The dab resembles the European plaice, but the lateral line curves over the pectoral fin.

Platichthys flesus - Flounder

The flounder usually occurs on hard bottoms. It is distinguished by the bony projections along the lateral line and on the gill cover.

The flounder is robust in appearance.

Distribution: In the eastern Atlantic from Gibraltar north to Norway and the White Sea. Black Sea and Mediterranean, Baltic. Absent around Iceland.

Description: Resembles the **European plaice**, ***Pleuronectes platessa*** and the **dab**, ***Limanda limanda***; like these, the flounder also can have red spots on the brown or greyish eyed side. However, the flounder has a more robust appearance, and is distinguished by rows of sharp bony spines along the lateral line, on the gill covers and along the base of the dorsal and anal fins. Around a third of flounder individuals are left-sided. Up to 50 cm in length and around 2.5 kg in weight

Habitat: Occurs both on soft and hard bottoms, from the shore down to around 100 metres depth. Tolerant of freshwater and appears in rivers and streams, particularly young, sexually immature individuals.

Biology: Spawning takes place between February and August. A large female spawns up to two million eggs. Eggs and larvae pelagic; hatching after 5 - 7 days. Larvae settle to the bottom at around 1 cm in length. Feeds on various bottom-dwelling organisms; mostly molluscs, bristle-worms, amphipods, isopods, shrimps and small fish.

Commercial value: Commercially fished in the North Sea and Baltic; popular as a smoked fish. Marketed fresh and frozen; can be steamed, fried, boiled, microwaved and baked.

Pleuronectes platessa - European plaice

A specimen of the European plaice displaying its typically striking reddish orange spots. ***Right inset****: Detail of pigment patch.*

Distribution: In the eastern Atlantic from Morocco north to Iceland, Norway and the White Sea. Black Sea and Mediterranean. Baltic.

Description: Perhaps the most well-known and highest priced of all our flatfish. Right-sided; eyed side smooth, except for the 4 - 7 characteristic bony projections extending from the eyes back to the gill covers and the anterior part of the lateral line. Eyed side brown or brownish-grey, with numerous reddish to orange spots, also on the fins. These are not unique to the European plaice, but also occur in the **dab**, ***Limanda limanda*** and the **flounder**, ***Platicthys flesus***. However, the European plaice is distinguished by the bony projections. Blind side white, although some individuals are pigmented on both sides. Left-sided individuals occasionally occur. Can reach up to 100 cm in length and 7 kg in weight. Up to 50-year-old individuals have been reported.

Habitat: Prefers sandy bottoms, from the shore down to around 250 metres depth. Usually the youngest individuals occur shallower than 10 metres, but also older specimens can be found close to the shore. Tolerant of brackish water.

Biology: After migrating to spawning grounds the European plaice breeds in southern North Sea in December-March (20-50 m depth), Irish Sea in February-April, and as late as June at the Murman coast. Eggs laid in depressions on the bottom, but float to the surface after fertilisation. Eggs hatch after around 20 days, depending on the temperature. Larvae live pelagically until they reach 12 - 17 mm in length. At this point, the larvae move to shallow water and remain there until the temperature drops during autumn, after which they move to deeper water, returning to the coast again the following summer. The small flatfish seen by swimmers along sandy beaches in summer usually are young European plaice. Feeds on various bottom-dwelling organisms, often burying bivalves, bristle-worms and various small crustaceans.

Commercial value: Highly prized as a food fish. Considerable fisheries in the North Sea. Over-fishing is a concern, and since the 1980s, total fishing bans have been established for defined periods, depending on the area. Hobby fishermen and sports divers should respect these bans!

Glyptocephalus cynoglossus - Witch

Distribution: Eastern Atlantic, from the Bay of Biscay to the Barents Sea, Iceland, Norway and the Kattegat.
Description: More elongate in shape than other flounders. Eyes large and situated close together. Mouth very small and lateral line almost straight. Eyed side pale brown, or reddish brown with numerous small black spots. Blind side also bears black spots. Pectoral fin on blind side black; that on the eyed side usually black at the tip. Anal spine present. Can reach over 60 cm in length and up to 2.5 kg in weight. Can reach at least 20 years of age.
Habitat: Usually found on soft bottoms, in deeper water than most other flounders. Seldom shallower than 40 metres depth; usually between 100 and 400 metres.
Biology: Spawning takes place between May and September. Eggs pelagic; hatching after 7 -8 days. Larvae live pelagically until settling to the bottom in relatively deep water at a length of 20 - 40 mm. Feeds on various bottom-dwelling organisms such as bristle-worms, bivalves, crustaceans and brittle-stars.
Commercial value: Mainly caught as by-catch in trawls. Popular food fish in countries including the UK. Seldom caught by sports fishermen except sporadically when using bait.

The witch seldom appears in shallow water, but here photographed at 15 m in southwest Norway.

The eyes of the witch are large and situated close together.

Microstomus kitt - Lemon sole

The lemon sole prefers hard substrates and has a fantastic ability to change colour.

Here amongst the sea squirt **Ciona intestinalis.**

Distribution: Eastern Atlantic, from the Bay of Biscay north to Iceland and the White Sea. Western Baltic.

Description: Right-sided, as in most flounders. Skin smooth with small scales. Anal spine absent. Blind side completely white. Head small; mouth with thick lips. Lateral line describing a slight curve over the pectoral fin. Up to 70 cm in length and 3 kg in weight. Females largest.

Habitat: Prefers hard bottoms close to the coast; found on soft bottoms in the North Sea. From a few metres depth down to approx. 250 metres; youngest individuals occurring shallowest. Moves to deeper water in winter. Often seen by divers amongst algae in shallow water.

Biology: Spawning takes place between April and September, latest farthest north. Eggs and larvae pelagic. Larvae settle to the bottom, often below 50 metres depth, at a length of 15 - 25 mm. Unlike most of our flatfish, the lemon sole is a specialist feeder, mainly on bristle-worms, but can also take other bottom-dwelling organisms.

Commercial value: Much sought-after in the UK, popular in restaurants. Little commercial importance in Scandinavia.

CHAMELEON OF THE SEA

With its often highly colourful marbled surface, which can change in both hue and intensity in a matter of seconds, the lemon sole is one of our most beautiful flatfish. It has a spectacular ability of change its colour to match the bottom conditions. The marbling can include various tones of red, yellow, orange, black, brown and green. The basal colour is brown, and when seen on brownish sandy bottoms, it is rather unremarkable. However, when the fish moves on to a more varied hard substrate, it will immediately change colour.

Hippoglossoides platessoides - American plaice

The American plaice has a large mouth opening. It can be confused with small Atlantic halibut, but distinguished by the convex, outwardly curving tail tip, which in the halibut is straight or concave.

Distribution: Northern distribution; in the eastern Atlantic from the English Channel north along the western and northern coasts of the British Isles, to Iceland, Norway and Spitsbergen, and east to the Kara Sea. Also western Baltic.

Description: In common with the **witch**, ***Glyptocephalus cynoglossus***, also elongate, but with an exceptionally large mouth. Mouth corners extend back to around mid-eye level. Head relatively large. Can be confused with small **Atlantic halibut**, ***Hippoglossus hippoglossus***, but distinguished by the tail fin, the end of which curves outwards. In the Atlantic halibut, the tail margin is straight or slightly concave. Up to 50 cm long in the northern part of its distribution area; rarely exceeding 35 cm farther south.

Habitat: Found on sandy bottoms or other soft substrates, usually between 50 - 250 metres depth. Can be seen as shallow as 10 metres depth in summer.

Biology: Spawning occurs between March and June, at depths between 100 - 200 metres. Eggs pelagic; hatching after 11 - 14 days. Newly-hatched larvae 4 - 6 mm in length. Settle to the bottom at 20 - 35 mm length, and adopt the adult lifestyle.

Commercial value: Not exploited in northern Europe, although it occurs in considerable quantities as by-catch in shrimp trawls. A prized food fish in North America.

Hippoglossus hippoglossus - Atlantic halibut

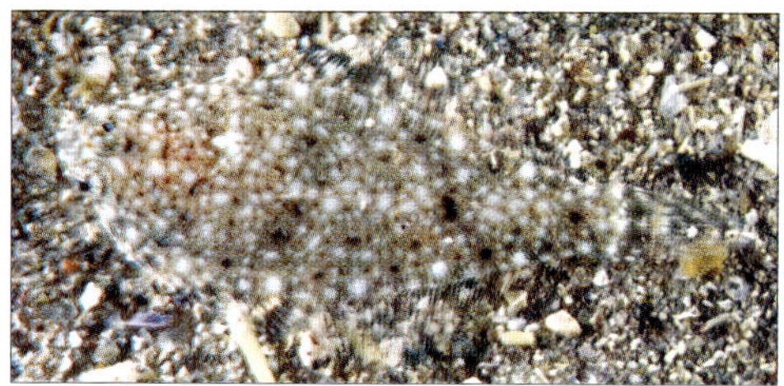

This young specimen is but 2 - 3 cm in length.

Distribution: Both sides of the north Atlantic. In the east, from the Bay of Biscay north to the Arctic Ocean, around Greenland, Iceland, Norway, Spitsbergen and Novaja Semlja. Western Baltic.

Description: This giant of a flatfish is recognisable by its size, elongate body and the relatively large mouth. Right-sided; eye side usually dark, but can be pale with dark spots in younger specimens. Blind side white. Anal spine present. Distinguished from the **Greenland halibut**, ***Reinhardtius hippoglossoides***, by the lateral line being markedly curved over the pectoral fin. Further, the blind side of the

Greenland halibut is dark in colour. Small Atlantic halibut can be confused with the **American flounder**, ***Hippoglossoides platessoides***, but can be distinguished by the shape of the tail. In the Atlantic halibut, the tip of the tail is straight or slightly concave, whereas that of the American halibut extends into a curved point.

Habitat: Small, young individuals occur in shallow coastal waters. Larger specimens usually found in deep water between 300 - 2000 metres depth. Unlike most other flatfish, it is a strong swimmer and often hunts in the water masses. On the bottom it occurs on a variety of substrates. Prefers temperatures not exceeding 10°C, or below freezing.

Biology: Spawning usually occurs in deep water, between 300 - 700 metres depth, in late winter. Although spawning is carried out in deep water, the fertilised eggs float to the surface as soon as they are spawned. Eggs hatch after 9 - 16 days; larvae live pelagically until reaching 45 - 70 mm in length. Young fish live on hard bottoms in coastal waters between 20 - 40 metres depth, until they are around two years of age-then around 20 - 30 cm in length. Although large individuals generally are found in deep waters, the occasional large halibut is caught in shallow areas. Feeds on various bottom-dwelling organisms, as well as various fish, which often are taken in the water masses.

Commercial value: The Atlantic halibut probably is the most highly-prized fish of all, at least in northern Scandinavia, and the price has always been high. The flesh is firm, white and extremely tasty. In southern Norway, it is known as the "holy fish", because it was permissible to eat it in during times of religious fasting. Because sexual maturation occurs first at 12 - 13 years of age, the stocks are at risk of over-exploitation; at present the stocks are so low that it is of little importance for fisheries. However, attempts at aquaculture of Atlantic halibut have been made.

DID YOU KNOW...

The Atlantic halibut is the largest of the bony fish permanently resident in the north Atlantic. Females can reach over 3.5 m in length and weigh almost 300 kg. Males reach up to 50 kg. These fish can become over 50 years old.

The Atlantic halibut is elongate, with a large mouth and strong tail.

Family Soleidae - Soles

Of the around 120 species within this family, five are recorded in northern European waters. The **scolenet**, **Buglossidium luteum** *are common around the British Isles and North Sea coast north to Denmark.* ***sand sole***, **Solea lascaris**, *and* ***thick-baked sole***, **Microchirus variegatus**, *are both distributed north to the west-coasts of the British Isles. A fifth species,* **Bathysolea profundicola**, *is distributed on deep water off the west-coast of Ireland.*

Solea vulgaris - Common sole

Distribution: From Senegal to the British Isles and mid-Norway.

Description: Right-sided. Characteristic oval shape, well developed pectoral fins and round head with curving snout. Mouth small and markedly curved. Dorsal fin beginning already at the snout. On the blind side, underside of jaws equipped with small feelers. Dorsal and anal fins continuous with the tail fin. Up to 60 cm in length and 3 kg in weight, but rarely larger than 40 cm in northern areas.

Habitat: Prefers sandy or clayey bottoms from the shore down to 150 metres depth. Youngest individuals shallowest. By day usually buried in the sand; hunts at night.

Biology: Migrates from the North Sea to the west coast of Jutland to spawn. Spawning migrations start early in the spring–tagging studies have shown that individuals can migrate up to 185 km. Spawning itself occurs between May and June, some hundreds of metres from the shore. Both eggs and larvae pelagic. Eggs hatch after approx. 10 days, but the larvae settle to the bottom first after 8 weeks in the free water masses, at which point they are 10 - 15 mm long. Feeds on various bottom-dwelling invertebrates, but also will take small fish such as gobies and fish larvae.

Commercial value: One of the most sought-after and highly priced on the market; caught with plaice nets and sole trawls. Has been over-exploited.

The sole perhaps is named after its characteristic shape.

Sports fishing records:

*Presented below is a list of world fishing records, according to The International Game Fish Association (IGFA); see **www.igfa.org**. Updated March 2004.*

Commen name:	Scientific name:	Weight (kg.):	Place and date of catch:
Allis shad	Alosa alosa	0.80	Netherlands, 21.08.98
American flounder	Hippoglossoides platessoides	0.99	Norway, 2003
Anglerfish	Lophius piscatorius	57.50	Høyanger, Norway, 07.04.96
Atlantic halibut	Hippoglossus hippoglossus	161.20	Valevågsbråtet, Norway, 20.10.97
Atlantic herring	Clupea harengus	0.48	New York, USA, 17.05.95
Atlantic mackerel	Scomber scombrus	1.20	Kråkvåg, Norway, 29.06.92
Ballan wrasse	Labrus berggylta	4.35	Ireland, 20.08.83
Blackbelly rosefish	Helicolenus dactylopterus	1.556	Langesund, Norway, 04.09.98
Blackmouth catshark	Galeus melastomus	1.37	Mausundvær, Norway, 17.09.94
Blue ling	Molva dypterygia	16.05	Trondheimsfjorden, Norway, 27.11.93
Blue whiting	Micromesistius poutassou	0.90	Mausundvær, Norway, 06.01.96
Cod	Gadus morhua	44.79	New Hampshire, USA, 08.06.69
Common skate	Raja batis	97.07	Orkney Islands, 16.07.68
Common sole	Solea vulgaris	1.96	Netherlands, 12.07.97
Conger eel	Conger conger	60.44	English Channel, 05.06.95
Dab	Limanda limanda	0.57	Germany, 20.09.98
European hake	Merluccius merluccius	7.08	Longva, Norway, 01.09.00
European seabass	Dicentrarchus labrax	9.40	France, 06.05.86
Flounder	Platichthys flesus	2.375	Fyksesund, Norway, 06.05.89
Garfish, garpike	Belone belone	1.18	France 2002
Greater argentine	Argentina silus	0.7	Hardangerfjorden, Norway
Greater weever	Trachinus draco	1.67	Høyanger, Norway, 07.04.96
Greenland halibut	Reinhardtius hippoglossoides	2.315	Andenes, Norway, 05.07.98
Grey gurnard	Eutrigla gurnardus	0.958	Kattegat, 20.05.98
Haddock	Melanogrammus aeglefinus	6.80	Saltstraumen, Norway, 15.08.97
Lemon sole	Microstomus kitt	1.27	Fedje, Norway, 20.05.99
Lesser spotted dogfish	Scyliorhinus caniculus	5.28	France, 2002
Ling	Molva molva	40.10	Shetland, 05.04.02
Megrim	Lepidorhombus whiffiagonis	2.36	Hardangerfjorden, Norway, 05.08.01
Northern wolffish	Anarhichas denticulatus	17.00	Greenland, 19.08.82
Oakettle	Somniosus microcephalus	775.00	Trondheimsfjorden, Norway, 18.10.87
Pollack/ lythe	Pollachius pollachius	12.80	English Channel, 16.01.86
Porbeagle	Lamna nasus	230.00	Scotland, 09.03.93
Rat-tail	Coryphaeonides rupestris	1.69	Trondheimsfjorden, Norway, 26.11.94
Sail ray	Dipturus lintea	11.25	Stavern, Norway, 13.06.99

Commen name:	Scientific name:	Weight (kg.):	Place and date of catch:
Saithe	Pollachius virens	22.70	Saltstraumen, Norway, 30.11.95
Spiny dogfish	Squalus acanthias	7.14	Ireland, 26.05.89
Spotted wolffish	Anarhichas minor	27.90	Vannøya, Norway, 29.05.00
Starry ray	Raja radiata	7.04	Hvasser, Norway, 10.10.82
Thicklip grey mullet	Chelon labrosus	4.414	Netherlands, 20.08.96
Thornback ray	Raja clavata	8.553	English Channel, 11.07.88
Tope shark	Galeorhinus galeus	44.67	California, USA 20.10.94
Tusk/ cusk	Brosme brosme	16.30	Langesund, Norway, 26.04.98
Twaite shad	Alosa fallax	0.71	Netherlands, 21.08.98
Red-fish	Sebastes marinus	8.440	Andenes, Norway, 05.07.98
Velvet-belly	Etmopterus spinax	0.85	Langesund, Norway, 07.10.00
Whiting	Merlangius merlangus	3.115	Romsdalsfjorden, Norway, 10.06.97
Wolffish	Anarhichas lupus	23.58	Massachusetts, USA, 11.06.86

Sources: www.villmarksinfo.no and www.igfa.org

The Norway redfish is common on Lophelia*-reefs.*

References

Abbott, R. Tucker, 1990. The Pocket Guide to Seashells of the Northern Hemisphere. Dragon's World, London. 176 pp.

Ackers, R. G., D. Moss, B. E. Picton & S. M. K. Stone. 1985. Sponges of the British Isles. Marine Conservation Society. Herefordshire. 199 pp.

Brattegard, T. & T. Holthe (editors). 1997: Distribution of marine, benthic macro-organisms in Norway. Research Report for DN 1997-1. Directorate for Nature Management.

Brown, J. m.fl. 1989. Waves, tides and shallow-water processes, Pergamon press. Oxford. 187 pp.

Brusca, R. C. & G. J. Brusca. 1990. Invertebrates. Sunderland, Sinauer Associates. 922 pp.

Carlgren, O. 1945. Polyppdyr (Coelenterata) III. Koraldyr. Danmarks fauna bind 51. G. E. C. Gads forlag, København. 168 pp.

Castro, P. & M. E. Huber. 1992. Marine Biology. Wm. C. Brown Publ. Dubuque. 592 pp.

Christensen, J. M., S. Larsen & B. O. Nystrøm. 1978. Muslinger. Cappelen, 125 pp.

Christiansen, M. E. 1972. Bestemmelsestabell over Crustacea Decapoda, Tifotkreps. Universitetsforlaget, Oslo. 71 pp.

Christiansen, M. E. 1972. Crustacea Decapoda Brachyura - Marine Invertebrates of Scandinavia, No. 2. Universitetsforlaget, Oslo. 143 pp.

DeHaas, W. & F. Knorr. 1990. Was lebt im Meer an Europas Kusten. Albert Muller Verlag, Wien. 390 pp.

Dons, C. 1927. Sjøen. Cappelen, Oslo. 108 pp.

Erwin, D. & B. E. Picton. 1987. A field guide to the Inshore Marine Life. Immel Publishing, London. 120 pp.

Froese, R. and D. Pauly. Editors. 2003. FishBase. World Wide Web electronic publication. www.fishbase.org, version 02 March 2004.

Fosså, J. H. 1995. Forvaltning av stortare - prioriterte forskningsoppgaver. Havforskningsinstituttet, Bergen. 102 pp.

Fosså, J. H. 1998. Invasjon av kolonimaneten *Apolemnia uvaria* langs norskekysten i 1997. Havets miljø 1998: 63-70.

Gulliksen, B. & E. Svensen. 2004. Svalbard and Life in Polar Oceans. Kom Forlag. 160 pp.

Hansson, H. G. 1998. Sydskandinaviska marina flercelliga evertebrater - utgåva 2. Länsstyrelsen Västre Götland 1998:4, Tjärnö Marinbiologiska Laboratorium.

Hayward, P.J., T. Nelson-Smith, & C. Shields. 1996. Sea Shore of Britain & Northern Europe. HarperCollins, London. 352 pp.

Hayward, P. J. & J. S. Ryland. (red.) 1995. Handbook of the Marine Fauna of North-West Europe. Oxford university press, Oxford. 800 pp.

Høisæter, T. 1998. Norske navn på marine evertebrater. Internet Sept. 1998 http://www.ifm.uib.no/nn/

Hågvar, E. 1995. Det Zoologiske Mangfoldet. Universitetsforlaget, Oslo. 383 pp.

Jonsson, B. & A. Semb-Johanssen 1992. Fiskene 2 saltvannsfisker- Norges Dyr. Cappelen, Oslo. 272 s.

Kaester, A. 1970. Invertebrate Zoology, volume III. Interscience Publishers, New York. 523 pp.

Jensen, AD. S. & R. Spärck. 1934. Bløddyr II. saltvandsmuslinger. Danmarks fauna bind 40. G. E. C. Gads forlag, København. 208 pp.

Jonsson, B. & A. Semb-Johansen (red.), 1992. Fiskene 1 krypdyr, amfibier, ferskvannsfisker - Norges dyr. Cappelen, Oslo. 199 pp.

Kramp, P. L. 1935. Polyppdyr (Coelenterata) I. Ferskvandspolypper og Goplepolypper. Danmarks fauna bind 41. G. E. C. Gads forlag, København. 208 pp.

Kramp, P. L. 1937. Polyppdyr (Coelenterata) II. Gopler. Danmarks fauna bind 43. G. E. C. Gads forlag, København. 223 pp.

Lawrence, E. 1989. Henderson's dictionary of Biological Terms. Tenth edition, Longman Scientific & Technical, 637 pp.

Lennmark, I. 1983. Kusthav. Platypus, Swede Books Int. 120 pp.

Lützen, J. G. 1967. Sækdyr. Danmarks fauna bind 75. G. E. C. Gads forlag, København. 267 pp.

Marcus, E. 1940. Mosdyr (Bryozòa eller Polyzòa). Danmarks fauna bind 46. G. E. C. Gads forlag, København. 401 pp.

Moen, F. E. & A. Eriksen. 1993. Taskekrabbe (*Cancer pagurus* L.); populasjonsstruktur, levesett og næringsvalg i et oppvekstområde ved trøndelagskysten - Hovedfagsoppgave. Trondhjem Biologiske Stasjon, Trondheim. 86 pp.

Mortensen, P. B. & M. Hovland. 1996. Korallrev i mørket: *Lophelia*-rev i Midt-Norge. Naturen Nr. 5: 247-254.

Mortensen, P. B. & M. Hovland. Brattegaard, T. & Farestveit, R. 1995. Deep water biotherms of the scleractinian coral *Lophelia pertusa* (L.) at 64° N on the Norwegian shelf: structure and associated megafauna. - Sarsia 80:145-158. Bergen.

Mortensen, S. & P. Eide. 1997. Et levende hav. KOM forlag, 184 pp.

Mortensen, T. H. 1924. Pighude (Echinodermer). Danmarks fauna bind 27. G. E. C. Gads forlag, København. 274 pp.

Muus, B. J. 1959. Skallus, Søtænder, Blæksprutter. Danmarks fauna bind 65. G. E. C. Gads forlag, København. 239 pp.

Muus, J. 1981. Våre saltvannsfisker. NKS - forlag, 244 pp.

Myklebust, B. 1979. Norske fisker i sjøen - Fiskeribiologi. Fabritius, Oslo. 261 pp.

Myklebust, B. 1991. Planter og dyr i sjøen. Gyldendal, Oslo. 91 pp.

Otterstrøm, C. V. 1912. Fisk I. Piggfinnefisk. Danmarks fauna bind 11. G. E. C. Gads forlag, København. 198 pp.

Pethon, P. 1989. Aschehougs store fiskebok. Aschehoug, Oslo. 447 pp.

Picton, B. E. 1993. A field guide to the shallow-water Echinoderms of the British Isles. Immel Publishing, London. 96 pp.

Picton, B. E. & C. C. Morrow. 1994. A field guide to the Nudibranchs of the British Isles. Immel Publishing, London. 143 s.

Rinde, E. m.fl. 1998, Kystøkologi. Universitetsforlaget, Oslo. 214 pp.

Semb-Johansson, A. (red.) 1988. Virvelløse dyr II - Verdens dyr bind 13. Cappelen. 232 pp.

Sneli, J. A. 1975. Mollusca Prosobranchia, Forgjellesnegler nordisk marine arter. Universitetsforlaget, Trondheim. 55 pp.

Southward, A. J. (red.) Barnacle Biology, i Crustacean Issues av Frederick R. Schram (red.). A.A.Balkema, Rotterdam. 443 pp.

Stephensen, K. 1910. Storkrebs I. Skjoldkreps. Danmarks Fauna bind 9. G. E. C. Gads forlag, København. 193 pp.

Stephensen, K. 1928. Storkrebs II. Ringkreps 1. Tanglopper (amfipoder). Danmarks Fauna bind 32. G. E. C. Gads forlag, København. 399 pp.

Stephensen, K. 1933. Havedderkopper og Rankefødder. Danmarks Fauna bind 38. G. E. C. Gads forlag, København. 158 pp.

Stephensen, K. 1948. Storkrebs IV. Ringkreps 3. Tanglus (marine isopoder) og Tanaider. Danmarks Fauna bind 53. G. E. C. Gads forlag, København. 187 pp.

Strømgren, T. 1970. Emergence of *Paramuricea placomus* (L.) and *Primnoa resedaeformis* (Gunn.) in the Inner Part of Trondheimsfjord (West Coast of Norway). Det Kongelige Norske Videnskabers Selskab skrifter No. 4: 6 pp.

Tendal, O. S. 1992. The North Atlantic distribution of the octocoral *Paragorgia arborea* (L., 1758) (Cnidaria, Anthozoa). - Sarsia 77: 213-217. Bergen.

Thompson, T. E., Brown, G. H. 1976, British Ophistobranch Molluscs in Synopses of the British Fauna No. 8, Academic press, London. 203 pp.

Ursing, B. 1971. Ryggradslosa djur P.A. Norstedt & Soners Forlag, Stockholm, 369 pp.

Whitehead, P.J.P., Bauchot, M.L. Hureau, J.C., Nielsen, J. & Tortonese, E. (Eds). 1986. Fishes of the North-eastern Atlantic and the Mediterranean. UNESCO. 1473 pp.

Åsen, P. A. 1980. Illustrert algeflora. Cappelen, Oslo. 64 pp.

Utbredelse til artene ved norskekysten er hentet fra **Brattegard & Holthe (1997)**. Systematisk inndeling og nomenklatur baserer seg i hovedsak på **Hansson (1998)**.

***Below**: Picture from 50 metres depth at Egersund, Norway*

Glossary of terms

Scientific terms are explained in the introduction of each chapter. Listed below is a selection of terms that occur frequently or which are not elsewhere explained.

abdomen – rear body part
alternation of generations – presence of alternating generations with sexual and asexual reproduction
amoeboid – resembling or moving like an amoeba
anoxic – without oxygen

basal – lowest plane or part of body
basal laceration – where a part of the organism is severed, and that part grows into a new individual
basis – at the bottom
benthic – living on or in the sea floor
bilateral symmetry – (of an organism) where only one plane allows a cut to produce two halves that are mirror images of each other
biodiversity – diversity of living organisms
biotope – defined area supporting a characteristic plant and animal community
brackish water – low salinity water (mixing of freshwater and seawater)
byssus threads – threads used by mussels to attach to the substrate

carapace – the outer 'shell' in crustaceans
cellular organ – unit within a cell
cellulose – a carbohydrate found within the cell wall of plants (polysaccharide consisting of long glucose chains)
cephalon – head region
cephalothorax – fused head and thoracic region in crustaceans
chloroplast – cellular organ within plants responsible for photosynthesis
chromatophores – organs containing colour pigments, located on the body surface; often can expand or contract to change the appearance of the organism
cilia – hair–like structures; often beating to create water current
circumpolar – occurring around the poles of the earth
clones – genetically identical individuals
commensalism – cohabitation between two species, where one benefits without harming the other
concave – curving inwards
concentric – having a common centre
convex – curving outwards
Coriolis effect – displacement of water due to the earth's rotation
cosmopolitan – occurring in all the worlds' oceans
cultivation – propagation of an organism; usually in artificial conditions, e.g. aquaculture
cuticle – outer layer of epithelium, or 'skin'
ecompose – to break down

detritus – organic 'litter'; particles or dead organisms
diffusion – passive transport of a substance from areas of high to low concentration
dorsal – back side (often the upper side of an organism)

ecosystem – defined area, or type of environment, giving consideration to the relationships between the animals, plants and their physical surroundings
ectoderm – outer cellular layer
ectoparasite – external parasite
exposed – (of water body) exposed to strong currents or waves
elliptical – not completely, but almost circular
embryo – foetus
encrusting fauna – animals growing as a layer on substrata
entoderm – inner cellular layer
enzyme – protein that speeds the rate of biochemical reactions
epifauna – animal living on the bottom, or attached to, hard substrates
epiphyte – plant living on plant/animal without directly feeding from it
epitokus stage – polychaete that is morphologically modified to leave the bottom and reproduce
excretion – removal of waste products from the body
exposed to – subjected to
extracellular digestion – breaking down of food outside the body cells

filter–feeders – organisms which feed by capturing particulate food matter from the water or by resuspending sediment
flagellum – whip–like structure; used to create water current or for locomotion

gametes – reproductive products; eggs/sperm
genetics – the study of genes and their function
globular – rounded in shape
gonad – reproductive organ

habitat – the natural living environment of an organism
habitus – holistic appearance of an organism
hermaphrodite – organism being of both sexes
holdfast – branching attachment organ of kelp
holoplankton – lives as plankton for entire life; contrast with meroplankton
homogenous – similar or not varying; e.g. of sediment structure

infauna – animals living buried within the sediment
invertebrates – animals without backbones
juvenile – young individual; not fully developed

kelp belt – depth zone dominated by kelp (laminarians)

Lamarckism – Jean Baptiste Lamarck (1744 – 1829) suggested that evolution had taken place, but that acquired characters could be inherited
lateral – side; e.g. of body
limnic – pertaining to freshwater
littoral pools – as rock pools

Medusoid stage – the planktonic stage of cnidarians
meroplankton – organisms living for a short period as plankton, usually in larval form, before adopting the adult lifestyle; contrast with holoplankton
metamorphosis – change; as in from larval to adult form
morphology– external shape and appearance
mutualism – mutually beneficial cohabitation between two species

Operculum – 'lid' covering the opening of snail shells or calcareous worm tubes
organic – (in present context) originating from living matter
osculum – exhalent opening in sponges
oviparous – animals that lay eggs
ovoviviparous – where eggs develop and hatch within the mother animal

Parasite – the benefiting party in parasitism
parasitism – cohabitation species that exploits another species, with negative consequences for the host
pathogenesis – development of an unfertilised egg cell
pelagic – free–living within the water masses
pheromones – scent substances
photopositive – attracted to light; opposite of photonegative
physiology – the study of organisms and their function
pigment patches – coloured patches or flecks on an organism
plankton – organisms drifting within the water masses; limited motility
planula larva – ciliated cnidarian larvae
polychaete – segmented worm bearing bristles
polyp stage – sedentary stage in cnidarians
predator – organism feeding on other live animals
proboscis – tube-shaped mouthpart in sea spiders, and the eversible mouthpart in many polychaetes
prokaryote (also **procaryote**) – organism lacking cell nucleus; genetic material contained within a single thread of DNA, e.g. bacteria and blue–green algae
protandric hermaphrodite – organism changing sex from male to female

Radial – arising from a common point and continuing outwards, like the radii of a circle
radial symmetry – (of an organism) where a vertical cut through two or more planes will produce two halves that are mirror images of each other (compare with bilateral symmetry)
regeneration – regrowth of a lost body part
respiration – breathing; process of oxygen uptake and release of carbon dioxide
respiratory organ – organ responsible for respiration; e.g. gills, lungs or tracheae
retractile – (of body part) able to be withdrawn into the body
rock pools – small bodies of water remaining after the tide has retreated
rostrum – thorn–like projection arising from between the eyes of many crustaceans
ROV – Remotely Operated Vehicle

Salinity – measure of salt content in water; given in parts per thousand
sculpting – (present context) surface formations; e.g. of shells
secondary infection – infection as a result of a weakened immune system, caused by a first (primary) infection
sediment – deposited particles; in present context e.g. sand or mud comprising the sea floor
segmented – divided into repeating body sections
separate sexes – (of species) where members are permanently either male or female
sexual dimorphism – where males and females of the same species have very different appearances
solitary – organism living independent of others; opposite of colonial
species complex – term of uncertainty for a group that may, or may not, include several species
statocyst – organ of balance, consisting of a sensory vesicle containing one or more small granules
statolith – granule within the statocyst; stimulates statocyst nerves and orients the organism of its horizontal/vertical position
substrate – (present context) matter comprising the sea floor

Taxon – a group of related organisms
terminal – at the tip
terrestrial – belonging to the land; land–living
thorax – 'chest' area; between the head and abdomen
triangular dredge – equipment with a triangular metal frame and net, dragged across the sea floor; commonly used to collect hard-bottom fauna
trochophore-larva – ciliated larva of certain molluscs and polychaetes
type specimen – specimen from the original description of the species that serves to fix the name

Umbilicus – distal (at the top) opening of a channel within the shell of certain gastropod snails
umbo – the oldest part of a bivalve shell; often knob-like
undulating – waved; curving

Veliger larva – shell-bearing pelagic mollusc larva
ventral – belly side
viviparous - where eggs hatch and larvae develop inside the mother organism

Zooid - individual within a colony of bryozoans and colonial ascidians
zoologist - researcher of animals
zoology - the study of animals

Systematic list

Only ranks used in the book are included. The taxonomy largely follows that of "Sydskandinaviska marina flercelliga evertebrater" by Hans. G. Hansson (2nd edition, 1998).

Abbreviations:				
	Subhp.	Sub-phylum	Ord.	Order
	Supercl.	Super-class	Subord.	Sub-order
	Cl.	Class	I.ord.	Infra-order
	Subcl	Sub-class	Superfam.	Super-family
	I.cl.	Infra-class	Fam.	Family
	Superord.	Super-order		

Phylum Porifera - sponges p. 42

Cl. Calcarea - calcareous sponges
Cl. Demospongia

Phylum Cnidaria - cnidarians p. 66

Supercl. Hydrozoa
Cl. Hydroida - hydroids
Subcl. Athecata
Subcl. Thecata
Subcl. Limnomedusae
Cl. Siphonophora - siphonophores
Supercl. Schyphozoa - jellyfishes
Ord. Stauromedusae
Ord. Semaeostomaeida
Ord. Rhizostomae
Subph. Anthozoa - corals
Cl. Octocorallia - octocorals
Ord. Alcyonacida - soft corals
Ord. Gorgonacida - horny corals
Ord. Pennatulacida - sea pens
Cl. Hexacorallia
Ord. Ceriantharia
Ord. Scleractinida - stony corals
Ord. Actiniarida - sea anemones

Phylum Ctenophora - comb jellies p. 146

Cl. Tentaculata

Phylum Plathelminthes - flatworms p. 149

Phylum Nemertea - ribbon worms p. 151

Phylum Priapulida p. 159

Phylum Annelida - segmented worms p. 161

Cl. Polychaeta - bristle-worms
Superfam. Aphroditoidea - scaleworms
Fam. Nereidae
Fam. Phyllodocidae
Fam. Hesionidae
Fam. Eunicidae
Fam. Pectinariidae
Fam. Flabelligeridae
Fam. Terebellidae
Fam. Cirratulidae
Fam. Serpulidae

Fam. Spirorbidae
Fam. Sabellidae - feather-duster worms
Fam. Arenicolidae
Cl. Clitellata
Subcl. Hirudinoidea - leeches

Phylum Echiura p. 193

Phylum Sipuncula p. 195

Phylum Arthropoda p. 197
Subph. Cheliceriformes
Supercl. Pycnogonida - sea spiders
Subph. Crustacea
Supercl. Maxillopoda
Cl. Copepoda - copepods
Cl. Thecostraca
Subcl. Cirripedia - barnacles
Supercl. Malacostraca -
Superord. Peracarida
Ord. Mysida - mysids
Ord. Amphipoda - amphipods
Subord. Gammarina
Subord. Hyperiidea
Subord. Caprellidea - skeleton shrimps
Ord. Isopoda - isopods
Superord. Eucarida
Ord. Euphausiacida - krill
Ord. Decapoda - decapods
Subord. Dendrobranchiatina
I.ord. Caridea - shrimps
I.ord. Astacidea
I.ord. Thalassinidea
I.ord. Anomura
Fam. Galatheidae - squat lobsters
Fam. Porcellanidae - porcelain crabs
Fam. Lithodidae - king crabs
Fam. Paguridae - hermit crabs
I.ord. Brachyura - crabs

Phylum Mollusca p. 282
Cl. Polyplacophora
Cl. Gastropoda - snails
Subcl. Prosobranchia
I.kl. Docoglossa
I.kl. Archaeogastropoda
I.kl. Caenogastropoda
Ord. Neotaenioglossa
Ord. Neogastropoda
Subcl. Heterobranchia
Superord. Ophistobranchia - sea-slugs, -hares & -butterflies
Ord. Cephalaspida
Ord. Gymnosomata
Ord. Aplysiomorpha

Ord. Sacoglossa
Ord. Nudibranchia - nudibranchs
Subord. Pleurobranchomorpha
Subord. Anthobranchia
Subord. Cladobranchia
Cl. Bivalvia - bivalves
Fam. Anomiidae - saddle oysters
Fam. Mytilidae - mussels
Fam. Ostreidae - oysters
Fam. Pectinidae - scallops
Fam. Limidae - file clams
Fam. Arcticidae
Fam. Cardiidae - cockles
Fam. Veneridae - venus clams
Fam. Tellinidae
Fam. Psammobiidae - sunset shells
Fam. Solenidae - razor shells
Fam. Myidae - gapers/ soft-shell clams
Fam. Hiatellidae
Cl. Cephalopoda
Ord. Sepiida - squids
Ord. Octopodida - octopi
Cl. Scaphopoda - tooth shells

Phylum Phoronida - phoronids p. 387
Phylum Brachiopoda - brachiopods p. 388
Cl. Inarticulata
Cl. Articulata

Phylum Bryozoa - bryozoans p. 393
Cl. Stenolaemata
Cl. Gymnolaemata
Phylum Echinodermata - echinoderms p. 405
Cl. Crinoidea - sea lilies
Cl. Asteroidea - starfishes
Cl. Ophiuroidea - brittle stars
Cl. Echinoidea - sea urchins
Ord. Cidaroida
Ord. Echinoida
Ord. Spatangoida
Cl. Holothurioidea - sea cucumbers
Subcl. Dendrochirotacea
Subcl. Aspidochirotacea

Phylum Chordata - ryggstrengdyr p. 452
Subph. Tunicata - kappedyr
Cl. Ascidiacea - sea squirts
Subph. Cephalochordata - lancelets
Subph. Vertebrata
Supercl. Agnatha - jawless fish
Supercl. Gnathostomata - jawed fish
Cl. Chondrichthyes - cartilaginous fishes

Superord. Selachimorpha - sharks
Fam. Squalidae - dogfish sharks
Fam. Carcharhinidae - requiem sharks
Fam. Scyliorhinidae - cat sharks
Ord. Rajiformes - skates and rays
Fam. Rajiidae - skates and rays
Ord. Chimaeriformes - ratfishes
Fam. Chimaeridae
Cl. Osteichthyes - bony fish
Ord. Clupeiformes - anchovies and herrings
Ord. Anguilliformes - eels
Or. Salmoniformes - salmons
Ord. Osmeriformes
Ord. Gadiformes - codfishes
Ord. Lophiiformes - anglerfishes
Ord. Beloniformes - needle fishes
Ord. Gobiesociformes - clingfishes
Ord. Syngnathiformes - pipe fishes and sea horses
Ord. Gasterosteiformes - sticklebacks
Ord. Scorpaniformes - scorpionfishes and flatheads
Ord. Perciformes - perch-likes
Fam. Carangidae - jacks and pompanos
Fam. Mullidae - goatfishes
Fam. Labridae - wrasses
Fam. Trachinidae - slimeheads
Fam. Blennidae - combtooth blennies
Fam. Pholididae - gunnels
Fam. Zoarcdae - eelpouts
Fam. Stichaeidae - pricklebacks
Fam. Anarhichadidae - wolffishes
Fam. Ammodytidae - sandeels/ sand lances
Fam. Gobiidae - gobies
Fam. Scombridae - mackerels, tunas
Fam. Callinymidae - dragonets
Ord. Pleuronectiformes - flatfishes, flounders & soles

Other photographers have contributed the following pictures:

Dr. Peter Glanvill: Page 162.
Nils Aukan: Page 85 (large).
Stein Johnsen: 116 (large), 244 (top).
Per Eide Studio: Page 258, 259.
Florian Graner: Page 91 (top).
Alf J. Nilsen: Page 201.
Cédric d'Udekem d'Acoz: Page 243.
Rudolf Svensen: Page 112 (small), 222 (large), 250 (large), 475 (amphipode), 495 (large), 498 (large), 506 (large), 512, 519 (both), 544 (both), 605.
Frank Emil Moen: Page 7, 58 (top), 59 (below left), 64 (bottom), 78 (top t.r.), 80, 88 (bottom right), 102 (bottom), 103 (bottom), 131 (small at top), 136 (large), 137 (both), 182 (small in middle), 184 (bottom), 205 (top), 233 (top), 257 (middle), 267 (two small), 275 (large), 276 (large), 367 (large), 399 (bottom right), 400 (two at bottom), 401 (bottom right), 409 (top right), 526 (eggs), 539 (middle).

INDEX SCIENTIFIC NAMES

INDEX COMMON NAMES

The book's photographer at work in Skarnsundet in Norway. The picture shows a Subal house with a Nikon F90x, macro port with Nikkor 60 2.8 micro, and two Hartenberger flashes. The photographic equipment alone weighs over 10 kg. Picture by the book photographer's brother, Rudolf Svensen.

How and why were the photographs taken?

It all started about 29 years ago. Life on the shore has fascinated me since childhood boat and beach trips together with my parents. I have always enjoyed rummaging around among seaweed in search of small creatures. The variety of marine life is enormous, and investigating a small crab or shell felt more exciting than the small creatures on land.

As I grew up, the craving became stronger to see animals that live more than half a metre below the water surface. A diving course was in order. Arnold Thomassen at Eigerøy Lighthouse had been diving for as long as I had known him as a little boy. It looked incredibly exciting. Bjørn Vigrestad and I signed up for a diving course and soon were proud owners of diving certificates. Now the exploration of unknown elements could begin.

Yours truly in action.

Foto: Rudolf Svensen

However, I soon became aware that something was missing. I lacked both documentation of what I had seen and an ability to explain it to others. Thor Petter Rasmussen had already begun to photograph life in the depths. I purchased a used camera from him and soon discovered that is was no easy affair to adjust the aperture and shutter speeds and handle a camera wearing thick gloves. The trickiest part, however, was the flash bulbs, which had to be retrieved and fitted snugly into the flash unit. Because these bulbs float, I became a big consumer.

However, progress was made. After 15 years with the Nikonos system, the same Thor Petter persuaded me to acquire a house for a single lens reflex camera. I decided on a Subal house, and after importing it myself from the manufacturers in Austria, I once more was ready. A new world opened up to me. Just the fact that 25 - 30 images out of a 36 exposure film were of good quality was a fantastic development. For the first time, I could see the composition through the finder, and the TTL system regulated the exposure. In addition, I could choose the composition even when using a macro lens. Previously, I had to choose in advance the size of animals to be photographed. All this is now obsolete.

There was an explosive development in technology for underwater photography during the beginning of the 1990s. The quality is fantastic and the possibilities immense. However, obtaining good pictures still presents certain challenges for both photographer and equipment. In the first case, around 50 kg diving gear must be worn, and diving skills must be mastered to such an extent that it feels like second nature. On top of this, photographic equipment needs to be understood and everything from light measurements, aperture and shutter speed adjustment to achieving a good composition needs to be carried out wearing thick gloves. Most importantly, all movement must be controlled such as to avoid disturbing settled particles, which cause 'snow' on the photographs. In addition, we have to control our enthusiasm, because we only have 36 photographs available for each dive. It is essential to avoid running out of film just as that unusual or spectacular animal appears.

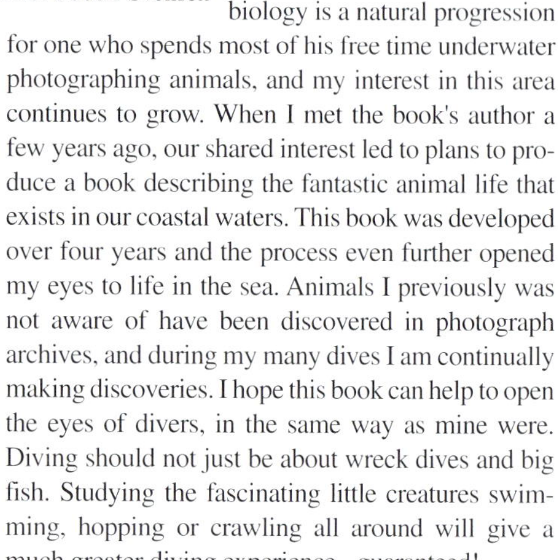

For macro-photography, it is important to know the animals; how and where they live, as well as what they eat. Study of biology is a natural progression for one who spends most of his free time underwater photographing animals, and my interest in this area continues to grow. When I met the book's author a few years ago, our shared interest led to plans to produce a book describing the fantastic animal life that exists in our coastal waters. This book was developed over four years and the process even further opened my eyes to life in the sea. Animals I previously was not aware of have been discovered in photograph archives, and during my many dives I am continually making discoveries. I hope this book can help to open the eyes of divers, in the same way as mine were. Diving should not just be about wreck dives and big fish. Studying the fascinating little creatures swimming, hopping or crawling all around will give a much greater diving experience - guaranteed!

Erling Svensen

Author's afterword

The sea has always fascinated me. I grew up in inland Norway, in the south-east, but fortunately most of my relatives are from the coast. And so it happened that summer holidays spent at my grandparents home in Vesterålen, northern Norway, were responsible for firing my interest in the mysteries of the sea. From the time I was a little boy, these holidays were spent either in a rowing boat or crawling amongst the rocks on the shore. After completing high school at Sortland, it was clear that I wanted to be a marine biologist. So university studies at Trondheim were completed, and it was first then that I obtained my diving certificate. It took only a short time before a need to document life in the sea made itself felt, and so underwater photography started. During my studies, it became clear that a good book illustrating the variety of marine life along our coast was lacking. Most available literature is based on line drawings, not photographs. Anyone who has attempted to identify collected animals, either for teaching, study or pure curiosity will know that such drawings are not always the easiest to use. When I went to work in Egersund, in south-west Norway, I knew that Erling Svensen lived there. He was easy to persuade-the idea of publishing underwater photographs in book form was already in his mind. And so it was simply a case of getting started.

There lies much impressive work in the literature used to write this book. I have particular respect for those researchers working in the 1800s and early 1900s, who devoted much of their lives to wide-ranging and thorough investigations of life in the sea. Even today, many works written in the early 1900s still rank among the most detailed and precise literature available.

We have written about animals living in places that are difficult for people to access. Therefore, our documentation of their geographic and depth distribution ranges, size and colour variations as well as habitat and way of life is sure to contain many errors. Even in the scientific literature, contradictory information on the same species can be found. This simply makes us aware that there is much we do not yet know. Previously published information on colour often appears to be based on observations of specimens brought to land, or fixed, and is often very different from that in their natural state. Similarly, recorded depth distributions are unreliable - we have seen many animals that are recorded as living below diveable depths, and we very often see species thought to be rare. I have drawn on my experience from teaching at school, college and university levels, and aimed to produce a useful aid, especially for field excursions. I hope you, the reader, enjoy using this book as much as we did making it.

Frank Emil Moen

Internet site: http://www.marinbi.com/

This web site aims to promote communication between the readers and authors of this book. We will provide updated information on the organisms presented in the book. We also invite readers to provide additional information on the species presented, particularly in situ observations. The book can also be ordered through this site. This book hopefully will have a long life, whereas internet addresses change, so search for the title or the authors if the above address no longer exists.

E-mail to the author:
moen@marinbi.com
and the photographer:
erling.svensen@marinbi.com.

Thanks and acknowledgements

Frank Emil Moen

Erling Svensen

Between the covers of this book lie the results of more than 20 years of underwater photography and study. Accompanying an underwater photographer during a dive can be a trying experience. Even worse, a biologically-minded photographer focusing on tiny, often barely visible animals can be a frustrating companion for a regular sports diver. We are therefore deeply grateful to our diving buddies over the years. In particular, we thank Rudolf Svensen, for his support in planning and carrying out challenging dives. We also thank him and the other photographers who have contributed photographs for this book.

The author diving near Egersund, Norway.

We aimed to present information that could withstand scientific critique. We therefore thank the various specialists we have consulted for information. In particular, we thank Torleiv Brattegard at the Institute for Fisheries and Marine Biology, University of Bergen in Norway for constructive criticism and input.

This book is previously published in Norwegian, and we thank everyone, scientists and hobby-biologists alike, who has given us useful information for amendments/additions to each of the three published editions. A special thanks to the translators Dr. Sabine Cochrane and Prof. Fredrik Pleijel, and for the final touch by Prof. Bjørn Gulliksen. We look forward to further input and will continue to update during subsequent editions.

Aside from printing and binding, this book is entirely produced by the author and photographer. This would not have been possible without access to high-quality computer graphics equipment, for which we are grateful to the newspaper Dalane Tidende in Egersund, Norway. We also thank the newspaper employees who have willingly helped us along the way.

Most of all, we give warm thanks to our families, who have patiently tolerated our extended periods of absence, both physically and mentally.

A genuine interest in nature does not just simply appear by itself. We are formed by those who have raised us, and in turn, we form our offspring. In an age where humans exploit, and often over-exploit, nature, it is particularly important to teach and inspire our children. We wish to thank in advance all parents, nursery-schools, teachers and others who will use this book in their activities, to encourage children and young people to become fascinated by life below the surface of the sea. In so doing, you contribute to an increasing awareness and respect for nature and the sea.